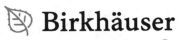

Géza Schay

Introduction to Probability
with Statistical Applications

Second Edition

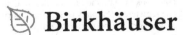 Birkhäuser

Géza Schay
Professor Emeritus
University of Massachusetts Boston
College of Science and Mathematics
Boston, MA, USA

ISBN 978-3-319-30618-6   ISBN 978-3-319-30620-9   (eBook)
DOI 10.1007/978-3-319-30620-9

Library of Congress Control Number: 2016938263

Mathematics Subject Classification (2010): 60-01, 62-01

Printed on acid-free paper

This book is published under the trade name Birkhäuser
The registered company is Springer International Publishing AG Switzerland
(www.birkhauser-science.com)

Preface

This book provides a calculus-based introduction to probability and statistics. It contains enough material for two semesters, but with judicious selection, it can be used as a textbook for a one-semester course, either in probability and statistics or in probability alone.

In each section, it contains many examples and exercises and, in the statistical sections, examples taken from current research journals.

The discussion is rigorous, with carefully motivated definitions, theorems, and proofs, but aimed for an audience, such as computer science students, whose mathematical background is not very strong and who do not need the detail and mathematical depth of similar books written for mathematics or statistics majors.

The use of linear algebra is avoided and the use of multivariable calculus is minimized as much as possible. The few concepts from the latter, like double integrals, that were unavoidable are explained in an informal manner, but triple or higher integrals are not used. The reader may find a few brief references to other more advanced concepts, but they can safely be ignored.

Some Distinctive Features

In Chapter 2, events are defined (following Kemeny and Snell, *Finite Mathematics*) as truth sets of statements. Venn diagrams are presented with numbered rather than shaded regions, making references to those regions much easier.

In Chapter 3, combinatorial principles involving all four arithmetic operations are mentioned, not just multiplication as in most books. Tree diagrams are emphasized. The oft-repeated mistake of presenting a limited version of the multiplication principle, in which the selections are from the same set in every stage and which makes it unsuitable for counting permutations, is avoided.

In Chapter 4, the axioms of probabilities are motivated by a brief discussion of relative frequency, and in the interest of correctness, measure-theoretical concepts are mentioned, though not explained.

In the combinatorial calculation of probabilities, evaluations with both ordered and unordered selections are given where possible.

De Méré's first paradox is carefully explained (in contrast to many books where it is mishandled).

Independence is defined before conditioning and is returned to in the context of conditional probabilities. Both concepts are illustrated by simple examples before stating the general definitions and more elaborate and interesting applications. Among the latter are a simple version of the gambler's ruin problem and Laplace's rule of succession as he applied it to computing the chances of the sun's rising the next day.

In Chapter 5, random variables are defined as functions on a sample space, and first, discrete ones are discussed through several examples, including the basic, named varieties.

The relationship between probability functions and distribution functions is stressed, and the properties of the latter are stated in a theorem, whose proof is relegated though to exercises with hints.

Histograms for probability functions are introduced as a vehicle for transitioning to density functions in the continuous case. The uniform and the exponential distribution are introduced next.

A section is then devoted to obtaining the distributions of functions of random variables, with several theorems of increasing complexity and nine detailed examples.

The next section deals with joint distributions, especially in two dimensions. The uniform distribution on various regions is explored and some simple double integrals are explained and evaluated. The notation $f(x, y)$ is used for the joint p.f. or density and $f_X(x)$ and $f_Y(y)$ for the marginals. This notation may be somewhat clumsy, but is much easier to remember than using different letters for the three functions, as is done in many books.

Section 5.5 deals with independence of random variables, mainly in two dimensions. Several theorems are given and some geometric examples are discussed.

In the last section of the chapter, conditional distributions are treated, both for discrete and for continuous random variables. Again, the notation $f_{X|Y}(x, y)$ is preferred over others that are widely used but less transparent.

In Chapter 6, expectation and its ramifications are discussed. The St. Petersburg paradox is explained in more detail than in most books, and the gambler's ruin problem is revisited using generating functions.

In Section 6.4 on covariance and correlation, following the basic material, the Schwarz inequality is proved and the regression line in scatter plots is discussed.

In the last section of the chapter, medians and quantiles are discussed.

In Chapter 7, the first section deals with the Poisson distribution and the Poisson process. The latter is not deduced from basic principles, because that would not be of interest to the intended audience, but is defined just by the distribution formula. Its various properties are derived though.

In Section 7.2, the normal distribution is discussed in detail, with proofs for its basic properties.

In the next section, the de Moivre-Laplace limit theorem is proved, and then used to prove the continuity correction to the normal approximation of the binomial, followed by two examples, one of them in a statistical setting. A rough outline of Lindeberg's proof of the Central Limit Theorem is given, followed by a couple of statistical examples of its use.

In Section 7.4, the negative binomial, the gamma, and beta random variables are introduced in a standard manner.

The last section of the chapter treats the bivariate normal distribution in a novel manner, which is rigorous, yet simple, and avoids complicated integrals and linear algebra. Multivariate normal distributions are just briefly described.

Chapter 8 deals with basic statistical issues. Section 8.1 begins with the method of maximum likelihood, which is then used to derive estimators in various settings. The method of moments for constructing estimators is also discussed. Confidence intervals for means of normal distributions are also introduced here.

Section 8.2 introduces the concepts of hypothesis testing and is then continued in the next section with a discussion of the power function.

In Section 8.4, the special statistical methods for normal populations are treated. The proof of the independence of the sample mean and variance and of the distribution of the sample variance is in part original. It was devised to avoid methods of linear algebra.

Sections 8.5, 8.6, and 8.7 describe chi-square tests, two-sample tests, and Kolmogorov-Smirnov tests.

Preface to the Second Edition

The organization of the book is the same as that of the first edition, except that Section 8.8 is new. It treats simple linear regression in some detail, pulling together and extending the partial strands of earlier discussions in Sections 6.4 and 7.5, which have also been expanded.

We made small improvements in many places to make the text clearer and more precise. This, of course, included the correction of all the known errors.

Many new examples have been added, especially more classical ones, such as the inclusion-exclusion principle, Montmort's problem, the ballot problem, the Monty Hall problem, Bertrand's paradox, Buffon's needle problem, and some new applications, e.g., the Maxwell-Boltzmann and the Bose-Einstein distributions in physics.

In Section 2.2 the previous treatment of the algebra of sets was quite superficial, because we assumed that this material was familiar to most students. Apparently, however, many students needed more, and so we have included a more detailed and rigorous discussion of set operations.

Section 5.3, Functions of Random Variables, was rewritten by adding more examples and omitting the theorems. It seemed to be adequate and pedagogically preferable just to provide brief suggestions for the necessary procedures and to use those in the examples always from scratch, instead of substituting into formulas of theorems.

The first edition had about 370 exercises; we have added about 30 more, especially in sections where their number was inadequate.

The students' online solution manual has been removed, since it was not very useful. Apparently, most students have not even looked at it, and now its removal has created a large number of new exercises available for homework. However, the appendix with brief answers and hints for selected odd-numbered exercises has been retained, and there is a complete online solution manual for instructors.

The author thanks his wife Maria and son Peter for their support and patience.

Boston, MA, USA   Géza Schay
January 2016

Contents

1. Introduction

Probability theory is a branch of mathematics that deals with repetitive events whose occurrence or nonoccurrence is subject to chance variation. Statistics is a related scientific discipline concerned with the gathering, representation, and interpretation of data and with methods for drawing inferences from them.

While the preceding statements are necessarily quite vague at this point, their meaning will be made precise and elaborated in the text. Here we can shed some light on them though by a few examples.

Suppose we toss a coin and observe whether it lands head (H) or tail (T) up. While the outcome may or may not be completely determined by the laws of physics and the conditions of the toss (such as the initial position of the coin in the tosser's hand, the kind of flick given the coin, the wind, the properties of the surface on which the coin lands, etc.), these conditions are usually not known anyway, and we cannot be sure which side the coin will fall on. We usually assign the number $1/2$ as the probability of either result. This can be interpreted and justified in several ways. First, it is just a convention that we take the numbers from 0 to 1 as probability values and the total probability for all the outcomes of an experiment to be 1. (We could use any other scale instead. For instance, when probabilities are expressed as percentages, we use the numbers from 0 to 100, and when we speak of odds, we use a scale from 0 to infinity.) Hence, the essential part of the probability assignment $1/2$ to both H and T is the equality of the probabilities of the two outcomes. Some people have explained this equality by a "principle of insufficient reason," that is, that the two probabilities should be equal because we have no reason to favor one outcome over the other, especially in view of the symmetrical shape of the coin. This reasoning does not stand up well in more complicated experiments. For instance, in the eighteenth century, several eminent mathematicians believed that in the tossing of two coins, there are three equally likely outcomes, HH, HT, and TT, each of which should have probability $1/3$. It was only through experimentation that people observed that when one coin shows H and the other T, then it makes a difference which coin shows which outcome, that is, that the four outcomes, HH, HT, TH, and TT, each show up about one fourth of the time, and so each should be assigned probability $1/4$. It is interesting to note, however,

© Springer International Publishing Switzerland 2016
G. Schay, *Introduction to Probability with Statistical Applications*,
DOI 10.1007/978-3-319-30620-9_1

that in modern physics, for elementary particles exactly the opposite situation holds, that is, they are, very strangely, indistinguishable from each other. Also, the laws of quantum theory directly give *probabilities* for the outcomes of measurements of various physical quantities, unlike the laws of classical physics, which predict the outcomes themselves.

The coin tossing examples above illustrate the generally accepted form of the frequency interpretation of probabilities: we assign probability values to the possible outcomes of an experiment so as to reflect the *proportions* of the occurrence of each outcome in a large number of repetitions of the experiment. Due to this interpretation, probability assignments and computations must follow certain simple rules, which are taken as axioms of the theory. The commonly used form of probability theory, which we present here, is based on this axiomatic approach. (There exist other approaches and interpretations of probability, but we are not going to discuss these. They are mostly incomplete and unsettled.) In this theory we are not concerned with the justification of probability assignments. We make them in some manner that corresponds to our experience, and we use probability theory only to compute other probabilities and related quantities. On the other hand, in the theory of statistics, we are very much concerned, among other things, with the determination of probabilities from repetitions of experiments.

An example of the kind of problem probability theory can answer is the following: Suppose we have a fair coin, that is, one that has probability $1/2$ for showing H and $1/2$ for T, and we toss it many times. I have 10 dollars and bet one dollar on each toss, playing against an infinitely rich adversary. What is the probability that I would lose all of my money within, say, 20 tosses (About 0.026.) Or, to ask for a quantity that is not a probability: For how many tosses can I expect my \$10 to last? (Infinitely many.) Similarly, how long can we expect a waiting line to grow, whether it involves people in a store or data in a computer? How long can a typical customer expect to wait?

Examples of the kinds of problems that statistical theory can answer are the following: Suppose I am playing the above game with a coin supplied by my opponent, and I suspect that he has doctored it, that is, that the probabilities of H and T are not equal. How many times do we have to toss to find out with reasonable certainty whether the coin is fair or unfair? What are reasonable assignments of the probabilities of H and T? Or in a different context, how many people need to be sampled in a preelection poll to predict the outcome with a certain degree of confidence? (Surprisingly, a sample of a few hundred people is usually enough, even though the election may involve millions.) How much confidence can we have in the effectiveness of a drug tested on a certain number of people? How do we conduct such tests?

Probability theory originated in the sixteenth century in problems of gambling, and even today, most people encounter it, if at all, only in that context. In this book we too shall frequently use gambling problems as illustrations, because of their rich history and because they can generally be described more simply than most other types of problems. Nevertheless we shall not lose sight of the fact that probability and statistics are used in many fields, such as insurance, public opinion polls, medical experiments, computer science, etc., and we shall present a wide-ranging set of real-life applications as well.

2. The Algebra of Events

2.1 Sample Spaces, Statements, and Events

Before discussing probabilities, we discuss the kinds of events whose probabilities we want to consider, make their meaning precise, and study various operations with them.

The events to be considered can be described by such statements as "a toss of a given coin results in head," "a card drawn at random from a regular 52-card deck is an Ace," or "this book is green."

What are the common characteristics of these examples?

First, associated with each statement, there is a set S of possibilities or possible outcomes.

Example 2.1.1. Tossing a Coin.

For a coin toss, S may be taken to consist of two possible outcomes, which we may abbreviate as H and T for head and tail. We say that H and T are the members, elements, or points of S, and write[1] $S = \{H, T\}$. Another choice might be $S' = \{HH, HT, TH, TT\}$, where we toss two coins, but ignore one of them. In this case, for instance, the outcome "the first coin shows H" is represented by the set $\{HH, HT\}$, that is, this statement is true if we obtain HH or HT and false if we obtain TH or TT. Here the toss of the first coin is regarded as a subexperiment of the tossing of two coins. ♦

Example 2.1.2. Drawing a Card.

For the drawing of a card from a 52-card deck, we can see a wide range of choices for S, depending on how much detail we want for the description of the possible outcomes. Thus, we may take S to be the set $\{A, \overline{A}\}$, where A stands for Ace and $\overline{A}$ for non-Ace. Or we may take S to be a set of 52 elements, each corresponding to the choice of a different card. Another choice might be $S = \{S, H, D, C\}$, where the letters stand for the suit of the card: spade, heart, diamond, and club. *Not every statement about drawing a card can be represented in every one of these sample spaces.* For example, the

---

[1] Recall that the usual notation for a set is a list of its members between braces, with the members separated by commas. More about this in the next section.

© Springer International Publishing Switzerland 2016
G. Schay, *Introduction to Probability with Statistical Applications*,
DOI 10.1007/978-3-319-30620-9_2

statement "an Ace is drawn" cannot be represented in the last sample space, but it corresponds to the simple set $\{A\}$ in the sample space $\{A, \overline{A}\}$. ◆

Example 2.1.3. Color of a Book.

In this example S may be taken to be the set $\{G, \overline{G}\}$, where G stands for green and $\overline{G}$ for not green. Or S may be the set $\{G, R, B, O\}$, where the letters stand for green, red, blue, and other. Another choice for S may be $\{LG, DG, \overline{G}\}$, where the letters stand for light green, dark green, and not green. ◆

Example 2.1.4. Tossing a Coin Until an H Is Obtained.

If we toss a coin until an H is obtained, we cannot say in advance how many tosses will be required, and so the natural sample space is $S = \{H, TH, TTH, TTTH, \ldots\}$, an infinite set. We can use, of course, many other sample spaces as well: for instance, we may be interested only in whether we had to toss more than twice or not, and then $S = \{1 \text{ or } 2, \text{ more than } 2\}$ is adequate. ◆

Example 2.1.5. Selecting a Number from an Interval.

Sometimes, we need an uncountable set for a sample space. For instance, if the experiment consists of choosing a random number between 0 and 1, we may use $S = \{x : 0 < x < 1\}$. ◆

As can be seen from these examples, many choices for S are possible in each case. In fact, infinitely many. This may seem confusing, but we must put every statement into some context, and while we have a choice over the context, we must make it definite; that is, we must specify a single, maximal set S whenever we want to assign probabilities. It would be very difficult to speak of the probability of an event if we did not know the alternatives.

The set S that consists of all the possible outcomes of an experiment is called the *universal set* or the *sample space* of the experiment. (The word "universal" refers to the fact that S is the largest set we want to consider in connection with the experiment; "sample" refers to the fact that in many applications the outcomes are statistical samples; and the word "space" is used in mathematics for certain types of sets.) The members of S are called the *possible outcomes* of the experiment or the (*sample*) *points* or *elements* of S. For instance, in Example 2.1.1 the points of S are H and T, and the points of S' are $\{HH, HT\}$ and $\{TH, TT\}$. Thus, what we call the possible outcomes of an experiment depend on the choice of the sample space; they correspond to the points of the sample space. For instance, in Example 2.1.2, for the drawing of a card, if we use $S = \{A, \overline{A}\}$, then we consider the possible outcomes to be those of obtaining an Ace or a non-Ace. On the other hand, if we use $S = \{S, H, D, C\}$, then we consider the possible outcomes to be those of obtaining a spade, a heart, a diamond, or a club[2].

---

[2] In the equation $S = \{S, H, D, C\}$, the S on the left denotes the sample space and the S on the right denotes "spade." We did not want to change these convenient notations to avoid the conflict, since the meaning should be obvious.

The second common characteristic of the examples is that the relevant statements are expressed as declarative sentences that are true (t) for some of the possible outcomes and false(f) for the others. For any given sample space, we do not want to consider statements whose truth or falsehood cannot be determined for each possible outcome, or conversely, once a statement is given, we must choose our sample space so that the statement will be t or f in each point.

For instance, in Example 2.1.2, the statement $p =$ "an Ace is drawn" is t for A and f for $\overline{A}$, if the first sample space is used. If we choose the more detailed sample space of 52 elements, then p is t for the four sample points AS, AH, AD, and AC (these stand for the drawings of the Ace of spades, hearts, diamonds, and clubs, respectively), and p is f for the other 48 possible outcomes. On the other hand, the sample space {black, red} is not suitable if we want to consider this statement, since we cannot determine whether this p is true or false if all we know is whether the card drawn is black or red.

All this can be summarized as follows:

We consider experiments that are described by:

1. The sample space S, i.e., the set of what we want to consider possible outcomes,
2. A statement or several statements which are true for certain outcomes in S and false for others. Such a statement is in effect a function from the set S to the two-element set $\{t, f\}$, that is, an assignment of t to the outcomes for which the given statement is true and f to the outcomes for which the statement is false. Only statements of this kind are considered.

Any performance of such an experiment results in one and only one point of S. Once the experiment has been performed, we can determine whether any allowed statements are t or f for this outcome. Thus, given S, the experiment we consider consists of selecting one point of the set S, and *we perform it only once*. If we want to model repetitions, then we make a single selection from a new sample space whose points represent the possible outcomes of the repetitions. For example, to model two tosses of a coin, we may use the sample space $S = \{HH, HT, TH, TT\}$ and the experiment consists of selecting exactly one of the four points HH, HT, TH, or TT, and we do this selection only once.

The set of sample points for which a statement p is t is called the *truth set* of p or the *event* described by or corresponding to p. For example, the event corresponding to the statement $p =$ "an Ace is drawn" is the set $P = \{AS, AH, AD, AC\}$ if the 52-element sample space is used. Thus, we use the word "event" to describe a subset[3] of the sample space. Actually, if S is a finite set, then we consider every subset of S to be an event. (If S

---

[3] Recall that a set A is said to be a subset of a set B if every element of A is also an element of B.

is infinite, some subsets may have to be excluded.) For example, if $S = \{LG, DG, \overline{G}\}$ is the sample space for the color of a book, then the event $P = \{LG, DG\}$ corresponds to the statement $p =$ "the book is green," and the event $Q = \{DG, \overline{G}\}$ corresponds to $q =$ "the book is dark green or not green" $=$ "the book is not light green." Incidentally, this example also shows that a statement can usually be phrased in several equivalent forms.

We say that an event P occurs if in a performance of the experiment the statement p corresponding to P turns out to be true.

Warning: As can be seen from the foregoing, when we make a statement such as $p =$ "a card drawn is an Ace," we do not imply that this is necessarily true, as is generally meant for statements in ordinary usage. Also, we must carefully distinguish the statement p from the statement $q =$ "p is true." In fact, even the latter statement may be false. Furthermore, we could have an infinite hierarchy of different statements based on this p. The next two would be $r =$ "q is true" and $s =$ "r is true."

In closing this section, let us mention that the events that consist of a single sample point are called *elementary events* or *simple events*. For instance, $\{LG\}, \{DG\}, and\{G\}$ are the elementary events in the sample space $\{LG, DG, G\}$. (The point LG and the set $\{LG\}$ are conceptually distinct, somewhat as the person who is the president is conceptually different from his or her role as president. More on this in the next section.)

Exercises

Exercise 2.1.1.

A coin is tossed twice. A sample space S can be described in an obvious manner as $\{HH, HT, TH, TT\}$:

a) What are the sample points and the elementary events of this S?
b) What is the event that corresponds to the statement "at least one tail is obtained"?
c) What event corresponds to "at most one tail is obtained"?

Exercise 2.1.2.

A coin is tossed three times. Consider the sample space $S = \{HHH, HHT, HTH, HTT, THH, THT, TTH, TTT\}$ for this experiment:

a) Is this S suitable to describe two tosses of a coin instead of the S in Exercise 2.1.1? Explain!

b) What events correspond in this S to the statements
   $x =$ "at least one head is obtained,"
   $y =$ "at least one head is obtained in the first two tosses,"
   $z =$ "exactly one head is obtained"?

Exercise 2.1.3.

a) List four different sample spaces to describe three tosses of a coin.
b) For each of your sample spaces in part a), give the event corresponding
   to the statement "at most one tail is obtained," if possible.
c) Is it possible to find an event corresponding to the above statement in
   every conceivable sample space for the tossing of three coins? Explain!

Exercise 2.1.4.

Describe three different sample spaces for the drawing of a card from a
52-card deck other than the ones mentioned in the text.

Exercise 2.1.5.

In the 52-element sample space for the drawing of a card:

a) Give the events corresponding to the statements $p =$ "an Ace or a red
   King is drawn" and $q =$ "the card drawn is neither red, nor odd, nor a
   face card."[4]
b) Give statements corresponding to the events
   $U = \{AH, KH, QH, JH\}$ and $V = \{2C, 4C, 6C, 8C, 10C, 2S, 4S, 6S, 8S, 10S\}$.
   (In each symbol the first letter or number denotes the rank of the card,
   and the last letter its suit.)

Exercise 2.1.6.

Three people are asked on a news show before an election whether they
prefer candidate A or B or have no preference. Give two sample spaces for
the possible answers.

Exercise 2.1.7.

The birth dates of a class of 20 students are recorded. Describe three
sample spaces for the possible birthday of one of these students chosen at
random.

---

[4] The face cards are J, Q, and K.

2.2 Operations with Sets

Before turning to a further examination of the relationships between statements and events, let us review the fundamentals of the algebra of sets.

As mentioned before, a common way of describing a set is by listing its members between braces. For example, $\{a, b, c\}$ is the set consisting of the three letters $a, b,$ and c. The order in which the members are listed is immaterial and so is any possible repetition in the list. Thus $\{a, b, c\}, \{b, c, a\}$, and $\{a, b, b, c, a\}$ each represent the same set. *Two sets are said to be equal if they have exactly the same members.* Thus, e.g., $\{a, b, c\} = \{a, b, b, c, a\}$.

Sometimes we just give a name to a set and refer to it by name. For example, we may call the above set A.

We use the symbol $\in$ to denote membership in a set. Thus $a \in A$ means that a is an element of A or a belongs to A. Similarly $d \notin A$ means that d is not a member of A.

Another common method of describing a set is that of using a descriptive statement, as in the following examples: Say S is the 52-element set that describes the drawing of a card. Then the set $\{AS, AH, AD, AC\}$ can also be written as $\{x | x \in S, x \text{ is an Ace}\}$ or as $\{x : x \in S, x \text{ is an Ace}\}$. These expressions we read as "the set of x's such that x belongs to S and x is an Ace." Also, if the context is clear, we just write this set as $\{x \text{ is an Ace}\}$.

Similarly, $\{x | 2 < x < 3\} = \{x : 2 < x < 3\} = \{2 < x < 3\}$ each denote the set of all real numbers strictly between 2 and 3. (This example also shows the real necessity of such a notation, since it would be impossible to list all the infinitely many numbers between 2 and 3.)

We say that a set A is a *subset* of a set B if every element of A is also an element of B and denote this relation by $A \subset B$. For instance, $\{a, b\} \subset \{a, b, c\}$. We may also read $A \subset B$ as "A is *contained* in B." Notice that by this definition, every set is a subset of itself, too. Thus $\{a, b, c\} \subset \{a, b, c\}$. While this usage may seem strange, it is just a convention, which one finds often useful in avoiding a discussion of "proper" subsets and the whole of a set separately. The notation $A \subset B$ can also be turned around and written as $B \supset A$ and read as "B is a *superset* of A."

Given two sets A and B, a new set, called the *intersection* of A and B, is defined as the set consisting of all the members common to both A and B and is denoted by $A \cap B$ or by AB. The name "intersection" comes from the case in which A and B are sets of points in the plane. In Figure 2.1, for instance, A and B are the sets of points inside the two circles, and AB is the set of points of the region labeled I.

Another example is $\{a, b, c, d\} \cap \{b, c, e\} = \{b, c\}$. See Figure 2.2. (Note the distinction of the notations in the figures: In Figure 2.1 the Roman numerals designate the regions, but in Figure 2.2 the letters are objects within those regions.) Such diagrams are called *Venn diagrams*.

Notice that the operation of intersection can be used to verify a relation of containment: $A \subset B$ if and only if $AB = A$.

For any two sets A and B, another useful set, called the *union* of A and B, is defined as the set whose members are all the members of A and B taken together and is denoted by $A \cup B$. Thus, in Figure 2.1 the regions I, II, and III together make up $A \cup B$.

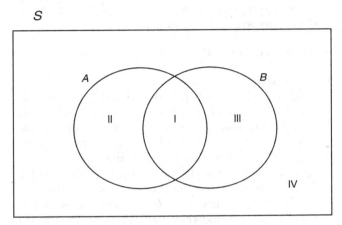

Fig. 2.1. A general Venn diagram for two sets

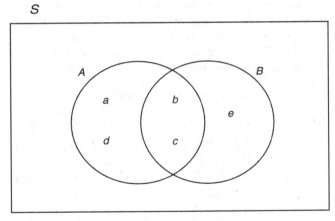

Fig. 2.2. A Venn diagram for two sets of letters

Also, $\{a,b,c,d\} \cup \{b,c,e\} = \{a,b,c,d,e\}$. The diagram of Figure 2.2 illustrates this relation as well.

Unions too can be used to verify a relation of containment: $A \subset B$ if and only if $A \cup B = B$.

A third important operation is the *subtraction* of sets: $A - B$ denotes the set of those points of A that do not belong to B. Thus in Figure 2.1 $A - B$ is region II and $B - A$ is region III.

If we subtract a set A from the universal set S, that is, consider $S - A$, the result is called the *complement* of A, and we denote it by $\overline{A}$. (There is no standard notation for this operation: some books use $\sim A, \tilde{A}, A'$ or A^c instead). In Figure 2.1, $\overline{A}$ consists of the regions III and IV, and $\overline{B}$ of II and IV.

Using both intersection and complement, we can represent each of the regions in Figure 2.1 in a very nice symmetrical manner as

$$\text{I} = AB, \quad \text{II} = A\overline{B}, \quad \text{III} = \overline{A}B, and \quad \text{IV} = \overline{AB}.$$

Also, we see that $A - B = A\overline{B}$ and $B - A = B\overline{A}$.

Here we end the list of set operations, but, in order to make these operations possible for all sets, we need to introduce a new set, the so-called *empty set*. The role of this set is similar to that of the number zero in operations with numbers: Instead of saying that we cannot subtract a number from itself, we say that the result of such a subtraction is zero. Similarly, if we form $A - A$ for any set A, we say that the result is the set with no elements, which we call the empty set and denote by $\emptyset$. We obtain $\emptyset$ in some other cases too: If A is contained in B, that is, $A \subset B$, then, $A - B = \emptyset$. Also, if A and B have no common element, then $AB = \emptyset$ and they are said to be *disjoint*. In view of this relation, $\emptyset$ is said to be a subset of every set A, that is, we extend the definition of $\subset$ to include $\emptyset \subset A$, for every A.

Warning: the empty set must not be confused with the number zero. While $\emptyset$ is a *set*, 0 is a number, and they are conceptually distinct from each other. (The empty set can also be used to illuminate the mentioned distinction between a one-member set and its single member: $\{\emptyset\}$ is a set with one element; and the one element is $\emptyset$, a set with no element.)

In the theorem below, we list the basic properties of the algebra of sets.

Theorem 2.2.1. *(Properties of the Basic Set Operations).*

For all subsets A, B, C of a sample space S, the following relations hold:

1. $A \cup B = B \cup A,$ $\qquad\qquad AB = BA$ $\qquad\qquad$ (*commutative rules*)
2. $A \cup (B \cup C) = (A \cup B) \cup C,$ $A(BC) = (AB)C$ $\qquad$ (*associative rules*)
3. $A \cup (BC) = (A \cup B)(A \cup C),$ $A(B \cup C) = AB \cup AC$ $\quad$ (*distributive rules*)
4. $A \cup A = A,$ $\qquad\qquad\qquad AA = A$ $\qquad\qquad$ (*idempotent rules*)
5. $A \cup \emptyset = A,$ $\qquad\qquad\qquad A\emptyset = \emptyset$ $\qquad\qquad$ (*rules for $\emptyset$*)
6. $A \cup S = S,$ $\qquad\qquad\qquad AS = A$ $\qquad\qquad$ (*rules for S*)
7. $A \cup \overline{A} = S,$ $\qquad\qquad\qquad A\overline{A} = \emptyset$ $\qquad\qquad$ (*rules for $\overline{A}$*)
8. $\overline{S} = \emptyset,$ $\qquad\qquad\qquad\qquad \overline{\emptyset} = S$ $\qquad$ (*rules for $\overline{S}$ and $\overline{\emptyset}$*)
9. $\overline{A \cap B} = \overline{A} \cup \overline{B},$ $\qquad\qquad \overline{A \cap B} = \overline{A} \cup \overline{B}$ $\quad$ (*deMorgan's laws*)
10. $\overline{\overline{A}} = A$ $\qquad\qquad\qquad\qquad$ (*rule of double complement*)

Proof. The proofs of these rules follow at once from the definitions of the operations. Here we just give two of them and several are left as exercises.

For example, the proof of the first commutative rule goes like this: $A \cup B$ is defined as the set whose members are all the members of A and B taken together, and $B \cup A$ as the set whose members are all the members of B and A taken together. Clearly, the order in which we put things together does not matter; we get the same set both ways.

DeMorgan's second law can be proved as follows: If $x \in \overline{A \cap B}$, then $x \notin A \cap B$, and so $x \notin A$ or $x \notin B$. Hence $x \in \overline{A}$ or $x \in \overline{B}$, that is, $x \in \overline{A} \cup \overline{B}$. Thus $\overline{A \cap B} \subset \overline{A} \cup \overline{B}$. Conversely, if $x \in \overline{A} \cup \overline{B}$, then $x \in \overline{A}$ or $x \in \overline{B}$, whence $x \notin A$ or $x \notin B$, and so $x \notin A \cap B$ and $x \in \overline{A \cap B}$. Thus $\overline{A} \cup \overline{B} \subset \overline{A \cap B}$. The two inclusions above imply $\overline{A \cap B} = \overline{A} \cup \overline{B}$. Alternatively, this law can be proved by using a Venn diagram. Referring to Figure 2.1, we have $\overline{A \cap B} = \overline{\{I\}} = \{II, III, IV\} = \{III, IV\} \cup \{II, IV\} = \overline{A} \cup \overline{B}$.   ■

The intersection of several sets $A, B, C, \ldots, Z$ is defined as the set of points that belong to each and is denoted by $A \cap B \cap C \cap \ldots \cap Z$ or by $ABC \cdots Z$. We can use this definition to eliminate parentheses in expressions with multiple consecutive intersections. For instance,[5] by Figure 2.3, $ABC = (AB)C$, since $ABC = \{1\}$ and $(AB)C = \{1, 4\} \cap \{1, 2, 3, 7\} = \{1\}$.

S

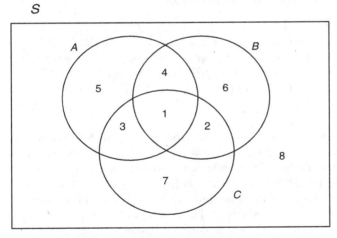

Fig. 2.3. A general Venn diagram for three sets

Similarly, the union $A \cup B \cup C \cup \ldots \cup Z$ of several sets is defined as the set consisting of all the points of the given sets taken together or, equivalently, as the set of all points that belong to at least one of the given sets. Again,

---

[5] Note that in Figure 2.3 the numerals designate the regions, but we prefer, somewhat against our conventions, to write braces and commas, e.g., $\{1, 4\}$ instead of $1 \cup 4$ and $\{1\}$ instead of 1, in order to emphasize the use of these numerals as labels and not as numbers.

as for intersections $A \cup B \cup C = (A \cup B) \cup C$, and so the parentheses are superfluous here and, similarly, in all expressions with multiple consecutive unions.

In expressions with multiple different operations, the *order of precedence* is intersection first, then union, and then subtraction, and the order of complementation is indicated by the placement of the bar. This convention often enables us to avoid using parentheses. For example, we write $AB \cup C$ for $(AB) \cup C$. Nevertheless, when in doubt, use parentheses; they do no harm if used correctly.

The rules of Theorem 2.2.1 can be used to simplify or compare expressions, as, for instance, in the following examples.

Example 2.2.1. Simplifying Expressions.

1. $A \cup AB = AS \cup AB = A(S \cup B) = AS = A.$
2. $(A \cup B)(A \cup C) = A(A \cup B) \cup C(A \cup B) = AA \cup AB \cup AC \cup BC = (A \cup AB) \cup AC \cup BC = (A \cup AC) \cup BC = A \cup BC.$
3. $(A \cup B)(A \cup C)(B \cup C) = (A \cup BC)(B \cup C) = (A \cup BC)B \cup (A \cup BC)C = AB \cup BBC \cup AC \cup BCC = AB \cup AC \cup BC.$
4. $(A - B) \cup (A - C) = A\overline{B} \cup A\overline{C} = A\left(\overline{B} \cup \overline{C}\right) = A\overline{BC}.$ ♦

Example 2.2.2. Comparing Two Expressions.

Under what condition is $A \cup (B - C) = (A \cup B) - C$?

We can reduce the left side as $A \cup (B - C) = A \cup B\overline{C} = (A \cup B)(A \cup \overline{C}) = A \cup AB \cup A\overline{C} \cup B\overline{C} = A \cup B\overline{C} = AC \cup A\overline{C} \cup B\overline{C}$, and for the right side we have $A \cup B - C = (A \cup B)\overline{C} = A\overline{C} \cup B\overline{C}$. Now, AC and $A\overline{C} \cup B\overline{C}$ are disjoint, since one is a subset of C and the other of $\overline{C}$. Thus the two sides are equal if and only if $AC = \emptyset$.   ♦

Let us mention that there exist objects other than sets for which we can build a similar algebra. For example, we have algebras of statements or propositions in mathematical logic, which will be discussed in the next section. Also, some electronic circuits with so-called logic gates follow analogous rules. Such sets with operations like those above, subject to the rules in Theorem 2.2.1, are called *Boolean algebras*. These rules are highly redundant though; about half of them are sufficient to define a Boolean algebra, and the other rules follow from those.

Notice that the regions in the Venn diagrams are all intersections of the given sets and their complements. For instance, in Figure 2.1 I $= AB$ and II $= A\overline{B}$, and in Figure 2.3, $1 = ABC$, $2 = \overline{A}BC$, etc. When dealing with complicated expressions, it is often helpful to reduce them to unions of such basic events, for instance, when we want to determine whether two expressions are equal. In fact, that is what we have done when we listed the region numbers in proofs. For more than three sets, we cannot draw Venn diagrams

with circles, and the high number of regions would also limit their applicability, but it is, however, still useful to write composite sets as such unions.
To this end, we first define the temporary notations $A^1 = A$ and $A^0 = \overline{A}$, for
all sets A. Then, for any distinct subsets $A_1, A_2, \ldots, A_n$ of a sample space
S, for $n = 2, 3, \ldots$, we define their *basic functions* as $A_1^{\epsilon_1} A_2^{\epsilon_2} \cdots A_n^{\epsilon_n}$, where
$\epsilon_i = 0$ or 1 for each i. Now, to list all of these for a given n, we consider the
ϵ_i values to be the digits of the binary representation $\epsilon_1 \epsilon_2 \ldots \epsilon_n$ of a number
k, for $k = 0, 1, \ldots, 2^n - 1$. (Clearly, k runs from 0 to $2^n - 1$, because that is
the range of n digit binary numbers.) Also, we write the shorthand symbol
$B_{n,k}$ for $A_1^{\epsilon_1} A_2^{\epsilon_2} \cdots A_n^{\epsilon_n}$.

Suppose a set A is built up from some given sets $A_1, A_2, \ldots, A_n$. A decomposition of A into a union of basic functions of $A_1, A_2, \ldots, A_n$ is called a
canonical representation of A. For such decompositions, we have the following
theorem.

Theorem 2.2.2. (Basic Decomposition). *Let $A_1, A_2, \ldots, A_n$ be distinct
subsets of a sample space S, for $n = 2, 3, \ldots$. Any set A built up from these
sets with intersections, unions, and complements can be written uniquely,
apart from order, as a union of basic functions, that is, it has a unique canonical representation.*

Proof. [6]First, we can see that

$$S = \bigcup_{k=1}^{2^n - 1} B_{n,k}, \tag{2.1}$$

since, on the one hand, by repeated application of the second distributive
rule,

$$\left(A_1 \cup \overline{A_1}\right)\left(A_2 \cup \overline{A_2}\right) \cdots \left(A_n \cup \overline{A_n}\right) = \cup_{k=1}^{2^n-1} B_{n,k}, \tag{2.2}$$

and on the other hand

$$\left(A_1 \cup \overline{A_1}\right)\left(A_2 \cup \overline{A_2}\right) \cdots \left(A_n \cup \overline{A_n}\right) = SS \cdots S = S, \tag{2.3}$$

by the definitions of complements and of multiple intersections and repeated
application of the second rule for S with $A = S$.

Second, if we intersect both sides of Equation 2.1 with A_i, for any i, then
we get

$$A_i = \bigcup_{k=1, \epsilon_k=1}^{2^n - 1} B_{n,k}, \tag{2.4}$$

because $A_i B_{n,k} = B_{n,k}$ if $\epsilon_k = 1$ and $A_i B_{n,k} = \emptyset$ if $\epsilon_k = 0$.

---

[6] This proof is taken, with some modifications, from Alfred Rényi: *Foundations of
Probability*, Holden-Day, San Francisco, 1970.

Next, let A and C be arbitrary sets that have canonical representations, that is, let

$$A = \bigcup_{k \in K} B_{n,k} \text{ and } C = \bigcup_{k \in L} B_{n,k} \tag{2.5}$$

for some sets of integers K and L. (Note that $\emptyset$ too has such a representation: $A = \emptyset$ if K is the empty set of integers.) Then, clearly

$$AC = \bigcup_{k \in KL} B_{n,k}, \tag{2.6}$$

$$A \cup C = \bigcup_{k \in K \cup L} B_{n,k}, \tag{2.7}$$

and

$$\overline{A} = \bigcup_{k \in \overline{K}} B_{n,k}, \tag{2.8}$$

where $\overline{K} = \{0, 1, \ldots, 2^n - 1\} - K$. Thus all sets built up from $A_1, A_2, \ldots, A_n$ by the three basic operations have canonical representations.

To prove the uniqueness of the representation, notice first that $B_{n,i} B_{n,k} = \emptyset$ if $i \neq k$, because then there must be a factor in $B_{n,i}$ that differs from the corresponding factor in $B_{n,k}$, say A_j and $\overline{A_j}$, and so $B_{n,i}$ and $B_{n,k}$ are subsets of the disjoint sets A_j and $\overline{A_j}$, respectively. Now, assume that a set A has two different canonical representations:

$$A = \bigcup_{k \in K} B_{n,k} \text{ and } A = \bigcup_{k \in L} B_{n,k}, \tag{2.9}$$

with $K \neq L$. In that case there must be a $B_{n,i}$ that occurs in one of the representations but not in the other. Let us say $i \in K$, but $i \notin L$. Then

$$B_{n,i} A = B_{n,i} \bigcup_{k \in K} B_{n,k} = \bigcup_{k \in K} B_{n,i} B_{n,k} = B_{n,i} \tag{2.10}$$

and

$$B_{n,i} A = B_{n,i} \bigcup_{k \in L} B_{n,k} = \bigcup_{k \in L} B_{n,i} B_{n,k} = \emptyset. \tag{2.11}$$

These two equations contradict each other, and so we cannot have two different canonical representations of A. ■

Example 2.2.3. A Canonical Form.

Find the canonical form of $A\overline{BC}$.
$$A\overline{BC} = A\left(\overline{B} \cup \overline{C}\right) = A\overline{B} \cup A\overline{C} = A\overline{B}\left(C \cup \overline{C}\right) \cup A\left(B \cup \overline{B}\right)\overline{C} = A\overline{B}C \cup$$
$$A\overline{B}\,\overline{C} \cup AB\overline{C} \cup A\overline{B}\,\overline{C} = A\overline{B}C \cup A\overline{B}\,\overline{C} \cup AB\overline{C}. \qquad\qquad\blacklozenge$$

Exercises

Exercise 2.2.1.

Use alternative notations to describe the following sets:

a) The set of odd numbers between 0 and 10,
b) {2, 4, 6, 8, 10},
c) The set of black face cards in a regular deck,
d) $\{x : -3 \leq x \leq 3 \text{ and } x^2 = 1, 4, \text{ or } 9\}$,
e) The set of all real numbers strictly between -1 and $+1$.

Exercise 2.2.2.

Referring to the Venn diagram in Figure 2.3, identify by numbers the regions corresponding to:

a) $(A \cup B) \cap C$,
b) $A \cap (B \cap C)$,
c) $\overline{A \cap (B \cap C)}$, (the complement of the set in part b)
d) $(A \cup B) \cup C$,
e) $\overline{A} \cap (\overline{B} \cap C)$,
f) $(A \cap B) \cap \overline{C}$,
g) $A - (B \cap C)$.

Exercise 2.2.3.

List all the subsets of $\{a, b, c\}$. (There are eight.)

Exercise 2.2.4.

Referring to Figure 2.1, prove DeMorgan's first law, by listing the regions corresponding to both sides of the equations, that is, prove that $\overline{A \cup B} = \overline{A} \cap \overline{B}$.

Exercise 2.2.5.

Prove the second associative rule, using Figure 2.3.

Exercise 2.2.6.

Show using Figure 2.3 that $A \cup B \cup C = A \cup (B \cup C) = (A \cup B) \cup C = (A \cup C) \cup B$.

Exercise 2.2.7.

Show using Figure 2.3 that in general:

a) $A \cap (B \cup C) \neq (A \cap B) \cup C$, but
b) $A \cap (B \cup C) = (A \cap B) \cup (A \cap C)$,
c) $(A \cap B) \cup C = (A \cup C) \cap (B \cup C)$.

Exercise 2.2.8.

Referring to Figure 2.3, express the following regions by using A, B, C and unions, intersections, and complements:

a) $\{8\}$
b) $\{3\}$
c) $\{1, 4, 5\}$
d) $\{1, 4, 5, 8\}$
e) $\{2, 6\}$
f) $\{2, 6, 7\}$

Exercise 2.2.9.

If $A \cap B = \emptyset$, what is $\overline{A} \cap \overline{B}$, and what is $\overline{A} \cup \overline{B}$? Illustrate by a Venn diagram.

Exercise 2.2.10.

We have $A = B$ if and only if $A \subset B$ and $B \subset A$. Use this equivalence to prove DeMorgan's first law, $\overline{A \cup B} = \overline{A} \cap \overline{B}$.

Exercise 2.2.11.

Prove that $A \subset B$ if and only if $A \cup B = B$.

Exercise 2.2.12.

Prove that $A \subset B$ if and only if $A \cap B = A$.

Exercise 2.2.13.

Show using Figure 2.3 that in general:

a) $(A - B) - C = (A - C) - (B - C)$
b) $A - (B \cup C) = (A - B) - C$,
c) $(AB) - C = (A - C)(B - C)$.

Exercise 2.2.14.

Show using the definition of subtraction and the rules of Theorem 2.2.1 that:

a) $A - BC = (A - B) \cup (A - C)$,
b) $(A - B) \cup C = ((A \cup C) - B) \cup BC$.

2.3 Relationships Between Compound Statements and Events

When dealing with statements, we often consider two or more at a time connected by words such as "and" and "or." This is also true when we want to discuss probabilities. For instance, we may want to know the probability that a card drawn is an Ace and red or that it is an Ace or a King. Often we are also interested in the negation of a statement, as in "the card drawn is not an Ace." We want to examine how these operations with statements are reflected in the corresponding events.

Example 2.3.1. Drawing a Card.

Consider the statements $p =$ "the card drawn is an Ace" and $q =$ "the card drawn is red." The corresponding sets are $P = \{AS, AH, AD, AC\}$ and $Q = \{2H, 2D, 3H, 3D, \ldots, AH, AD\}$. Now the statement "$p$ and q" can be abbreviated to "the card drawn is an Ace and red" (which is short for "the card drawn is an Ace and the card drawn is red"). This is obviously true for exactly those outcomes of the drawing for which p and q are both true, that is, for those sample points that belong to both P and Q. The set of these sample points is exactly $P \cap Q = \{AH, AD\}$. Thus, the truth set of "p and q," that is, the event corresponding to this compound statement, is $P \cap Q$.

Similarly, "p or q" is true for those outcomes for which p is true or q is true, that is, for the points of P and of Q put together.[7] This is by definition the union of the two sets. Thus the truth set of "p or q" is $P \cup Q$. In our case "p or q" = "the card drawn is an Ace or red" has the 28-element truth set $P \cup Q = \{AS, AC, 2H, 2D, 3H, 3D, \ldots, AH, AD\}$.

Furthermore, the statement "not p" = "the card drawn is not an Ace" is obviously true whenever any of the 48 cards other than one of the Aces is drawn. The set consisting of the 48 outcomes not in P is by definition the complement of P. Thus the event corresponding to "not p" is $\overline{P}$.   ♦

The arguments used in the above example obviously apply to arbitrary statements, too, not just to these specific ones. Thus we can state the following general result.

Theorem 2.3.1. *Correspondence between Logical Connectives and Set Operations.* *If P and Q are the events that correspond to any given statements p and q, then the events that correspond to "p and q," "p or q," and "not p" are $P \cap Q$, $P \cup Q$, and $\overline{P}$, respectively.*

Some other, less important connectives for statements will be mentioned in the next example and in the exercises.

Example 2.3.2. *Choosing a Letter.*

Let $S = \{a, b, c, d, e\}$, $A = \{a, b, c, d\}$, and $B = \{b, c, e\}$. (See Figure 2.2.) Thus S corresponds to our choosing one of these five letters. Let us name the statements corresponding to A and B, p and q. In other words, let $p = $ "$a, b, c,$ or d is chosen" and $q = $ "$b, c,$ or e is chosen." Then $A - B = \{a, d\}$ obviously corresponds to the statement "p but not q" = "$a, b, c,$ or d, but not $b, c,$ or e is chosen." (As we know, we can also write $A\overline{B}$ for $A - B$.) Similarly $B - A = \{e\}$ corresponds to "q but not p," and $(A - B) \cup (B - A) = \{a, d, e\}$ corresponds to "either p or q (but not both)." (The set $A \triangle B = (A - B) \cup (B - A)$ is called the *symmetric difference* of A and B, and the corresponding "or" used here is called the *exclusive or*.)   ♦

Example 2.3.3. *Two Dice.*

Two dice are thrown, say, a black one and a white one. Let b stand for the number obtained on the black die and w for the number on the white die. A convenient diagram for S is shown in Figure 2.4. The possible outcomes are pairs of numbers such as (2, 3) or (6, 6). (We write such pairs within *parentheses*, rather than braces, and call them ordered pairs, because, unlike in sets, the order of the numbers is significant: the first number stands for

---

[7] In mathematics, we use "or" in the inclusive sense, that is, including tacitly the possibility "or both."

the result of the throw of one die, say the black one, and the second number for the other die.) The set S can be written as $S = \{(b, w) : b = 1, 2, \ldots, 6$ and $w = 1, 2, \ldots, 6\}$.

Let $p = $ "$b + w = 7$," that is, $p = $ "the sum of the numbers thrown is 7," and $q = $ "$w \leq 3$." The corresponding truth sets $P = \{(b, w) : b + w = 7\}$ and $Q = \{(b, w) : w \leq 3\}$ are shown shaded in Figure 2.4. The event corresponding to "p and q" = "the sum of the numbers thrown is 7 and the white die shows no more than 3" is the doubly shaded set $P \cap Q = \{(4, 3), (5, 2), (6, 1)\}$. The event corresponding to "p or q" is represented by the $18 + 3 = 21$ shaded squares in Figure 2.4; it is $P \cup Q = \{(b, w) : b + w = 7 \text{ or } w \leq 3\}$. The 15 unshaded squares represent the event $\overline{P} \cap \overline{Q}$, which corresponds to "neither p nor q."   ♦.

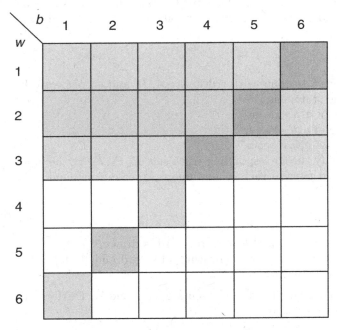

Fig. 2.4. Throwing two dice

Exercises

Exercise 2.3.1.

Consider the throw of two dice as in Example 2.3.3. Let S, p, and q be the same as there, and let $r = $ "b is 4 or 5." Describe and illustrate as in Figure 2.4 the events corresponding to the statements:

a) r,
b) q or r,
c) r but not q,
d) p and q and r,
e) q and r, but not p.

Exercise 2.3.2.

Let a, b, c be statements with truth sets A, B, and C, respectively. Consider the following statements:

$p_1 = $ "exactly one of a, b, c occurs"
$p_2 = $ "at least one of a, b, c occurs"
$p_3 = $ "at most one of a, b, c occurs"

In Figure 2.3 identify the corresponding truth sets P_1, P_2, P_3 by the numbers of the regions, and express them using unions, intersections, and complements of A, B, and C.

Exercise 2.3.3.

Again, let a, b, c be statements with truth sets A, B, and C, respectively. Consider the following statements:

$p_4 = $ "exactly two of a, b, c occur"
$p_5 = $ "at most two of a, b, c occur"
$p_6 = $ "at least two of a, b, c occur"

In Figure 2.3 identify the corresponding truth sets P_4, P_5, P_6 by the numbers of the regions, and express them using unions, intersections, and complements of A, B, and C.

Exercise 2.3.4.

Let $a = $ "an Ace is drawn" and $b = $ "a red card is drawn," and let S be our usual 52-point sample space for the drawing of a card and A and B the events corresponding to a and b:

  i. What logical relations for these statements correspond to DeMorgan's laws (Part 9, Theorem 2.2.1)?
 ii. What statement does S correspond to?

Exercise 2.3.5.

Suppose A and B are two subsets of a sample space S, such that $A \cup B = S$. If A and B correspond to some statements a and b, what can you say about the latter?

Exercise 2.3.6.

Again, let A and B be events corresponding to statements a and b. How are a and b related if $A \cap B = \emptyset$?

Exercise 2.3.7.

For any two events A and B, the expression $A \triangle B = A\overline{B} \cup \overline{A}B$ is called their *symmetric difference* and corresponds to the "exclusive or" of the corresponding statements, that is, to "one or the other but not both." (See Example 2.3.2.) Prove, using Figure 2.3:

1. That it is associative, that is, $(A \triangle B) \triangle C = A \triangle (B \triangle C)$,
2. That intersection is distributive over symmetric difference, that is, $A(B \triangle C) = AB \triangle AC$.

Exercise 2.3.8.

Is union distributive over symmetric difference, that is, is $A \cup (B \triangle C) = (A \cup B) \triangle (A \cup C)$?

3. Combinatorial Problems

3.1 The Addition Principle

As mentioned in the Introduction, if we assume that the elementary events of an experiment with finitely many possible outcomes are equally likely, then the assignment of probabilities is quite simple and straightforward.[1] For example, if we want the probability of drawing an Ace, when the experiment consists of the drawing of a card under the assumption that any card is as likely to be drawn as any other, then we can say that $\frac{1}{52}$ is the probability of drawing any of the 52 cards, and $\frac{4}{52} = \frac{1}{13}$ is the probability of drawing an Ace, since there are four Aces in the deck. We obtain the probability by taking the number of outcomes making up the event that an Ace is drawn and dividing it by the total number of outcomes in the sample space. Thus the assignment of probabilities is based on the counting of numbers of outcomes, if these are equally likely. Now the counting was very simple in the above example, but in many others it can become quite involved. For example, the probability of drawing two Aces if we draw two cards at random (this means "with equal probabilities for all possible outcomes") from our deck is $\frac{4\cdot3}{52\cdot51} = .0045$, since, as we shall see in the next section, $4 \cdot 3 = 12$ is the number of ways in which two Aces can be drawn and $52 \cdot 51 = 2652$ is the total number of possible outcomes, that is, of possible pairs of cards.

Since the counting of cases can become quite complicated, we are going to present a systematic discussion of the methods required for the most important counting problems that occur in the applications of the theory. Such counting problems are called combinatorial problems, because we count the numbers of ways in which different possible outcomes can be combined.

The first question we ask is: What do our basic set operations do to the numbers of elements of the sets involved? In other words if we let $n(X)$ denote the number of elements of the set X for any X, then how are $n(A), n(B)$, $n(A \cup B), n(AB), n(\overline{A}), n(A - B)$, etc. related to each other?

We can obtain several relations from the following obvious special case:

---

[1] In this chapter every set will be assumed to be finite.

© Springer International Publishing Switzerland 2016
G. Schay, *Introduction to Probability with Statistical Applications*,
DOI 10.1007/978-3-319-30620-9_3

Addition Principle:

If $A \cap B = \emptyset$, then $n(A \cup B) = n(A) + n(B)$. $\qquad\qquad$ (3.1)

We can restate this as follows: if A and B do not overlap, then the number of elements in their union equals the sum of the number of elements of A and of B. Basically this is nothing else but the definition of addition: the sum of two natural numbers has been defined by putting two piles together.

When two sets do not overlap, that is, $A \cap B = \emptyset$, then we call them *disjoint or mutually exclusive*. Similarly, we call any number of sets disjoint or mutually exclusive if no two of them have a point in common. For three sets, $A, B, and C$, for instance, we require that $A \cap B = \emptyset$, $A \cap C = \emptyset$, and $B \cap C = \emptyset$, if we want them to be disjoint. Notice that it is not enough to require $A \cap B \cap C = \emptyset$. While the latter does follow from the former equations, we do not have it the other way around, and obviously we need the first three conditions if we want to extend the addition principle to A, B, and C. By repeated application of the addition principle, we can generalize it to any finite number of sets:

Theorem 3.1.1. (Additivity of Several Disjoint Sets). If $A_1, A_2, \ldots,$ A_k are k disjoint sets, then

$$n(A_1 \cup A_2 \cup \cdots \cup A_k) = n(A_1) + n(A_2) + \cdots + n(A_k). \qquad (3.2)$$

We leave the proof as an exercise.

If the sets involved in a union are not necessarily disjoint, then the addition principle leads to:

Theorem 3.1.2. (Size of the Union of Two Arbitrary Sets). For any two sets A and B,

$$n(A \cup B) = n(A) + n(B) - n(A \cap B). \qquad (3.3)$$

Proof. We have $A \cup B = A \cup \overline{A}B$ with A and $\overline{A}B$ disjoint and $B = AB \cup \overline{A}B$, with AB and $\overline{A}B$ disjoint (see Figure 3.1).Thus, by the addition principle,

$$n(A \cup B) = n(A) + n(\overline{A}B) \qquad (3.4)$$

and

$$n(B) = n(AB) + n(\overline{A}B). \qquad (3.5)$$

Subtracting, we get

$$n(A \cup B) - n(B) = n(A) - n(AB), \qquad (3.6)$$

and adding $n(B)$ to both sides, we get the formula of the theorem. ∎

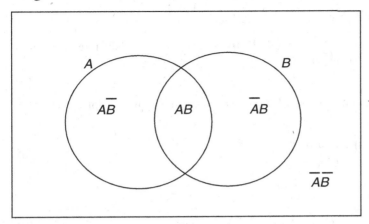

Fig. 3.1. General Venn diagram for two sets with basic decomposition

Example 3.1.1. Survey of Drinkers and Smokers.

In a survey, 100 people are asked whether they drink or smoke or do both or neither. The results are 60 drink, 30 smoke, 20 do both, and 30 do neither. Are these numbers compatible with each other?

If we let A denote the set of drinkers, B the set of smokers, N the set of those who do neither, and S the set of all those surveyed, then the data translate to $n(A) = 60, n(B) = 30, n(AB) = 20, n(N) = 30$, and $n(S) = 100$. Also, $A \cup B \cup N = S$, and $A \cup B$ and N are disjoint. So we must have $n(A \cup B) + n(N) = n(S)$, that is, $n(A \cup B) + 30 = 100$. By Theorem 3.1.2, $n(A \cup B) = n(A) + n(B) - n(AB)$. Therefore in our case $n(A \cup B) = 60 + 30 - 20 = 70$, and $n(A \cup B) + 30 = 70 + 30$ is indeed 100, which shows that the data are compatible. ◆

Let us mention that we could have argued less formally that Theorem 3.1.2 must be true because, if we form $n(A)+n(B)$, we count all the points of $A \cup B$, but those in AB are then counted twice (once as part of $n(A)$ and once as part of $n(B)$). So, in forming $n(A) + n(B) - n(AB)$, the subtraction undoes the double counting, and each point in $A \cup B$ is counted exactly once.

Theorem 3.1.2 can be generalized to unions of three or more sets. For three sets we have

$$n(A \cup B \cup C) = n(A) + n(B) + n(C) - n(AB) - n(AC) - n(BC) + n(ABC). \quad (3.7)$$

We leave the proof of this equation by using a Venn diagram as Exercise 3.1.5. It is also a special case of the theorem below.

Theorem 3.1.3. (Inclusion-Exclusion Theorem). *For any positive integer N and arbitrary sets $A_1, A_2, \ldots, A_N$,*

$$n\left(\bigcup_{i=1}^{N} A_i\right) = \sum_{1 \leq i \leq N} n(A_i) - \sum_{1 \leq i < j \leq N} n(A_i A_j)$$

$$+ \sum_{1 \leq i < j < k \leq N} n(A_i A_j A_k) - \cdots + (-1)^{N-1} n(A_1 A_2 \cdots A_N).$$

$$(3.8)$$

Proof. The *indicator function*[2] I_A of an event A in any sample space S is defined by

$$I_A(s) = \begin{cases} 1 \text{ if } s \in A \\ 0 \text{ if } s \in \overline{A}. \end{cases} \tag{3.9}$$

Let $A = \bigcup_{i=1}^{N} A_i$. Then

$$(I_A - I_{A_1})(I_A - I_{A_2}) \cdots (I_A - I_{A_N}) = 0, \tag{3.10}$$

because if $s \in \overline{A}$, then every factor is $0 - 0$, and if $s \in A$, then also $s \in A_k$ for some k, and for such an s, the factor $I_A - I_{A_k} = 1 - 1 = 0$. Now we expand the product and use the results of Exercise 3.1.7. Then one of the terms will be $I_A^N = I_A$. Also, we get terms in which we choose $-I_{A_i}$ from one factor and I_A from the others; these terms yield $-\sum_{1 \leq i \leq N} I_{A_i} I_A^{N-1} = -\sum_{1 \leq i \leq N} I_{A_i A} = -\sum_{1 \leq i \leq N} I_{A_i}$. Similarly, we get terms with two I_{A_i} factors, then with three, and so on. If we put the I_A term on the left and collect all other terms on the right, then the expansion results in

$$I_A = \sum_{1 \leq i \leq N} I_{A_i} - \sum_{1 \leq i < j \leq N} I_{A_i A_{ji}} + \cdots + (-1)^{N-1} I_{A_1 A_2 \cdots A_N}. \tag{3.11}$$

Now, if we sum both sides over all $s \in S$, then we get the statement of the theorem. ∎

Example 3.1.2. Counting the Number of Integers with Three Properties.

How many positive integers ≤ 1000 are there that are not divisible[3] by 6, 7, and 8?

We use Equation 3.7 with $S = \{1, \ldots, 1000\}$, $A = \{$multiples of 6 in $S\}$, $B = \{$multiples of 7 in $S\}$, and $C = \{$multiples of 8 in $S\}$. Then[4] $n(A) = \lfloor 1000/6 \rfloor = 166$, $n(B) = \lfloor 1000/7 \rfloor = 142$, $n(C) = \lfloor 1000/8 \rfloor = 125$, $n(AB) =$

---

[2] In other branches of mathematics, I_A is called the characteristic function of A, but in probability theory, that name is reserved for a different function.

[3] Recall that an integer m is called divisible by an integer k if m/k is an integer. In that case m is also called a multiple of k.

[4] Here $\lfloor x \rfloor$ denotes the greatest integer or floor function, that is, for any x, the integer such that $\lfloor x \rfloor \leq x < \lfloor x \rfloor + 1$.

$\lfloor 1000/42 \rfloor = 23$, $n(AC) = \lfloor 1000/24 \rfloor = 41$, $n(BC) = \lfloor 1000/56 \rfloor = 17$, and $n(ABC) = \lfloor 1000/168 \rfloor = 5$. Thus,

$$n(\overline{A \cup B \cup C}) = 1000 - 166 - 142 - 125 + 23 + 41 + 17 - 5 = 643. \quad (3.12)$$

♦

From Theorem 3.1.2 it is easy to see that in general

$$\cdot \ n(B - A) = n(B) - n(AB) \tag{3.13}$$

and

$$n(B - A) = n(B) - n(A) \text{ if and only if } A \subset B. \tag{3.14}$$

(This relation is sometimes called the *subtraction principle*.) Substituting S for B, we get

$$n(\overline{A}) = n(S) - n(A). \tag{3.15}$$

Exercises

Exercise 3.1.1.

If in a survey of 100 people, 65 people drink, 28 smoke, and 30 do neither, then how many do both?

Exercise 3.1.2.

Give an example of three pairwise nondisjoint sets A, B, and C such that $A \cap B \cap C = \emptyset$.

Exercise 3.1.3.

Prove that any one of the conditions $A \cap B = \emptyset$, $A \cap C = \emptyset$, or $B \cap C = \emptyset$ implies $A \cap B \cap C = \emptyset$.

Exercise 3.1.4.

Prove Theorem 3.1.1:

a)   For $k = 3$,
b)   For arbitrary k.

Exercise 3.1.5.

Prove the formula given in Equation 3.7 for $n(A \cup B \cup C)$ by using the Venn diagram of Figure 2.3 on page 13.

Exercise 3.1.6.

How many cards are there in a deck of 52 that are:

a) Aces or spades,
b) Neither Aces nor spades
c) Neither Aces nor spades nor face cards (J,Q,K)?

Exercise 3.1.7.

Prove that for any indicator functions:

1. $I_{AB} = I_A I_B$.
2. $I_{A_1 A_2 \cdots A_n} = I_{A_1} I_{A_2} \cdots I_{A_n}$ for any $n \geq 2$.
3. $I_{A \cup B} = I_A + I_B$ if A and B are disjoint.
4. $I_{\overline{A}} = 1 - I_A$.

Exercise 3.1.8.

How many positive integers ≤ 1000 are there that are divisible by 3, 6, or 8?

3.2 Tree Diagrams and the Multiplication Principle

In the previous section, we worked with fixed sample spaces and counted the number of points in single events. Here we are going to consider the construction of new sample spaces and events from previously given ones and count the number of possibilities in the new sets. For example, we throw a die three times and want to relate the number of elements of a sample space for this experiment to the three six-element sample spaces for the individual throws. Or we draw two cards from a deck and want to find the number of ways in which the two drawings both result in Aces, by reasoning from the separate counts in the two drawings.

The best way to approach such multistep problems is by drawing a so-called tree diagram. In such diagrams we first list the possible outcomes of the first step and then draw lines from each of those to the elements in a list of the possible outcomes that can occur in the second step depending on the outcome in the first step. We continue likewise for the subsequent steps, if any.

The above description may be unclear at this point; let us clarify it by some examples.

Example 3.2.1. Drawing Two Aces.

Let us illustrate the possible ways of successively drawing two Aces from a deck of cards (we do not replace the first one before drawing the second). In the first step, we can obtain AS, AH, AD, AC, but in the second step, we can only draw an Ace that has not been drawn before. This is shown in Figure 3.2.

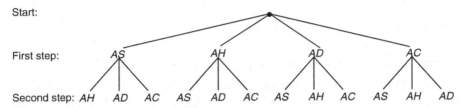

Fig. 3.2. Tree diagram for dealing two Aces without replacement

As we see, for each choice in the first step, there are three possible choices in the second step; thus altogether there are $4 \cdot 3 = 12$ choices for the two Aces. In the figure, for the sake of completeness, we included a harmless extra point on the top, labeled "Start," so that the four choices in the first step would not hang loose. We could turn the diagram upside down (or sideways, too), and then it would resemble a tree: this is the reason for the name. The number 12 shows up two ways in the diagram: first, it is the number of branches from the Start to the bottom, and second, it is the number of branch tips, that is, entries in the bottom row, whether they are distinct or not.   ◆

Example 3.2.2. Primary Elections.

Before primary elections, voters are polled about their preferences in a certain state. There are two Republican candidates R_1 and R_2 and three Democratic candidates D_1, D_2, and D_3. The voters are first asked whether they are registered Republicans (R), Democrats (D), or independents (I) and, second, which candidate they prefer. The independents are allowed to vote in either primary, so in effect they can choose any of the five candidates. The possible responses are shown in the tree of Figure 3.3.

Notice that the total number of branches in the second step is 10, which can be obtained by using the addition principle: we add the three branches through D, the two through R, and the five through I. The branches correspond to mutually exclusive events in the 10-element compound sample space $\{DD_1, DD_2, DD_3, RR_1, RR_2, ID_1, ID_2, ID_3, IR_1, IR_2\}$. This is the new sample space built up in a complicated manner from the simpler ones $\{D, R, I\}$, $\{D_1, D_2, D_3\}$, and $\{R_1, R_2\}$.   ◆

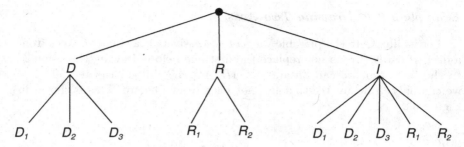

Fig. 3.3. Tree diagram for Example 3.2.2

Example 3.2.3. Tennis Match.

In a tennis match, two players, A and B, play several sets until one of them wins three sets. (The rules allow no ties.) The possible sequence of winners is shown in Figure 3.4.

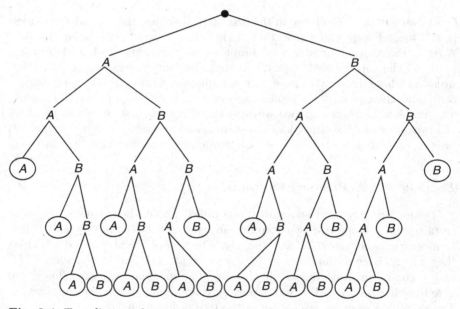

Fig. 3.4. Tree diagram for tennis match

The circled letters indicate the ends of the 20 possible sequences. As can be seen, the branches have different lengths, and this makes the counting more difficult than in the previous examples. Here, by repeated use of the sample space $\{A, B\}$, we built up the 20-element sample space $\{AAA, AABA, AABBA, AABBB, ABAA, ABABA, ABABB, \ldots, BBB\}$.

Notice that if we look upon these strings of A's and B's as words, then they are arranged in alphabetical order (e.g., AAA before $AABA$). *Arranging selections in alphabetical or numerical order is often very helpful in making counts accurate, since it helps 1) to avoid unwanted repetitions and 2) to ensure that everything is listed.*   ♦

We discussed in Example 3.2.2 how the addition principle was applicable there. Now, it is easy to see that it is applicable in Example 3.2.1 and Example 3.2.3 as well. The latter was intended to illustrate branches of various lengths, and we cannot extract any important regularity from it. In Example 3.2.1, however, we see the operation of multiplication showing up for the first time. The four choices in the first step fan out into three branches each, and so, by the addition principle, we obtain the total number of branches for the second step if we add 3 to itself four times. This operation, however, is the same as multiplication of 3 by 4. In general, since multiplication by a natural number is repeated addition, if we have n_1 choices in the first step of an experiment and each of those gives rise to n_2 choices in the second step, then the number of possible outcomes for both steps together, that is, the number of paths from top to bottom of the corresponding two-step tree, is $n_1 n_2$.

We can easily generalize this statement to experiments with several steps and call it a new principle:

The Multiplication Principle: *If an experiment is performed in m steps, and there are n_1 choices in the first step, and for each of those there are n_2 choices in the second step, and so on, with n_m choices in the last step for each of the previous choices, then the number of possible outcomes, for all the steps together, is given by the product $n_1 n_2 n_3 \cdots n_m$.*

Example 3.2.4. Three Coin Tosses.

Toss a coin three times. Then the number of steps is $m = 3$, and in each step we have two possibilities H or T; hence, $n_1 = n_2 = n_3 = 2$. Thus the total number of possible outcomes, that is, of different triples of H's and T's, is $2 \cdot 2 \cdot 2 = 2^3 = 8$. Similarly in m tosses, we have 2^m possible sequences of H's and T's.   ♦

Example 3.2.5. Number of Subsets.

The number of subsets of a set of m elements is 2^m. This can be seen by considering any subset as being built up in m steps: We take in turn each of the m elements of the given set and decide whether it belongs to the desired subset or not. Thus we have m steps and in each step two choices, namely, yes or no, to the question of whether the element belongs to the desired subset. The 2^m subsets include $\emptyset$ and the whole set. (Why?)   ♦

Example 3.2.6. Drawing Three Cards.

The number of ways three cards can be drawn one after the other from a regular deck is 52^3 if we replace each card before the next one is drawn and $52 \cdot 51 \cdot 50$ ways if we do not replace them. For, obviously, we have three steps in both cases, i.e., $m = 3$; and with replacement we can pick any of the 52 cards in each step, that is, $n_1 = n_2 = n_3 = 52$; and without replacement we can pick any of the $n_1 = 52$ cards in the first step, but for the second step, only $n_2 = 51$ cards remain to be drawn from, and for the third step only $n_3 = 50$.   ◆

Example 3.2.7. Seating People.

There are four seats and three people in a car, but only two can drive. In how many ways can they be seated if one is to drive?

For the driver's seat, we have two choices and for the next seat three, because either of the remaining two people can sit there or it can remain empty. For the third seat, we have two possibilities in each case; if the second seat was left empty, then either of the remaining two people can be placed there, and if the second seat was occupied, then the third one can either be occupied by the remaining person or be left empty. The use of the fourth seat is uniquely determined by the use of the others. Consequently, the solution is $2 \cdot 3 \cdot 2 \cdot 1 = 12$.

Alternatively, once the driver has been selected in two possible ways, the second person can take any one of three seats and the third person one of the remaining two seats. Naturally, we get the same result: $2 \cdot 3 \cdot 2 = 12$.

Notice that in this problem, we had to start our counting with the driver, but then had a choice whether to assign people to seats or seats to people. Such considerations are typical in counting problems, and often the nature of the problem favors one choice over another.   ◆

Example 3.2.8. Counting Numbers with Odd Digits.

How many natural numbers are there under 1000 whose digits are odd?

Since all such numbers have either one, two, or three digits, we count those cases separately and add up the three results. First, there are five single-digit odd numbers. Second, there are 5^2 numbers with two odd digits, since each of the two digits can be chosen in five ways. Third, we can form 5^3 three-digit numbers with odd digits only. Thus the solution is $5 + 5^2 + 5^3 = 155$.   ◆

Exercises

Exercise 3.2.1.

a) What sample space does Figure 3.2 illustrate?
b) What are the four mutually exclusive events in this sample space that correspond to the drawing of AS, AH, AD, AC, respectively, in the first step?
c) What is the event corresponding to the statement "one of the two cards drawn is AH"?

Exercise 3.2.2.

In a survey, voters are classified according to sex (M or F), party affiliation ($D, R,$ or I), and educational level (say $A, B,$ or C). Illustrate the possible classifications by a tree diagram! How many are there?

Exercise 3.2.3.

In an urn there are two black and four white balls. (It is traditional to call the containers urns in such problems.) Two players alternate drawing a ball until one of them has two white ones. Draw a tree to show the possible sequences of drawings.

Exercise 3.2.4.

In a restaurant a complete dinner is offered for a fixed price in which a choice of one of three appetizers, one of three entrees, and one of two desserts is given. Draw a tree for the possible complete dinners. How many are there?

Exercise 3.2.5.

Three different prizes are simultaneously given to students from a class of 30 students. In how many ways can the prizes be awarded:

a) If no student can receive more than one prize,
b) If more than one prize can go to a student?

Exercise 3.2.6.

How many positive integers are there under 5000 that:

a) Are odd,
b) End in 3 or 4,

c) Consist of only 3's and/or 4's,

d) Do not contain 3's or 4's?

(Hint: In some of these cases, it is best to write these numbers with four digits, for instance, 15 as 0015, to choose the four digits separately and use the multiplication and addition principles.)

Exercise 3.2.7.

In the Morse code, characters are represented by code words made up of dashes and dots:

a) How many characters can be represented with three or fewer dashes and/or dots?

b) With four or fewer?

Exercise 3.2.8.

A car has six seats including the driver's, which must be occupied by a driver. In how many ways is it possible to seat:

a) Six people if only two can drive,

b) Five people if only two can drive,

c) Four people if each can drive?

Exercise 3.2.9.

How many sets can be built up from given distinct subsets $A_1, A_2, \ldots, A_n$ of a sample space S, for $n = 2, 3, \ldots$, with intersections, unions, and complements? The collection of such sets is called the *Boolean algebra generated by* $A_1, A_2, \ldots, A_n$, and so the question can also be phrased as "how many members does the Boolean algebra generated by n sets have?" (*Hint*: Count the number of different canonical representations as in Theorem 2.2.1.)

3.3 Permutations and Combinations

Certain counting problems recur so frequently in applications that we have special names and symbols associated with them. These will now be discussed.

Any arrangement of things in a row is called a permutation of those things. We denote the number of permutations of r different things out of n different ones by $_nP_r$. This number can be obtained by the multiplication principle. For example, $_8P_3 = 8 \cdot 7 \cdot 6 = 336$, because we have $r = 3$ places to fill in a row, out of $n = 8$ objects. The first place can be filled in eight ways and the second place seven ways, since one object has been used up, and for the third place, six objects remain. Because all these selections are performed one after the other, $_8P_3$ is the product of the three numbers 8, 7, and 6.

In general, $_nP_r$ can be obtained by counting backwards r numbers starting with n and multiplying these r factors together. If we want to write a formula for $_nP_r$ (which we need not use, we may just follow the above procedure instead), we must give some thought to what the expression for the last factor will be: In place 1 we can put n objects, which we can write as $n - 1 + 1$; in place 2 we can put $n - 1 = n - 2 + 1$ objects; and so on. Thus the rth factor will be $n - r + 1$, and so, for any[5] positive integers n and $r \le n$,

$$_nP_r = n(n - 1)(n - 2) \cdots (n - r + 1). \tag{3.16}$$

We can check that for our example, in which $n = 8$ and $r = 3$, we obtain $n - r + 1 = 8 - 3 + 1 = 6$, which was indeed the last factor in $_8P_3$.

For the product that gives $_nP_n$, we have a special name and a symbol. We call it n-*factorial* and write it as $n!$. Thus, for any positive integer n,

$$n! = n(n - 1)(n - 2) \cdots 3 \cdot 2 \cdot 1. \tag{3.17}$$

The symbol $n!$ is just a convenient abbreviation for the above product, that is, for the product of all natural numbers from 1 to n (the order does not really matter). For example, $1! = 1$, $2! = 2 \cdot 1 = 2$, $3! = 3 \cdot 2 \cdot 1 = 6$, $4! = 4 \cdot 3 \cdot 2 \cdot 1 = 24$.

As we have said, the number of permutations of n things out of n is $_nP_n = n!$.

From the definitions of $n!$, $(n - r)!$, and $_nP_r$, we can obtain the following relation: $n! = [n(n - 1)(n - 2) \ldots (n - r + 1)][(n - r)(n - r - 1) \ldots 2 \cdot 1] = _nP_r \cdot (n - r)!$, and so

$$_nP_r = \frac{n!}{(n - r)!}. \tag{3.18}$$

Formulas 3.16 and 3.17 defined $_nP_r$ and $n!$ for all positive integer values of n and $r \le n$. The above formula, however, becomes meaningless for $r = n$, since then $n - r = 0$, and we have not defined $0!$. To preserve the validity of this formula for the case of $r = n$, we define $0! = 1$.

Then, for $r = n$, Formula 3.18 becomes

$$_nP_n = \frac{n!}{0!} = n!, \tag{3.19}$$

---

[5] Note that the product on the right of Equation 3.16 does not have to be taken literally as containing at least four factors. This expression is the usual way of indicating that the factors should start with n and go down in steps of 1 to $n - r + 1$. For instance, if $r = 1$, then $n - r + 1 = n$, and the product should start and end with n, that is, $_nP_1 = n$. The obvious analog of this convention is generally used for any sums or products in which a pattern is indicated, for example, in Equation 3.17 as well.

as it should be by Equation 3.16. We shall see later that by this definition many other formulas also become meaningful whenever 0! appears. We can also extend the definition of $_nP_r$ to the case of $r = 0$, by setting

$$_nP_0 = 1, \tag{3.20}$$

as required by Equation 3.18, and we can further extend the definition to $n = 0$, by defining $_0P_0 = 1$ as well.

Example 3.3.1. Dealing Three Cards.

In how many ways can three cards be dealt from a regular deck of 52 cards?

The answer is $_{52}P_3 = 52 \cdot 51 \cdot 50 = 132,600$. Notice that in this answer, the order in which the cards are dealt is taken into consideration, not only the result of the deal. Thus a deal of AS, AH, KH is counted as a case different from AH, KH, AS.   ♦

In many problems, as in the above example, for instance, it is unnatural to concern ourselves with the order in which things are selected, and we want to count only the number of different possible selections without regard to order. The number of possible *unordered* selections of r different things out of n different ones is denoted by $_nC_r$, and each such selection is called a *combination* of the given things.

To obtain a formula for $_nC_r$, we can argue the following way. If we select r things out of n without regard to order, then, as we have just said, this can be done in $_nC_r$ ways. In each case we have r things which can be ordered $r!$ ways. Thus, by the multiplication principle, the number of ordered selections is $_nC_r \cdot r!$. On the other hand, this number has been denoted by $_nP_r$. Therefore $_nC_r \cdot r! = {}_nP_r$, and so

$$_nC_r = \frac{_nP_r}{r!} = \frac{n!}{r!(n-r)!}. \tag{3.21}$$

The quantity on the right is usually abbreviated as $\binom{n}{r}$ and is called a *binomial coefficient*, for reasons that will be explained in the next section. We have, for example, $\binom{3}{2} = \frac{3!}{2!(3-2)!} = \frac{6}{2 \cdot 1} = 3$ and $\binom{7}{3} = \frac{7!}{3!4!} = \frac{7 \cdot 6 \cdot 5}{3 \cdot 2 \cdot 1} = 35$.

In the latter example, the 4! could be cancelled, and we could similarly cancel $(n-r)!$ in the general formula, as we did for $_nP_r$. Thus, for any positive integer n and $r = 1, 2, \ldots, n$,

$$_nC_r = \binom{n}{r} = \frac{n(n-1)(n-2)\cdots(n-r+1)}{r!}. \tag{3.22}$$

For $r = 0$ the cancellation, together with $0! = 1$, gives

$$_nC_0 = \binom{n}{0} = \frac{n!}{0!(n-0)!} = 1, \tag{3.23}$$

and we extend the validity of this formula to $n = 0$ as well.

The formula

$$\binom{n}{r} = \frac{n!}{r!(n-r)!} \tag{3.24}$$

remains unchanged if we replace r by $n - r$, and so

$$\binom{n}{n-r} = \binom{n}{r}. \tag{3.25}$$

This formula says that the number of combinations of $n - r$ things out of n equals the number of combinations of r things out of n. We can easily see that this must be true, since whenever we make a particular selection of $n - r$ things out of n, we are also selecting the r things that remain unselected, that is, we are splitting the n things into two sets of $n - r$ and r things simultaneously.

Example 3.3.2. Selecting Letters.

Let us illustrate the relationship between permutations and combinations, that is, between ordered and unordered selections, on a simple example, in which all cases can easily be enumerated. Say we have four letters A, B, C, D and want to select two. If order counts, then the possible selections are

$AB, AC, AD, BC, BD, CD,$

$BA, CA, DA, CB, DB, DC.$

Their number is $_4P_2 = 4 \cdot 3 = 12$. If we want to disregard the order in which the letters are selected, then AB and BA stand for the same combination, also AC and CA for another single combination, and so on. Thus the number of selections written in the first row above, that is, 6, gives us $_4C_2$. Indeed, $\binom{4}{2} = \frac{4 \cdot 3}{2 \cdot 1} = 6$. In this case, the argument we used for obtaining $_nC_r$ amounts to saying that each unordered selection gives rise to two ordered selections, and there are 12 of the latter; hence $2 \cdot {_4C_2} = 12$, and so $_4C_2 = \frac{12}{2} = 6$.

We can also look at this slightly differently: We have 12 permutations. To make them into combinations, we must identify pairs such as AB and BA with each other. Thus, the number of combinations is the number of unordered pairs into which a set of 12 objects can be partitioned, and this is, by the definition of division, $\frac{12}{2}$.   ◆

The argument above can be generalized as follows.

Division Principle: *If we have m things and k is a divisor[6] of m, then we can divide the set of m elements into m/k subsets of k elements each.*

Applied to permutations and combinations, this principle says that $m = {_nP_r}$ permutations can be grouped into subsets with $k = r!$ elements, with those permutations that have the same letters making up each subset, and

---

[6] This means that m/k is a whole number.

the number of these subsets is $\frac{nP_r}{r!}$. Since these subsets represent all the combinations, their number is, on the other hand, nC_r. Thus, the division principle can give us directly the previously obtained relationship $nC_r = \frac{nP_r}{r!}$.

Example 3.3.3. Three-Card Hands.

The number of different three-card hands from a deck of 52 cards is $_{52}C_3 = \binom{52}{3} = \frac{52 \cdot 51 \cdot 50}{3 \cdot 2 \cdot 1} = 22,100$.   ♦

Example 3.3.4. Committee Selection.

In a class there are 30 men and 20 women. In how many ways can a committee of two men and two women be chosen?

We have to choose 2 men out of 30 and 2 women out of 20. These choices can be done in $\binom{30}{2}$ and $\binom{20}{2}$ ways, respectively. By the multiplication principle, the whole committee can be selected in $\binom{30}{2} \cdot \binom{20}{2} = \frac{30 \cdot 29}{2 \cdot 1} \cdot \frac{20 \cdot 19}{2 \cdot 1} = 15 \cdot 29 \cdot 10 \cdot 19 = 82,650$ ways.   ♦

Exercises

Exercise 3.3.1.

Evaluate $_5P_2$, $_6P_3$, $_8P_1$, $_5P_0$, $_6P_6$.

Exercise 3.3.2.

How many three-letter "words" can be formed, without repetition of any letter, from the letters of the word "symbol"? (We call any permutation of letters a word.)

Exercise 3.3.3.

Prove that $n! = n \cdot (n-1)!$.

Exercise 3.3.4.

Evaluate $_5C_2$, $_6C_3$, $_8C_1$, $_5C_0$, $_6C_6$.

Exercise 3.3.5.

List all permutations of three letters taken at a time from the letters A, B, C, D, mark the groups whose members must be identified to obtain the *combinations* of three letters out of the given four, and explain how the division principle would give the number of combinations in this case.

Exercise 3.3.6.

In how many ways can a committee of 4 be formed from 10 men and 12 women if it is to have:

a) Two men and two women,
b) One man and three women,
c) Four men,
d) Four people regardless of sex?

Exercise 3.3.7.

A salesman has to visit any four of the cities A, B, C, D, E, F, starting and ending in his home city, which is other than these six. In how many ways can he schedule his trip?

Exercise 3.3.8.

A die is thrown until a 6 comes up, but only five times if no 6 comes up in five throws. How many possible sequences of numbers can come up?

Exercise 3.3.9.

In how many ways can five people be seated on five chairs around a round table if:

a) Only their positions relative to each other count (i.e., the arrangements obtained from each other by rotation of everybody are considered to be the same),
b) Only who sits next to whom counts, but not on which side (rotations and reflections do not change the arrangement)?

Exercise 3.3.10.

Answer the same questions as in Exercise 3.3.9, but for five people and seven chairs.

Exercise 3.3.11.

How many positive integers are there under 5000 that are:

a) Multiples of 3,
b) Multiples of 4,
c) Multiples of both 3 and 4,
d) Not multiples of either 3 or 4?

   (Hint: Use the division principle adjusted for divisions with remainder!)

3.4 Some Properties of Binomial Coefficients and the Binomial Theorem

The binomial coefficients have many interesting properties, and some of these will be useful to us later, so we describe them now.

If we write the binomial coefficients in a triangular array, so that $\binom{0}{0}$ goes into the first row; $\binom{1}{0}$ and $\binom{1}{1}$ into the second row; $\binom{2}{0}$, $\binom{2}{1}$, and $\binom{2}{2}$ into the third row; and so on, then we obtain the following table, called Pascal's triangle:

$$
\begin{array}{ccccccccc}
 & & & & 1 & & & & \\
 & & & 1 & & 1 & & & \\
 & & 1 & & 2 & & 1 & & \\
 & 1 & & 3 & & 3 & & 1 & \\
1 & & 4 & & 6 & & 4 & & 1 \\
\end{array}
$$

$$
\begin{array}{ccccccccccc}
1 & & 5 & & 10 & & 10 & & 5 & & 1
\end{array}
$$

$$\cdots$$

It is easy to see that each entry other than 1 is the sum of the two nearest entries in the row immediately above it; for example, the 6 in the fifth row is the sum of the two threes in the fourth row. In general, we have the following theorem.

Theorem 3.4.1. *Sums of Adjacent Binomial Coefficients.*

For any positive integers r and $n > r$,

$$\binom{n-1}{r-1} + \binom{n-1}{r} = \binom{n}{r}. \tag{3.26}$$

Proof. We give two proofs.

To prove this formula algebraically, we only have to substitute the expressions for the binomial coefficients and simplify. For $r = 1$ the left side becomes

$$\binom{n-1}{0} + \binom{n-1}{1} = 1 + (n-1) = n = \binom{n}{1}, \tag{3.27}$$

and for $r > 1$

$$
\frac{(n-1)(n-2)\ldots(n-r+1)}{(r-1)!} + \frac{(n-1)(n-2)\ldots(n-r+1)\,(n-r)}{r!}
$$

$$
= \frac{[(n-1)(n-2)\ldots(n-r+1)]\cdot r}{r\cdot(r-1)!} + \frac{[(n-1)(n-2)\ldots(n-r+1)]\cdot(n-r)}{r!}
$$

$$
= \frac{[(n-1)(n-2)\ldots(n-r+1)]\cdot(r+n-r)}{r!}
$$

$$
= \frac{(n-1)(n-2)\ldots(n-r+1)\cdot n}{r!} = \binom{n}{r}. \tag{3.28}
$$

An alternative, so-called combinatorial proof of Equation 3.26, is as follows: $\binom{n}{r}$ equals the number of ways of choosing r objects out of n. Let x denote one of the n objects. (It does not matter which one.) Then the selected r objects will either contain x or will not. The number of ways of selecting r objects with x is $\binom{n-1}{r-1} \cdot 1$, because there are $n-1$ objects other than x, and we must choose $r-1$ of those, in addition to x, which we can choose in just one way. On the other hand, the number of ways of selecting r objects without x is $\binom{n-1}{r}$, because there are $n-1$ objects other than x, and we must choose r of those. Using the addition principle for these two ways of choosing r objects out of n completes the proof.   ∎

The next topic we want to discuss is the binomial theorem.

An expression that consists of two terms is called a binomial, and the binomial theorem gives a formula for the powers of such expressions. The binomial coefficients are the coefficients in that formula, and this circumstance explains their name. Let us first see how they show up in some simple cases.

We know that

$$(a+b)^2 = a^2 + 2ab + b^2 \tag{3.29}$$

and

$$(a+b)^3 = a^3 + 3a^2b + 3ab^2 + b^3. \tag{3.30}$$

The coefficients on the right sides are 1, 2, 1 and 1, 3, 3, 1, and these are the numbers in the rows for $n = 2$ and 3 in Pascal's triangle. In general we have:

Theorem 3.4.2. *The Binomial Theorem.* *For any natural number[7] n and any numbers a, b,*

$$(a+b)^n = \binom{n}{0} a^n + \binom{n}{1} a^{n-1}b + \binom{n}{2} a^{n-2}b^2 + \cdots + \binom{n}{n} b^n = \sum_{k=0}^{n} \binom{n}{k} a^k b^{n-k}.$$

Proof. Let us first illustrate the proof for $n = 3$. Then

$$(a+b)^3 = (a+b)(a+b)(a+b), \tag{3.31}$$

and we can perform the multiplication in one fell swoop, instead of obtaining $(a+b)^2$ first and then multiplying that by $(a+b)$. When we do both multiplications simultaneously, then we have to multiply each letter in each pair of parentheses by each letter in the other pairs of parentheses and add up these

---

[7] In fact, the theorem can be extended to arbitrary real exponents as discussed in calculus courses, but then the combinatorial meaning shown in the present proof, which is what we need, is lost.

products of three factors each. Thus the products we add up are obtained by multiplying one letter from each expression in parentheses in every possible way. Since we choose from two letters three times, we have $2^3 = 8$ products such as aaa, aab, etc. to add up. Now, some of these products are equal to each other, for example, $aab = aba = baa = a^2b$. The number of ways in which we can choose the three a's from the three $(a + b)$'s is one. Thus, we have one a^3 in the result. The number of a^2b terms is $\binom{3}{1} = 3$, since we can choose the one $(a + b)$ from which the factor b comes in $\binom{3}{1}$ ways. Similarly, the number of ab^2 terms is $\binom{3}{2} = 3$, since we can choose the two $(a + b)$'s from which the two b's come in $\binom{3}{2}$ ways. Finally, we have just one b^3 term. Thus,

$$(a + b)^3 = a^3 + \binom{3}{1}a^2b + \binom{3}{2}ab^2 + b^3. \tag{3.32}$$

To make each term conform to the general pattern, we could write the first and last terms as $\binom{3}{0}a^3b^0$ and $\binom{3}{3}a^0b^3$ and write b^1 for b and a^1 for a in the second and third terms. Then, for instance, $\binom{3}{0}b^0 = 1$ means that there is only one way to select zero b's, and the product with no b is the same as the one multiplied by b^0.

In the general case of $(a + b)^n$, the result will have all possible kinds of terms in which a total of n a's and b's are multiplied together: one letter from each of the n factors $(a + b)$. If the number of a's chosen is k, then the number of b's must be $n - k$, since a total of n letters must be multiplied for each term of the result. Furthermore, the coefficient of a^kb^{n-k} must be $\binom{n}{k}$, since we can select the k factors $(a + b)$ from which we take the a's in exactly that many ways. Thus the expansion of $(a + b)^n$ must consist of terms of the form $\binom{n}{k}a^kb^{n-k}$, with k taking all possible values from 0 to n. ∎

We can of course use the binomial theorem for the expansion of binomials with all kinds of expressions in place of a and b, as in the next example.

Example 3.4.1. A Binomial Expansion.

$$\begin{aligned}
(3x - 2)^4 &= (3x + (-2))^4 \\
&= (3x)^4 + \binom{4}{1}(3x)^3(-2) + \binom{4}{2}(3x)^2(-2)^2 \\
&\quad + \binom{4}{3}(3x)(-2)^3 + (-2)^4 \\
&= 3^4x^4 - 4 \cdot 3^3 \cdot 2x^3 + 6 \cdot 3^2 \cdot 2^2x^2 - 4 \cdot 3 \cdot 2^3x + 2^4 \\
&= 81x^4 - 216x^3 + 216x^2 - 96x + 16. \tag{3.33}
\end{aligned}$$

♦

Example 3.4.2. Counting Subsets.

If we put $a = b = 1$ in the binomial theorem, then it gives

$$\binom{n}{0} + \binom{n}{1} + \binom{n}{2} + \ldots + \binom{n}{n} = (1+1)^n = 2^n. \tag{3.34}$$

This can also be seen directly from the combinatorial interpretations of the quantities involved: If we have a set of n elements, then $\binom{n}{0}$ is the number of its zero-element subsets, $\binom{n}{1}$ is the number of its one-element subsets, and so on, and the sum of these is the total number of subsets of the set of n elements, which is 2^n, as we know from Example 3.2.5. ♦

Example 3.4.3. Alternating Sum of Binomial Coefficients.

Putting $a = 1$ and $b = -1$ in the binomial theorem, we obtain

$$\binom{n}{0} - \binom{n}{1} + \binom{n}{2} - \cdots \pm \binom{n}{n} = (1-1)^n = 0. \tag{3.35}$$

This would be more difficult to interpret combinatorially; we do not do it here (but see Exercise 3.4.6). ♦

There is one other property of binomial coefficients that is important for us; we approach it by an example.

Example 3.4.4. Counting Ways for a Committee.

In Exercise 3.3.6 we asked a question about forming a committee of four people out of 10 men and 12 women. Such a committee can have either zero men and four women, or one man and three women, or two men and two women, or three men and one woman, or four men and zero women. Since these are the disjoint possibilities that make up the possible choices for the committee, regardless of sex, we can count their number on the one hand by using the addition and multiplication principles and on the other hand directly without considering the split by sex. Thus

$$\binom{10}{0}\binom{12}{4} + \binom{10}{1}\binom{12}{3} + \binom{10}{2}\binom{12}{2} + \binom{10}{3}\binom{12}{1} + \binom{10}{4}\binom{12}{0} = \binom{22}{4}. \tag{3.36}$$

♦

We can generalize this example as follows: If we have n_1 objects of one kind and n_2 objects of another kind and take a sample of r objects from these, with $r \le n_1$ and $r \le n_2$, then the number of choices can be evaluated in two ways, and we get

$$\binom{n_1}{0}\binom{n_2}{r} + \binom{n_1}{1}\binom{n_2}{r-1} + \cdots + \binom{n_1}{r}\binom{n_2}{0} = \binom{n_1+n_2}{r}. \tag{3.37}$$

Exercises

Exercise 3.4.1.

Write down Pascal's triangle to the row with $n = 10$.

Exercise 3.4.2.

Use Pascal's triangle and the binomial theorem to expand $(a + b)^6$.

Exercise 3.4.3.

Expand $(1 + x)^5$.

Exercise 3.4.4.

Expand $(2x - 3)^5$.

Exercise 3.4.5.

What would be the coefficient of x^8 in the expansion of $(1 + x)^{10}$?

Exercise 3.4.6.

Explain the formula $\binom{3}{0} - \binom{3}{1} + \binom{3}{2} - \binom{3}{3} = 0$ by using the expansion of $n(A \cup B \cup C)$ from Equation 3.7.

Exercise 3.4.7.

Use the binomial theorem to evaluate:

a) $\sum_{k=0}^{n} \binom{n}{k} 4^k$,
b) $\sum_{k=0}^{n} \binom{n}{k} x^k$ for any $x \neq 0$.

Exercise 3.4.8.

In how many ways can a committee of 4 be formed from 10 men (including Bob) and 12 women (including Alice and Claire) if it is to have two men and two women but:
a) Alice refuses to serve with Bob,
b) Alice refuses to serve with Claire,
c) Alice will serve only if Claire does, too,
d) Alice will serve only if Bob does, too?

Exercise 3.4.9.

How many subsets does a set of $n > 4$ elements have that contain:

a)   At least two elements,
b)   At most four elements?

Exercise 3.4.10.

Generalize Theorem 3.4.1 by considering two special objects x and y instead of the single object x in the combinatorial proof.

3.5 Permutations with Repetitions

Until now, we have discussed permutations of objects different from each other, except for some special cases to which we will return below. In this section, we consider permutations of objects, some of which may be identical or which amounts to the same thing of different objects that may be repeated in the permutations.

The special cases we have already encountered are the following: First, the number of possible permutations of length n out of r different objects with an arbitrary number of repetitions, that is, with any one of the r things in any one of the n places, is r^n. (e.g., the number of two letter "words" made up of a, b, or c is 3^2: $aa, ab, ac, ba, bb, bc, ca, cb, cc$.)

The second case we have seen in a disguise is that of the permutations of length n of two objects, with r of the first object and $n - r$ of the second objects chosen. The number of such permutations is obviously $_nC_r$, since to obtain any one of them, we may just select the r places out of n for the first object.

In general, if we have k different objects and we consider permutations of length n, with the first object occurring n_1 times, the second n_2 times, and so on, with the kth object occurring n_k times, then we must have $n_1 + n_2 + \cdots + n_k = n$, and the number of such permutations is

$$\frac{n!}{n_1! n_2! \cdots n_k!}. \tag{3.38}$$

This follows at once from our previous counts for permutations and the division principle. Since, if all the n objects were different, then the number of their permutations would be $n!$. When, however, we identify the n_1 objects of the first kind with each other, then we are grouping the permutations into sets with $n_1!$ members in each; and so we must divide the $n!$ by $n_1!$ to account for the indistinguishability of the objects of the first kind. Similarly, we must divide the count by $n_2!$ to reflect the indistinguishability of the n_2 objects of the second kind and so on.

The quantity above is called a *multinomial coefficient* and is sometimes denoted by the symbol

$$\binom{n}{n_1,\ n_2,\ \cdots,\ n_k}. \tag{3.39}$$

Note that for $k = 2$, the multinomial coefficient equals the corresponding binomial coefficient, that is,

$$\binom{n}{n_1,\ n_2} = \binom{n}{n_1} = \binom{n}{n_2}. \tag{3.40}$$

The reason for this relation is that when we have n_1 objects of one kind and n_2 objects of another kind, then the number of ways of arranging them in a row is the same as the number of ways of selecting the n_1 spaces for the first type from the total of $n_1 + n_2 = n$ spaces or the number of ways of selecting the n_2 spaces for the second type from the same total.

Example 3.5.1. Number of Words.

How many seven-letter words can be made up of two a's, two b's, and three c's?

Here $n = 7, k = 3, n_1 = 2, n_2 = 2$, and $n_3 = n_k = 3$. Thus the answer is

$$\binom{7}{2,\ 2,\ 3} = \frac{7!}{2! \cdot 2! \cdot 3!} = 210. \tag{3.41}$$

♦

The reason for calling the quantities above multinomial coefficients is that they occur as coefficients in a formula giving the nth power of expressions of several terms, called multinomials:

Theorem 3.5.1. Multinomial Theorem. *For any real numbers $x_1, x_2, \ldots, x_k$ and any natural number n,*

$$(x_1 + x_2 + \ldots + x_k)^n = \sum \binom{n}{n_1,\ n_2,\ \cdots,\ n_k} x_1^{n_1} x_2^{n_2} \cdots x_k^{n_k}, \tag{3.42}$$

with the sum taken over all nonnegative integer values $n_1, n_2, \ldots, n_k$ such that $n_1 + n_2 + \cdots + n_k = n$.

The proof of this theorem is omitted; it would go much like the proof of the binomial theorem.

Example 3.5.2. A Multinomial Expansion.

$$(x + y + z)^4 = x^4 + y^4 + z^4 + 4(x^3y + xy^3 + x^3z + xz^3 + y^3z + yz^3)$$
$$+ 6(x^2y^2 + x^2z^2 + y^2z^2) + 12(x^2yz + xy^2z + xyz^2), \quad (3.43)$$

since

$$\binom{4}{4,\ 0,\ 0} = \frac{4!}{4! \cdot 0! \cdot 0!} = 1, \quad \binom{4}{3,\ 1,\ 0} = \frac{4!}{3! \cdot 1! \cdot 0!} = 4,$$

$$\binom{4}{2,\ 2,\ 0} = \frac{4!}{2! \cdot 2! \cdot 0!} = 6, \quad \binom{4}{2,\ 1,\ 1} = \frac{4!}{2! \cdot 1! \cdot 1!} = 12, \qquad (3.44)$$

and permuting the numbers in the lower row in any multinomial coefficient leaves the latter unchanged.   ♦

In closing this section, let us consider a problem that can be reduced to one of counting permutations with two kinds of indistinguishable objects:

Example 3.5.3. Placing Indistinguishable Balls Into Distinguishable Boxes.

In how many ways can k indistinguishable[8] balls be distributed into n different boxes?

If there are $k = 2$ balls and $n = 3$ boxes, then the possible distributions can be listed as ordered triples of nonnegative whole numbers that give the numbers of balls in the boxes. The numbers of each triple must add up to two, since we are distributing two balls. Thus the possible distributions are $(2, 0, 0)$, $(0, 2, 0)$, $(0, 0, 2)$, $(1, 1, 0)$, $(1, 0, 1)$, and $(0, 1, 1)$, and so in this case, the answer is 6.

In the general case, the problem can be solved by the following trick:

Each distribution can be represented by a sequence of circles and bars, with the circles representing the balls and the bars the walls of the boxes (we put only one bar as a wall between two boxes). For instance, Figure 3.5 shows the distribution $(0, 3, 1, 2, 0, 2)$ of $k = 8$ balls into $n = 6$ boxes arranged in a row.

Fig. 3.5. Distribution of 8 balls in 6 boxes

---

[8] Actually, the balls may be distinguishable, but we may not want to distinguish them. In some applications, for instance, involving distribution of money to people, all we care about is how many dollars someone gets, not which dollar bills.

Now, if there are six boxes, then we have seven bars. Two of those must be fixed at the ends, and the remaining five can have various positions among the eight balls. Thus, out of $5 + 8 = 13$ positions for balls and bars together, we must choose 5. We can do this in $\binom{13}{5} = 1287$ ways.

In general, if we have n boxes, then we can choose the positions of $n - 1$ bars freely. Thus, the problem becomes that of counting the number of permutations of $n - 1$ bars and k circles. We know that the number of such permutations is $\binom{n-1+k}{n-1,\ k} = \binom{n-1+k}{k} = \binom{n-1+k}{n-1}$. This expression is the answer to our question. If $k = 2$ and $n = 3$, then it becomes $\binom{3-1+2}{2} = 6$, as we have seen above by a direct enumeration.   ♦

Exercises

Exercise 3.5.1.

In how many ways can we form six-letter words:

a) From a's and/or b's,
b) From two a's and four b's,
c) From two a's, one b, and three c's,
d) From two a's and four letters each of which may be b or c?

Exercise 3.5.2.

On how many paths can a rook move from the lower left corner of a chessboard to the diagonally opposite corner by moving only up or to the right at each step?

Exercise 3.5.3.

a) How many permutations are there of the letters of the word *"success"*?
b) How many of the above have exactly three s's together (Hint: Consider *sss* as if it were a single letter.)
c) How many have two or three s's together? (Hint: Regard *ss* as a single letter)
d) How many have exactly two s's together?

Exercise 3.5.4.

Prove, both algebraically and combinatorially (i.e., in terms of selections) that if $n_1 + n_2 + n_3 = n$, then $\binom{n}{n_1,\ n_2,\ n_3} = \binom{n}{n_1}\binom{n-n_1}{n_2}$.

Exercise 3.5.5.

What is the coefficient of:

a) $a^2b^3c^2$ in the expansion of $(a+b+c+d)^7$,
b) $a^2b^3c^2$ in the expansion of $(2a-3b+c-d)^7$?

Exercise 3.5.6.

Expand $(2+3+1)^4$ by the multinomial theorem, and show that the terms add up to $6^4 = 1,296$.

Exercise 3.5.7.

a) In how many ways can 10 cents be distributed among three children? (All that matters is how much each child gets and not which coins, that is, cents are considered indistinguishable.)
b) In how many ways if each child is to get at least one cent? (*Hint*: From the spaces between circles, choose some for bars, or first give 1 cent to each and then distribute the remaining 7 cents.)

Exercise 3.5.8.

In how many ways can k indistinguishable balls be distributed into $n \leq k$ different boxes if each box is to get at least one ball? (Hint: From the spaces between circles, choose some for bars.)

Exercise 3.5.9.

In how many ways can k indistinguishable balls be distributed into $n \geq k$ different boxes if no box is to get more than one ball?

Exercise 3.5.10.

How many distinct terms are there in the multinomial expansions of:

a) $(a+b+c)^6$
b) $(a+b+c+d)^5$?

Explain!
    (*Hint*: Use Example 3.5.3.)

4. Probabilities

4.1 Relative Frequency and the Axioms of Probabilities

We begin our discussion of probabilities with the definition of relative frequency, because this notion is very concrete and probabilities are, in a sense, idealizations of relative frequencies.

Definition 4.1.1. *Relative Frequency*. *If we perform an experiment n times (each performance is called a trial) and the event A occurs in n_A trials, then the ratio $\frac{n_A}{n}$ is called the* relative frequency *of A in the n trials and will be denoted by f_A.*

For example, if we toss a coin $n = 100$ times and observe heads $n_H = 46$ times, then the relative frequency of heads in those trials is $f_H = \frac{n_H}{n} = \frac{46}{100} = .46$.

For two mutually exclusive events A and B the relative frequency of $A \cup B$ in n trials turns out to be the sum of the relative frequencies of A and B, because $n_{A \cup B} = n_A + n_B$, by the addition principle, and so $f_{A \cup B} = f_A + f_B$.

As mentioned in the Introduction, we assign probabilities to events in such a way that the relative frequency of an event in a large number of trials should approximate the probability of that event. We can expect this to happen only if we define probabilities so that they have the same basic properties that relative frequencies have. Thus we make the following definition.

Definition 4.1.2. *Probabilities*. *Given a sample space S and a certain collection $\mathcal{F}$ of its subsets, called events,[1] an assignment P of a number $P(A)$ to each event A in $\mathcal{F}$ is called a* probability measure *and $P(A)$ the* probability *of A, if P has the following properties:*

1. $P(A) \geq 0$ for every A,

---

[1] If S is a finite set, then the collection $\mathcal{F}$ of events is taken to be the collection of all subsets of S. If S is infinite, then $\mathcal{F}$ must be a so-called sigma-field, which we do not discuss here.

© Springer International Publishing Switzerland 2016
G. Schay, *Introduction to Probability with Statistical Applications*,
DOI 10.1007/978-3-319-30620-9_4

2. $P(S) = 1$,
3. $P(A_1 \cup A_2 \cup \cdots) = P(A_1) + P(A_2) + \cdots$ *for any finite or countably infinite set of mutually exclusive events* $A_1, A_2, \ldots$.

The sample space S together with $\mathcal{F}$ and P is called a *probability space*.

The properties of P in the definition are also called the *axioms* of the theory. Furthermore, if Axiom 3 were stated for merely two sets, then from that form, it could be proved for an arbitrary finite number of sets (finite additivity) by mathematical induction but not for an infinite number (countable additivity), which we also need.

From this definition, several other important properties of probabilities follow rather easily, which we give as theorems. *In each of these theorems, an underlying arbitrary probability space will be tacitly understood.*

Theorem 4.1.1. *The Probability of the Empty Set Is 0.*

In every probability space, $P(\emptyset) = 0$.

Proof. Consider an event A. Then $A \cup \emptyset = A$, and A and $\emptyset$ are mutually exclusive, since $A \cap \emptyset = \emptyset$. Hence $P(A \cup \emptyset) = P(A)$ on the one hand, and on the other, by Property 3 applied to $A_1 = A$ and $A_2 = \emptyset$, $P(A \cup \emptyset) = P(A) + P(\emptyset)$. Thus $P(A) = P(A) + P(\emptyset)$, and so $P(\emptyset) = 0$. ∎

Note, however, that the empty set need not be the only set with zero probability, that is, in some probability spaces we have events $A \neq \emptyset$ for which $P(A) = 0$. There is nothing in the axioms that would prevent such an occurrence. In fact, such events need not be impossible. For instance, if the experiment consists of picking a point at random (that is, with equal probabilities) from the interval $[0, 1]$ of real numbers, then each number must have zero probability, because otherwise Axiom 3 would imply that the sum of the probabilities of an infinite sequence of such numbers is infinite, in contradiction to Axiom 2. (Why?) To make useful probability statements in this case, we assign probabilities to subintervals of nonzero length of $[0, 1]$, rather than to single numbers. Details will be discussed in Chapter 5 and thereafter.

Theorem 4.1.2. *The Probability of the Union of Two Events.*

For any two events A and B,

$$P(A \cup B) = P(A) + P(B) - P(AB). \tag{4.1}$$

Proof. $A \cup B = A \cup \overline{A}B$ with A and $\overline{A}B$ disjoint.(See Figure 3.1.) Thus, by Axiom 3,

$$P(A \cup B) = P(A) + P(\overline{A}B). \tag{4.2}$$

Similarly,

$$P(\overline{A}B) + P(AB) = P(B) \tag{4.3}$$

Adding the two equations and canceling $P\left(\overline{A}B\right)$, we get

$$P(A \cup B) + P(AB) = P(A) + P(B), \tag{4.4}$$

which is equivalent to the formula of the theorem. ∎

Theorem 4.1.3. *The Probability of the Union of Three Events.*

For any three events,

$$P(A \cup B \cup C) = P(A) + P(B) + P(C) - P(AB) - P(AC) - P(BC) + P(ABC).$$

Proof. We apply Theorem 4.1.2 three times:

$$
\begin{aligned}
P(A &\cup B \cup C)\\
&= P(A \cup (B \cup C)) = P(A) + P(B \cup C)) - P(A(B \cup C))\\
&= P(A) + P(B) + P(C) - P(BC) - P(AB \cup AC)\\
&= P(A) + P(B) + P(C) - P(BC) - [P(AB) + P(AC) - P(ABAC)]\\
&= P(A) + P(B) + P(C) - P(AB) - P(AC) - P(BC) + P(ABC).
\end{aligned}
\tag{4.5}
$$

∎

Theorem 4.1.4. *Probability of Complements.*

For any event A,

$$P(\overline{A}) = 1 - P(A). \tag{4.6}$$

Proof. $\overline{A} \cap A = \emptyset$ and $\overline{A} \cup A = S$ by the definition of $\overline{A}$. Thus, by Axiom 3, $P(S) = P\left(\overline{A} \cup A\right) = P(\overline{A}) + P(A)$. Now, Axiom 2 says that $P(S) = 1$, and so, comparing these two values of $P(S)$, we obtain $P(\overline{A}) + P(A) = 1$. ∎

Theorem 4.1.5. *Probability of Subsets.*

If two events A and B satisfy $A \subset B$, then $P(A) \leq P(B)$.

Proof. If $A \subset B$, then $B = A \cup (B \cap \overline{A})$, with A and $B \cap \overline{A}$ being disjoint. Thus, by Property 3, $P(B) = P(A) + P(B \cap \overline{A})$, and by Property 1, $P(B \cap \overline{A}) \geq 0$. Therefore $P(B)$ is $P(A)$ plus a nonnegative quantity, and so greater than or equal to $P(A)$. ∎

Corollary 4.1.1. $P(A) \leq 1$ *for all events A.*

Proof. In Theorem 4.1.5, take $B = S$. Since $A \subset S$ for every event A and $P(S) = 1$ by Axiom 2, Theorem 4.1.5 gives $P(A) \leq 1$.   ∎

Example 4.1.1. Drawing a Card.

For the drawing of a card at random from a deck of 52 cards, we consider the sample space S made up of the 52 elementary events corresponding to the 52 possible choices of drawing any one of the cards. We assign $\frac{1}{52}$ as the probability of each of the elementary events, and for any compound event A, we define its probability $P(A)$ as the number $n(A)$ of the elementary events that make up A times $\frac{1}{52}$, that is, as

$$P(A) = n(A) \cdot \frac{1}{52}. \tag{4.7}$$

For example, the probability of drawing a spade is $13 \cdot \frac{1}{52} = \frac{1}{4}$, since there are 13 spades and the drawing of each spade is an elementary event, the 13 of which make up the event $A = \{$a spade is drawn$\}$.

It is easy to verify our axioms for this case:

1. Obviously, the assignment, Equation 4.7, makes every $P(A)$ nonnegative.
2. $P(S) = 1$, since S is made up of all the 52 elementary events, and so $P(S) = 52 \cdot \frac{1}{52} = 1$.
3. By Theorem 3.1.1, for k pairwise disjoint sets $A_1, A_2, \ldots, A_k$, Equation 3.2 gives

$$\frac{n(A_1 \cup A_2 \cup \cdots \cup A_k)}{52} = \frac{n(A_1)}{52} + \frac{n(A_2)}{52} + \cdots + \frac{n(A_k)}{52} \tag{4.8}$$

and, by Equation 4.7, Equation 4.8 becomes

$$P(A_1 \cup A_2 \cup \cdots \cup A_k) = P(A_1) + P(A_2) + \cdots + P(A_k). \tag{4.9}$$

◆

Much as we did in the special case of the above example, we can prove

Theorem 4.1.6. *Assignment of Probabilities in a Finite Sample Space.*

In a finite sample space, we obtain a probability measure by assigning nonnegative numbers whose sum is 1 as probabilities of the elementary events and, for general A, by taking the sum of the probabilities of the elementary events that make up A as $P(A)$.

Example 4.1.2. An Assignment of Unequal Probabilities.

Let $S = \{s_1, s_2, s_3, s_4\}$ and assign probabilities to the elementary events[2] as $P(s_1) = \frac{1}{2}$, $P(s_2) = \frac{1}{3}$, $P(s_3) = \frac{1}{6}$, $P(s_4) = 0$, and, for general A, take $P(A)$ as the sum of the probabilities of the elementary events that make up A. For instance, if $A = \{s_1, s_2\}$, then take $P(A) = \frac{1}{2} + \frac{1}{3} = \frac{5}{6}$. We could easily verify the axioms for this assignment.

How could we realize an experiment that corresponds to this probability space? One way of doing this would be to consider picking a number at random from the interval $[0, 1]$ of real numbers (as random number generators do on computers, more or less) and letting $s_1 = [0, \frac{1}{2})$, $s_2 = [\frac{1}{2}, \frac{5}{6})$, $s_3 = [\frac{5}{6}, 1)$, $s_4 = \{1\}$. ◆

Theorem 4.1.6 has a very important special case, which we state as a corollary:

Corollary 4.1.2. *If a probability space consists of n elementary events of equal probability, then this common probability is $\frac{1}{n}$ and, if an event A is the union of k elementary events, then $P(A) = \frac{k}{n}$.*

If the elementary events have equal probability, then we say that we are choosing one of them *at random*. Also, it is customary to call the k outcomes that make up A the outcomes *favorable* to A and to call n the *total* number of possible outcomes. Thus, for equiprobable elementary events, the assignment can be summarized as

$$P(A) = \frac{favorable}{total}. \tag{4.10}$$

For a long time, this formula was considered to be the definition of $P(A)$ and is still called the classical definition of probabilities. Example 4.1.1 provided an illustration of this: The probability of drawing a spade from a deck of 52 cards, if one card is drawn at random (i.e., with equal probability for each card), is $\frac{13}{52}$, since $k = 13$ and $n = 52$.

Note, however, that the probability of drawing a spade is not $\frac{13}{52}$ under all conditions. Corollary 4.1.2 ensures this value only if all cards have the same probability of being drawn, which will not be true if the deck is not well shuffled or we use some special method of drawing. In fact, there is no way of proving that all cards must have the same probability of being drawn, no matter how we do the shuffling and drawing. The equal probabilities in this case are assignments based on our experience. In every case, some probabilities must somehow be assigned, and the theory is only intended to show how to calculate certain probabilities and related quantities from others (also see the Introduction). For instance, Theorem 4.1.6 and its corollary tell us

---

[2] It is customary to omit the braces in writing the probabilities of the elementary events, such as writing $P(s_1)$ instead of the correct, but clumsy, $P(\{s_1\})$.

how to calculate the probabilities of compound events from those of the elementary events. Thus, the so-called classical definition of probabilities is *not really a definition* by present-day standards, but a very useful formula for the calculation of probabilities in many cases.

Example 4.1.3. (The Monty Hall Problem).

The Monty Hall problem is a brain teaser, in the form of a probability puzzle, loosely based on the American television game show Let's Make a Deal and named after its original host, Monty Hall. The problem was originally posed in a letter by Steve Selvin to the *American Statistician* in 1975. It became famous as a question from a reader's letter quoted in Marilyn vos Savant's "Ask Marilyn" column in Parade magazine in 1990 and generated thousands of letters to her and a huge literature because of the counterintuitive nature of the solution.[3]

Suppose you're on a game show, and you're given the choice of three closed doors: Behind one door is a car; behind the others are goats. You pick a door, say No. 1, and the host, who knows what's behind the doors, opens another door, say No. 3, which has a goat. He then says to you, "Do you want to switch your pick to door No. 2?" What's behind your final choice is yours.

Is it to your advantage to switch your choice?

It is tacitly assumed that the host will always open a door with a goat, and if your pick hides the car, then he will open one of the other doors with probability 1/2 for each. (We are going to return to this problem in Example 4.4.4 and in Exercise 4.5.15 with the conditions modified.)

Vos Savant's response was that the player should switch to the other door. Under the standard assumptions, players who switch have a 2/3 chance of winning the car, while players who stick to their choice have only a 1/3 chance. This is the correct answer as we shall explain below. However, most of the readers' comments were wrong and some very rude, like this one:

> "You blew it, and you blew it big! Since you seem to have difficulty grasping the basic principle at work here, I'll explain. After the host reveals a goat, you now have a one-in-two chance of being correct. Whether you change your selection or not, the odds are the same. There is enough mathematical illiteracy in this country, and we don't need the world's highest IQ propagating more. Shame!" – Scott Smith, Ph.D., University of Florida

Many people were hard to convince that the quoted argument is wrong: Phrased more precisely, it says: when the decision must be made, there are two closed doors, and each one is just as likely as the other to conceal the car. So, it does not matter whether you switch or not; either way P(win car) = 1/2.

---

[3] See https://en.wikipedia.org/wiki/Monty_Hall_problem

The fallacy lies in the assumption that after seeing a goat behind door No. 3, the car is just as likely to be behind door No. 1 as behind door No. 2.

But that would only be true if the host were to go first in opening door No. 3. The fact that the player goes first to select a door skews the chances because the host then cannot open the player's selection. The host is left with the choice of opening one of the two doors 2 or 3. In one case, when door 1 hides the car, he has the choice of two doors with goats, and in two cases, that is, when the car is behind door 2 or door 3, he has the choice of one door with a goat and one with the car. Since he must reveal a goat, in the latter two out of the three equally likely cases (of the car being behind doors 1, 2, or 3), he has to leave a door with the car to the player. Thus, we see that switching doors leads the player to the car with probability $2/3$, while staying with the original choice gets the car only with probability $1/3$.   ♦

Exercises

Exercise 4.1.1.

We draw a card at random from a deck of 52. Let $A = \{$the card drawn is a spade$\}$, $B = \{$the card drawn is a face card$\}$, $C = \{$the card drawn is a King$\}$. Find:

a) $P(A)$,
b) $P(B)$,
c) $P(C)$,
d) $P(A \cap B)$,
e) $P(A \cup B)$,
f) $P(B \cap C)$,
g) $P(\overline{B} \cap C)$,
h) $P(\overline{B} \cup \overline{C})$.

Exercise 4.1.2.

We throw two dice as in Example 2.3.3, a black one and a white one. If b denotes the result of the throw of the black die and w that of the white die, then let $A = \{b + w = 7\}, B = \{b \leq 3\}, C = \{w > 4\}$. Find the eight probabilities listed in Exercise 4.1.1 above, but with this assignment of A, B, and C.

Exercise 4.1.3.

Let A, B, C be arbitrary events in a probability space. Use the Venn diagram, Figure 2.3, to prove that the probability that exactly one of them occurs is $P(A) + P(B) + P(C) - 2[P(AB) + P(AC) + P(BC)] + 3P(ABC)$.

Exercise 4.1.4.

Prove that, for any two events A and B, $P(AB) \geq P(A) + P(B) - 1$.

Exercise 4.1.5.

When is $P(A - B) = P(A) - P(B)$? Prove your answer, paying attention to events with zero probability other than $\emptyset$ (see, e.g., Example 4.1.2.)

Exercise 4.1.6.

As we have seen in Exercise 2.3.7, the expression $A \triangle B = A\overline{B} \cup \overline{A}B$ is called the symmetric difference of A and B and corresponds to the "exclusive or" of the corresponding statements, that is, to "one or the other but not both."

1. Find an expression for $P(A \triangle B)$ in terms of $P(A), P(B)$, and $P(AB)$ and prove it.
2. Prove that this operation satisfies a "triangle inequality": For any three events, $P(A \triangle B) \leq P(A \triangle C) + P(C \triangle B)$.

Exercise 4.1.7.

Prove, for arbitrary events and any integer $n > 1$,

a) $P(A \cup B) \leq P(A) + P(B)$,
b) $P(A \cup B \cup C) \leq P(A) + P(B) + P(C)$,
c) $P(\bigcup_{i=1}^{n} A_i) \leq \sum_{i=1}^{n} P(A_i)$ for $n = 2, 3, \ldots$. (Boole's inequality)

Exercise 4.1.8.

Consider the sample space $S = \{a, b, c, d\}$ and assign probabilities to the elementary events as $P(\{a\}) = \frac{1}{7}, P(\{b\}) = \frac{2}{7}, P(\{c\}) = \frac{4}{7}, P(\{d\}) = 0$.

a) Compute the probabilities of all compound events, as described in Theorem 4.1.6.
b) Find two nonempty sets A and B from the S above such that $AB \neq B$, but $P(AB) = P(B)$.

4.2 Probability Assignments by Combinatorial Methods

In this section, we consider several examples of probability assignments to complex events, under the assumption that the elementary events are equiprobable. In such problems, with coins, cards, and dice, it is always assumed that all elementary events are equally likely. Thus, we use the classical definition and, because of the complexity of the problems, the combinatorial methods developed in Chapter 3.

Example 4.2.1. Probability of Drawing Two Given Cards.

We draw two cards from a deck of 52 without replacement. What is the probability of drawing a King and an Ace without regard to order?

We solve this problem in two ways, by using two different sample spaces.

First, we use S = set of all ordered pairs of Aces and Kings. Then the total number of ways of drawing two cards with regard to order is $52 \cdot 51$, and there are 4^2 ways of drawing a King first and an Ace second, and another 4^2 ways of drawing an Ace first and a King second. Thus,

$$P(K \text{ and } A) = \frac{2 \cdot 4^2}{52 \cdot 51}. \tag{4.11}$$

The other way to solve this problem is to start by disregarding the order from the outset, that is, by using S = set of all unordered pairs of Aces and Kings. Then the total number of possible outcomes is $\binom{52}{2}$, which are again equally likely, and the number of ways of choosing one King out of four is $\binom{4}{1}$ and of one Ace out of four also $\binom{4}{1}$. Thus,

$$P(K \text{ and } A) = \frac{\binom{4}{1}\binom{4}{1}}{\binom{52}{2}} = \frac{4^2}{(52 \cdot 51)/(2 \cdot 1)} = \frac{2 \cdot 4^2}{52 \cdot 51}, \tag{4.12}$$

the same as before.   ◆

Example 4.2.2. Probability of Head and Tail.

We toss two coins. What is the probability of obtaining one head and one tail?

If we denote the outcomes of the toss of the first coin by H_1 and T_1, those of the second by H_2 and T_2, then the possible outcomes of the toss of both are the sets $\{H_1, H_2\}$, $\{H_1, T_2\}$, $\{T_1, H_2\}$, $\{T_1, T_2\}$. These outcomes are equally probable, and the second and third ones are the favorable ones. Thus P(one H and one T) $= \frac{2}{4} = \frac{1}{2}$.

For two successive tosses of a single coin instead of simultaneous tosses of two coins, the possible outcomes could be listed exactly the same way, with H_1 and T_1 denoting the result of the first toss, and H_2 and T_2 that of the second toss, or more simply as HH, HT, TH, and TT, and so the probabilities remain the same.

Notice, that we cannot solve this problem by the alternate method of combining the ordered pairs into unordered ones, as in Example 4.2.1, since $\{HH\}, \{HT, TH\}$, and $\{TT\}$ are not equally likely. Their probabilities are $\frac{1}{4}, \frac{1}{2}$, and $\frac{1}{4}$, respectively. By ignoring the inequality of the probabilities of the elementary events, we would get P(one H and one T) $= \frac{1}{3}$, which is incorrect.
♦

Example 4.2.3. Six Throws of a Die.

A die is thrown six times. What is the probability of obtaining at least one six?

It is easiest to calculate this probability by using Theorem 4.1.4, that is, from P(at least one six) = 1 - P(no six). Now the total number of possible (ordered) outcomes is 6^6, and since on each throw there are 5 ways of obtaining something other than six, in six throws, we can get numbers other than six in 5^6 ways. Thus, P(at least one six) $= 1 - 5^6/6^6 \approx .665$.

Notice, that in this problem, as in the previous one, we must use ordered outcomes, because the unordered ones would not be equally likely, which is a prerequisite for computing probabilities by the classical definition.   ♦

Example 4.2.4. Sampling Good and Bad Items without Replacement.

In a batch of N manufactured items, there are N_1 good ones and N_2 defective ones, with $N_1 + N_2 = N$. We choose a random sample of n items *without replacement*, that is, once an item is chosen, we take it out of the pool from which the next items are picked. Here n is called the *size of the sample and N the size of the population*. We ask: what is the probability of the sample having n_1 good items and n_2 bad ones, where $n_1 + n_2 = n$?

We solve this problem with unordered selections. (It could be done with ordered selections as well, see Exercise 4.2.20.) The total number of equally probable ways of choosing n items out of N different ones is $\binom{N}{n}$; the number of ways of choosing n_1 good ones out of N_1 is $\binom{N_1}{n_1}$, and that of n_2 defectives out of N_2 is $\binom{N_2}{n_2}$. Thus, the required probability is given by

$$P(n_1; n, N_1, N_2) = \frac{\binom{N_1}{n_1}\binom{N_2}{n_2}}{\binom{N}{n}}. \tag{4.13}$$

We have used the notation $P(n_1; n, N_1, N_2)$ for this probability, since it is the probability of the sample containing n_1 good items under the given experimental data of sample size n, and N_1 and N_2 good and bad items in the total population (n_2 is given by $n - n_1$). The variable n_1 can take on any nonnegative integer value satisfying $n_1 \leq n$, $n_1 \leq N_1$ and $n - n_1 \leq N_2$, that is, $\max(0, n - N_2) \leq n_1 \leq \min(n, N_1)$. (See Exercise 4.2.21.) Since the events described by the different values of n_1 are mutually exclusive and their union is the sure event, the above probabilities sum to 1 as n_1 varies from max$(0, n - N_2)$ to min(n, N_1), for any fixed values of n, N_1, and N_2. (See also Equation 3.37.) Thus, the above formula describes how the total probability 1 is distributed over the events corresponding to the various values of n_1.

Whenever we give the probabilities of disjoint events whose union is the whole sample space, we call such an assignment of probabilities a *probability distribution*. The distribution just given is called the *hypergeometric distribution* with parameters n, N_1, and N_2.   ♦

Example 4.2.5. Sampling with Replacement.

Let us modify the previous problem by asking what the probability of obtaining n_1 good items is if we choose a random sample of n items *with replacement* from N_1 good and N_2 bad ones, that is, we choose one item at a time, note whether it is good or bad, and replace it in the population before choosing the next one.

Since in each of the n steps of the sampling we have $N = N_1 + N_2$ items to choose from, the total number of equally probable elementary events is N^n. Next, we have to count how many of these are favorable, that is, how many elementary events have n_1 good items and n_2 bad ones. Now, at each of the n steps of the sampling, we can choose either a good or a bad item, but in n_1 of them, we must choose a good one. We can choose these n_1 steps in $\binom{n}{n_1}$ ways. Then at each of these n_1 steps, we have a choice of N_1 items and at each of the remaining $n_2 = n - n_1$ steps a choice of N_2 items, for a total of $N_1^{n_1} \cdot N_2^{n_2}$ choices. Thus the required probability is

$$f(n_1; n, N_1, N_2) = \frac{\binom{n}{n_1} N_1^{n_1} N_2^{n_2}}{N^n}. \tag{4.14}$$

If we write $N^n = N^{n_1 + n_2} = N^{n_1} N^{n_2}$ and replace n_2 by $n - n_1$, then we can write the above formula as

$$f(n_1; n, N_1, N_2) = \binom{n}{n_1} \cdot \left(\frac{N_1}{N}\right)^{n_1} \left(\frac{N_2}{N}\right)^{n - n_1}. \tag{4.15}$$

Here N_1/N is the probability of choosing a good item at any given step and N_2/N is that of choosing a bad item. It is customary to denote these probabilities by p and q (with $p + q = 1$), and then the required probability of obtaining n_1 good items can be written as

$$f(n_1; n, p) = \binom{n}{n_1} p^{n_1} q^{n - n_1}. \tag{4.16}$$

Since these probabilities are the terms of the expansion of $(p + q)^n$ by the binomial theorem and they are the probabilities of disjoint events (the different values of n_1) whose union is the sure event, they are said to describe the so-called *binomial distribution* with parameters n and p.

It is easy to check that, indeed,

$$\sum_{n_1=0}^{n} \binom{n}{n_1} p^{n_1} q^{n-n_1} = (p+q)^n = 1. \tag{4.17}$$

♦

Example 4.2.6. The Birthday Problem.

What is the probability that at least two people, out of a given set of n persons, for $1 \leq n \leq 365$, have the same birthday? Disregard February 29 and assume that the 365^n possible birthday combinations are equally likely.

If all the n persons had different birthdays, then there would be 365 choices for the birthday of the first person, 364 for that of the second, and so on. Thus,

$$\text{P(at least two have same birthday)} = 1 - \frac{365P_n}{365^n}.$$

It is interesting and very surprising that this probability is about 0.5 for as few as 23 people and about 0.99 for 60 people. ♦

Example 4.2.7. The Ballot Problem.

Consider an election in which two candidates A and B run for an office and A wins with a total of m votes over B's $n < m$ votes. What is the probability that A is ahead throughout the counting of the ballots? Although this problem had been posed and solved earlier by W. A. Whitworth and J. Bertrand, we present a simplified version of the solution given by D. André from 1887, utilizing the so-called *reflection principle*.

Let us illustrate the vote-counting process by a polygonal path in the xy coordinate system that starts at the origin and ends at (m, n), with a unit step to the right for each vote for A and a unit step up for each vote for B. (See Figure 4.1.) We assume that these paths are all equally likely.

A path for which A is ahead throughout the count is under the $y = x$ line, except for its starting point. We shall call such paths good and all others bad. Clearly, all good paths from the origin O must go through the point $(1, 0)$, and those that start through $(0, 1)$ are all bad and they must meet the $y = x$ line somewhere, because $(0, 1)$ is above the $y = x$ line and the path ends at (m, n) under that line. On the other hand, a bad path through $(1, 0)$ must also touch or cross the $y = x$ line. Call the first point where this happens P. Every path through $(0, 1)$ can be paired one-to-one with a bad path through $(1, 0)$ by reflecting its section between O and P across the $y = x$ line. Thus,

$$\text{P (path is bad and starts through } (1, 0)) = \text{P (path starts through } (0, 1))$$

$$= \frac{n}{m+n}. \tag{4.18}$$

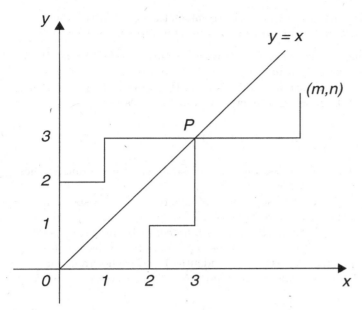

Fig. 4.1. Each bad path from O to (m, n) that starts horizontally must meet the $y = x$ line at some point P and can be paired with a path through P that starts vertically.

This relation is called the reflection principle.

Now, since the bad paths from the origin must start either north or east and those that start north are all bad,

$$P\,(\text{bad}) = P\,(\text{path starts through }(0, 1))$$

$$+ P\,(\text{path is bad and starts through }(1, 0)) = \frac{2n}{m + n}. \quad (4.19)$$

Hence the probability that A is ahead throughout the counting of the ballot is

$$P\,(\text{good}) = 1 - \frac{2n}{m + n} = \frac{m - n}{m + n} = \frac{1 - (n/m)}{1 + (n/m)}.$$

Notice that, interestingly, this probability depends only on the ratio n/m and not on the sizes of m and n separately. Thus, even if a million votes are cast, the probability that A is ahead of B throughout the counting of the ballots is the same as if there were only a hundred ballots, as long as A wins with the same ratio.   ◆

Example 4.2.8. Particle Distributions in Physics.

In statistical physics, the problem of randomly distributing k particles into n distinguishable cells has been considered under three different assumptions about the nature of the particles. The cells are the possible states of the system such as the various modes of electromagnetic waves in the theory of heat radiation or the quantum states of electrons in atoms.

1. Historically, the first case considered was the *Maxwell-Boltzmann* distribution (or "statistics" as physicists call it), in which the particles are assumed to be distinguishable. In this case, the probability of obtaining $k_1, k_2, \ldots, k_n \geq 0$ particles, respectively, in $n \geq 1$ cells is

$$\frac{k!}{k_1! k_2! \cdots k_n!} \cdot \frac{1}{n^k}, \tag{4.20}$$

   where $k_1 + k_2 + \ldots + k_n = k$. This distribution, as it turned out, applies only to macroscopic particles.

2. To everybody's great surprise, it was discovered that the spectrum of heat radiation can only be explained by the *Bose-Einstein* distribution (statistics), in which the particles are assumed to be indistinguishable (roughly, because the particles may be represented by waves and when those waves fuse the particles lose their identity), and each arrangement of $k > 0$ particles in $n > 0$ cells has the same probability (see Example 3.5.3)

$$\frac{1}{\binom{k+n-1}{k}}. \tag{4.21}$$

   This distribution applies to photons and atoms or ions that contain an even number of elementary particles.

3. Experiments have lead physicists to consider a third type of distribution as well, the *Fermi-Dirac* distribution (statistics), in which the particles are again assumed to be indistinguishable but no more than one particle can occupy a cell. So there are $\binom{n}{k}$ possible arrangements for $k \leq n$, which are assumed to be equiprobable, that is, each one having probability

$$\frac{1}{\binom{n}{k}}. \tag{4.22}$$

   This distribution applies to electrons, protons, neutrons, and atoms or ions that contain an odd number of elementary particles.

Consider, for instance, the arrangement shown in Figure 4.2 as in

Fig. 4.2. Distribution of 8 balls in 6 boxes

Example 3.5.3, where we have distributed $k = 8$ particles into $n = 6$ cells. Its probability is

$$\frac{8!}{0!3!1!2!0!2!} \cdot \frac{1}{6^8} \approx 10^{-3} \tag{4.23}$$

with the Maxwell-Boltzmann distribution, and

$$\frac{1}{\binom{8+6-1}{8}} = \frac{1}{1287} \approx 7.77 \times 10^{-4} \tag{4.24}$$

with the Bose-Einstein distribution. It is impossible with the Fermi-Dirac distribution. ♦

Example 4.2.9. Seating Men and Women.

m men and n women are seated at random in a row on $m+n$ chairs, with $m \leq n$. What is the probability that no men sit next to each other?

The total number of possible arrangements is $(m + n)!$ and the number of favorable arrangements can be obtained as follows.

Consider any arrangement of the n women in a row. Then there are $n+1$ spaces between or around them, from which we must choose m for the men. Thus, we have $\binom{n+1}{m}$ choices for the seats of the men once the women's order is set. For any of the just counted choices, the men can be ordered in $m!$ ways and the women in $n!$ ways, and so the number of favorable arrangements is $\binom{n+1}{m}m!n!$. Hence

$$P(\text{no men sit next to each other}) = \frac{\binom{n+1}{m}m!n!}{(m+n)!}. \tag{4.25}$$

♦

Example 4.2.10. Four of a Kind in Poker.

In a variant of the game of poker, players bet on the value of a five-card hand dealt to them from a standard 52-card deck. The value of the hand is determined by the type of combination of cards. In playing the game, it is helpful to know the probabilities of various combinations. In "four of a kind," the player's hand consists of all four cards of a certain kind, say all four Aces, plus one other card. The probability of being dealt four of a kind can be computed with both ordered and unordered selection, because the unordered selections are equiprobable, each consisting of $5!$ ordered selections.

With ordered selection, the total number of possible hands is $_{52}P_5$, and the number of favorable hands is $13 \cdot 48 \cdot 5!$, because the four like cards can be chosen 13 ways, the odd card can be any one of the remaining 48 cards, and any of the $5!$ orders of dealing the same cards results in the same hand. Thus,

$$P(\text{four of a kind}) = \frac{13 \cdot 48 \cdot 5!}{_{52}P_5} \approx 0.00024. \tag{4.26}$$

With unordered selection, the total number of possible hands is $\binom{52}{5}$ (these hands are now equally likely), and the number of favorable hands is $13 \cdot 48$, because the four like cards can be chosen 13 ways and the odd card can be any one of the remaining 48 cards. (Now we don't multiply by 5! because the order does not matter.) Thus

$$\text{P(four of a kind)} = \frac{13 \cdot 48}{\binom{52}{5}}. \tag{4.27}$$

$\blacklozenge$

Example 4.2.11. Two Pairs in Poker Dice.

The game of poker dice is similar to poker but uses dice instead of cards. We want to find the probability of obtaining two pairs with five dice, that is, a combination of the type x, x, y, y, z in any order, with x, y, z being distinct numbers from one to six.

Now the total number of possible outcomes is 6^5. (For this problem, we must use ordered outcomes, because the unordered ones would not be equally likely.) For the favorable cases, the numbers x and y can be chosen $\binom{6}{2} = 15$ ways and the number z four ways. Furthermore, the number of ways x, x, y, y, z can be ordered is $\binom{5}{2,\,2,\,1} = 30$. Thus,

$$\text{P(two pairs)} = \frac{15 \cdot 4 \cdot 30}{6^5} \approx 0.23. \tag{4.28}$$

$\blacklozenge$

An analog of the inclusion-exclusion principle (Theorem 3.1.3) holds for probabilities:

Theorem 4.2.1. *(Inclusion-Exclusion Theorem for Probabilities).* *For any positive integer n and arbitrary nonnull events $A_1, A_2, \ldots, A_n$,*

$$\text{P}\left(\bigcup_{i=1}^{n} A_i\right) = \sum_{1 \leq i \leq n} \text{P}(A_i) - \sum_{1 \leq i < j \leq n} \text{P}(A_i A_j)$$

$$+ \sum_{1 \leq i < j < k \leq n} \text{P}(A_i A_j A_k) - \cdots + (-1)^{n-1} \text{P}(A_1 A_2 \cdots A_n). \tag{4.29}$$

Proof. By Theorem 2.2.2, the union on the left can be written as the probability of the union of the disjoint sets $B = A_{i_1} \ldots A_{i_k} \overline{A_{i_{k+1}}}, \ldots, \overline{A_{i_n}}$ for every k in $\{1, \ldots, n\}$ and every pair of combinations $\{i_1, \ldots, i_k\}$ and $\{i_{k+1}, \ldots, i_n\}$. We can show that $\text{P}(B)$ is counted exactly once on the right, too, for every such B. Indeed, for a fixed k, such a B is a subset of each of $A_{i_1}, \ldots, A_{i_k}$ and so it will contribute $\text{P}(B)$ to k terms of the first sum. Also, B is a subset of the intersection of each pair of $A_{i_1}, \ldots, A_{i_k}$ and so it will contribute $\text{P}(B)$ to $\binom{k}{2}$ terms of the second sum,..., to $\binom{k}{k}$ terms of the kth sum, and nothing to the others. Thus the total contribution of B on the right is $\left[\binom{k}{1} - \binom{k}{2} + \cdots + (-1)^{k-1}\binom{k}{k}\right]P(B)$. The sum in the brackets is 1, because, by the binomial theorem, $(1-1)^k = 1 - \left[\binom{k}{1} - \binom{k}{2} + \cdots + (-1)^{k-1}\binom{k}{k}\right] = 0$.

While the proof above is valid in arbitrary probability spaces, we give an additional, alternate proof for finite spaces. Thus, assume that our sample space S is finite and the probabilities $P(s)$ of the elementary events are given. Consider Equation 3.11:

$$I_A = \sum_{1 \le i \le n} I_{A_i} - \sum_{1 \le i < j \le n} I_{A_i A_{ji}} + \cdots + (-1)^{n-1} I_{A_1 A_2 \cdots A_n}. \tag{4.30}$$

Then, by Theorem 4.1.6,

$$P(A) = \sum_{s \in S} I_A(s) P(s) \tag{4.31}$$

for any event A. Thus, multiplying Equation 4.30 by $P(s)$ for each s and summing over all s, we get the statement of the theorem.   ∎

This theorem has an important special case in which the probability of an intersection depends only on the number of sets in the intersection and not on the sets themselves.

Corollary 4.2.1. *If for every k in $\{1,\ldots,n\}$, there is a p_k such that*

$$P(A_{i_1} \ldots A_{i_k}) = p_k \tag{4.32}$$

for all permutations $(i_1,\ldots,i_k)$ of k elements of $\{1,\ldots,n\}$, then

$$P\left(\bigcup_{i=1}^{n} A_i\right) = \sum_{k=1}^{n} (-1)^{k-1}\binom{n}{k} p_k. \tag{4.33}$$

Proof. The number of terms in the kth sum in Equation 4.29 is $\binom{n}{k}$ for each k and the probability of each term there is p_k.   ∎

Example 4.2.12. (Montmort's Matching Problem).

In a book on the mathematics of games of chance, P. Montmort published the following problem and its solution in 1708.

Suppose we have two identical decks of cards with the cards numbered from 1 to n in each deck. The decks are shuffled and the order of the cards compared. If, for any i, the ith card is the same in both decks, we say that a match has occurred. What is the probability of k matches, for $k = 0, 1, \ldots, n$? This problem can clearly be equivalently stated as asking for the probability of k matches in a random permutation $(i_1, \ldots, i_n)$ of the $n$tuple $(1, \ldots, n)$, where a match occurs at the jth place if $i_j = j$. Instead of cards, the problem is frequently stated with hats and then it is called the *hatcheck problem*:

Suppose n people leave their hats in a checkroom and receive them back in a random order. What is then the probability that exactly k people get their own hats, for $k = 0, 1, \ldots, n$?

Let A_j be the set of permutations with a match at the jth place. Then $p_1 = \mathrm{P}(A_j) = (n-1)!/n! = 1/n$ since the total number of permutations is $n!$ and the favorable number of permutations is $(n-1)!$ for each j because j is fixed and the remaining $n-1$ numbers can be arbitrarily permuted. Similarly, the probability of a match at the ith and the jth place, for any i, j, is $p_2 = \mathrm{P}(A_i A_j) = (n-2)!/n! = 1/n(n-1)$, and $p_k = (n-k)!/n!$ for any k. Substituting these p_k values into Equation 4.33, we obtain for the probability of obtaining at least one match:

$$\mathrm{P}\left(\bigcup_{i=1}^{n} A_i\right) = \sum_{k=1}^{n} (-1)^{k-1} \binom{n}{k} \frac{(n-k)!}{n!} \tag{4.34}$$

$$= \sum_{k=1}^{n} (-1)^{k-1} \frac{n!}{k!\,(n-k)!} \frac{(n-k)!}{n!} = \sum_{k=1}^{n} (-1)^{k-1} \frac{1}{k!}. \tag{4.35}$$

A permutation with no matches is called a *derangement*. Hence the probability that a random permutation of $(1, \ldots, n)$ is a derangement is

$$p_{0,n} = 1 - \sum_{k=1}^{n} (-1)^{k-1} \frac{1}{k!} = \sum_{k=0}^{n} (-1)^{k} \frac{1}{k!}. \tag{4.36}$$

Notice that the sum on the right is a truncation of the Maclaurin series of e^{-1} and its value is about 0.37 for every $n > 2$. Thus, very surprisingly, the probability that nobody gets his hat back is about the same whether there are three hats or three million.

The probability of k matches, for $k > 0$, will be discussed in Example 4.2.13, but first we need to generalize the inclusion-exclusion theorem for probabilities (Theorem 4.2.1.).   ◆

Theorem 4.2.2. (Inclusion-Exclusion Theorem for the Probability of k Events). *For any positive integers n and k, with $1 \leq k \leq n$, and arbitrary nonnull events $A_1, A_2, \ldots, A_n$, the probability of exactly k events occurring is given by*

$$\mathrm{P}\left(\bigcup A_{i_1} \ldots A_{i_k} \overline{A_{i_{k+1}}}, \ldots, \overline{A_{i_n}}\right)$$

$$= \sum \mathrm{P}(A_{i_1} \ldots A_{i_k}) - \binom{k+1}{k} \sum \mathrm{P}(A_{i_1} \ldots A_{i_{k+1}})$$

$$+ \binom{k+2}{k} \sum \mathrm{P}(A_{i_1} \ldots A_{i_{k+2}}) - \cdots \pm \binom{n}{k} \mathrm{P}(A_1 A_2 \cdots A_n), \tag{4.37}$$

where the union is taken over all combinations of k numbers $i_1, \ldots, i_k$ from the set $\{1, \ldots, n\}$ and the sums are taken over all combinations of $k, k+1, \ldots, n$ numbers, respectively.

Proof. The union on the left is the probability of the union of the disjoint sets $A_{i_1} \ldots A_{i_k} \overline{A_{i_{k+1}}}, \ldots, \overline{A_{i_n}}$ for every combination $\{i_1, \ldots, i_k\}$ from $\{1, \ldots, n\}$.

On the right, however, the individual terms include sets of the form $B = A_{i_1} \ldots A_{i_j} \overline{A_{i_{j+1}}}, \ldots, \overline{A_{i_n}}$ with $j \neq k$ as well (for instance, $A_1 A_2 A_3 \overline{A_4}$ is a subset of $A_1 A_2 A_3, A_1 A_2, A_1 A_3, A_2 A_3, A_1, A_2, A_3$), and we must show that their total contribution is 0, while those with $j = k$ contribute just once.

Now P(B) does not contribute to any sum on the right if $j < k$, because each term there is the probability of a subset of A_{i_k}, but when $j < k$, then B is a subset of $\overline{A_{i_k}}$.

If $j = k$, then P(B) is counted exactly once on the right, just in the first sum, for every such B.

However, if $j > k$, then P(B) is counted in every sum up to the jth one. Thus, the coefficient of P(B) on the right side is

$$\binom{j}{k} - \binom{k+1}{k}\binom{j}{k+1} + \binom{k+2}{k}\binom{j}{k+2} - \cdots$$

$$= \frac{j!}{k!\,(j-k)!} - \frac{j!}{k!\,(j-k-1)!} + \frac{j!}{2!k!\,(j-k-2)!} - \cdots$$

$$= \binom{j}{k}\left[1 - (j-k) + \frac{(j-k)(j-k-1)}{2!} - \cdots \right]$$

$$= \binom{j}{k}\left[\binom{j-k}{0} - \binom{j-k}{1} + \binom{j-k}{2} - \cdots \pm \binom{j-k}{j-k} \right]$$

$$= \binom{j}{k}(1-1)^{j-k} = 0. \tag{4.38}$$

∎

Example 4.2.13. (*Montmort's Matching Problem Continued*).

We are now able to solve Montmort's matching problem for any number of matches.

As in Example 4.2.12, $P(A_{i_1} \ldots A_{i_k}) = (n-k)!/n!$ for all permutations $(i_1, \ldots, i_k)$ of k elements of $\{1, \ldots, n\}$ for every k. Substituting this into Equation 4.37, we get

P (exactly k matches)

$$= \binom{n}{k}\frac{(n-k)!}{n!} - \binom{k+1}{k}\binom{n}{k+1}\frac{(n-k-1)!}{n!}$$

$$+ \binom{k+2}{k}\binom{n}{k+2}\frac{(n-k-2)!}{n!} \cdots \pm \binom{n}{k}\frac{1}{n!}$$

$$= \frac{1}{k!}\left[1 - 1 + \frac{1}{2!} - \frac{1}{3!} + \cdots \pm \frac{1}{(n-k)!} \right] = \frac{1}{k!}p_{0,n-k}, \text{ for } k = 0, 1, 2, \ldots, n.$$

$$\tag{4.39}$$

Note that, in terms of hats, if there are k matches, then there are k "good" hats and $n - k$ "bad" hats. Then the result is the product of the probability $1/k!$ that the k good hats all go to their owners and of the probability $p_{0,n-k}$ that none of the $n - k$ bad hats are matched.   ◆

Exercises

Exercise 4.2.1.

From a deck of cards, use only AS, AH, KS, KH and choose two of these cards without replacement.

a) List all possible ordered pair outcomes.
b) Using the above, find the probability of obtaining an Ace and a King in either order.
c) Find the same probability by using unordered pairs.
d) Explain why the unordered pairs have equal probabilities unlike those in Example 4.2.2

Exercise 4.2.2.

If in Exercise 4.2.1 the drawing is done with replacement, find the probability of obtaining an Ace and a King. Can you find this probability by counting unordered pairs? Explain.

Exercise 4.2.3.

Explain why in Example 4.2.3 we did not get P(at least one six) = 1, in spite of the fact that on each throw the probability of getting a six is $\frac{1}{6}$ and 6 times $\frac{1}{6}$ is 1.

Exercise 4.2.4.

What is the probability that a 13-card hand dealt from a deck of 52 cards will contain:

a) The Queen of spades,
b) Five spades and 8 cards from other suits,
c) Five spades, five hearts, two diamonds, and the Ace of clubs?

Exercise 4.2.5.

Three dice are rolled. What is the probability that they show different numbers?

Exercise 4.2.6.

m men and n women are seated at random in a row on $m + n$ chairs. What is the probability that all the men sit next to each other?

Exercise 4.2.7.

m men and n women are seated at random around a round table on $m+n$ chairs. What is the probability that all the men sit next to each other?

Exercise 4.2.8.

An elevator in a building starts with six people and stops at eight floors. Assuming that all permutations of the passengers getting off at various floors are equally likely, find the probability that at least two of them get off on the same floor.

Exercise 4.2.9.

In the Massachusetts Megabucks game, a player selects six distinct numbers from 1 to 42 on a ticket, and the Lottery Commission draws six distinct numbers at random from 1 to 42. If all the player's numbers match the drawn ones, then she/he wins the jackpot and if five numbers match, then a smaller prize. Find the probability of each event.

Exercise 4.2.10.

A random sample of size 10 is chosen from a population of 100 without replacement. If A and B are two individuals among the 100, what is the probability that the sample will contain

a) Both,
b) Neither,
c) A,
d) Either A or B, but not both?

Simplify the answers.

Exercise 4.2.11.

Three integer digits $(0, 1, \ldots, 9)$ are chosen at random with repetitions allowed. What is the probability that

a) Exactly one digit will be even,
b) Exactly one digit will be less than three,

c) Exactly two digits will be divisible by three?

Exercise 4.2.12.

Two cards are dealt from n decks of 52 cards mixed together. (Mixing several decks is common in the game of 21 in casinos.) Find the probability of getting a pair, that is, two cards of the same denomination, for $n = 1, 2, 4, 6, 8$.

Exercise 4.2.13.

Assuming Bose-Einstein statistics (Part 2 of Example 4.2.8) with n cells and k particles, the probability that a given cell contains exactly m particles, for $0 \leq m \leq k$, is

$$\binom{k+n-m-2}{k-m} \bigg/ \binom{k+n-1}{k}. \tag{4.40}$$

Explain why (see Exercise 3.5.8.)

Exercise 4.2.14.

Assuming Bose-Einstein statistics (Part 2 of Example 4.2.8) with n cells and k particles, the probability that exactly m cells remain empty, for $0 \leq m < n$, is

$$\binom{n}{m}\binom{k-1}{n-m-1} \bigg/ \binom{k+n-1}{k}. \tag{4.41}$$

Explain why (see Exercise 3.5.8.)

Exercise 4.2.15.

Compute the probability that a poker hand dealt from a deck of 52 cards contains five different denominations (that is, no more than one of each kind: no more than one ace, one 2, etc.).

Exercise 4.2.16.

Compute the probability that a poker hand dealt from a deck of 52 cards contains two pairs.

Exercise 4.2.17.

Compute the probability that a poker hand dealt from a deck of 52 cards is a full house, that is, contains a pair and a triple (that is, x, x, y, y, y).

Exercise 4.2.18.

Compute the probability that in poker dice we get four of a kind.

Exercise 4.2.19.

Compute the probability that in poker dice we get a full house.

Exercise 4.2.20.

Show combinatorially that the probability in Example 4.2.4 can be obtained by using ordered selections as

$$P(n_1; n, N_1, N_2) = \frac{\binom{n}{n_1, \, n_2} {}_{N_1}P_{n_1} \, {}_{N_2}P_{n_2}}{{}_{N}P_{n}}, \tag{4.42}$$

and show algebraically that this quantity equals the one obtained in Equation 4.13.

Exercise 4.2.21.

Prove that in Example 4.2.4, the four inequalities $0 \le n_1 \le n$, $n_1 \le N_1$ and $n - n_1 \le N_2$, together, are equivalent to the double inequality $\max(0, n - N_2) \le n_1 \le \min(n, N_1)$.

Exercise 4.2.22.

A 13-card hand is dealt from a standard deck of 52 cards. What is the probability that

a) It contains exactly three spades and all four aces,
b) At least three of each suit?

4.3 Independence

The calculation of certain probabilities is greatly facilitated by the knowledge of any relationships, or lack thereof, between the events under consideration. In this section, we want to examine the latter case, that is, the case in which the occurrence of one event has no influence on the probability of the other's occurrence. We want to call such events independent of each other, and want to see how this is reflected in the probabilities. We begin with two examples.

Example 4.3.1. Repeated Tosses of Two Coins.

Suppose we toss two coins repeatedly. We describe this experiment by the sample space $S = \{HH, HT, TH, TT\}$, and want to estimate the relative frequency of HH. Of course, we know that it should be about $\frac{1}{4}$, but we want to look at this in a novel way. We can argue that the first coin shows H in about $\frac{1}{2}$ of the trials, and since the outcome of the first coin's toss does not influence that of the second, the second coin shows H in not only about $\frac{1}{2}$ of all trials but also among those in which the first coin turned up H. Thus, HH occurs in about $\frac{1}{2}$ of $\frac{1}{2}$, that is, in about $\frac{1}{4}$ of the trials. So P(HH) = P(the first coin shows H) $\cdot$ P(the second coin shows H), that is, the probability of both events occurring equals the product of the probabilities of the separate events. If we denote the event {the first coins shows H} = $\{HH, HT\}$ by A, and the event {the second coin shows H} = $\{HH, TH\}$ by B, then $\{HH\} = A \cap B$, and the above result can be written as P(AB) = P(A) $\cdot$ P(B). ◆

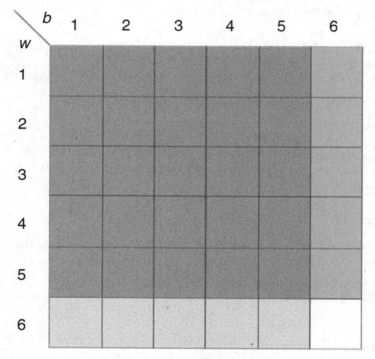

Fig. 4.3. Two dice with neither of them showing 6

Example 4.3.2. Two Dice.

We throw two dice, a black and a white one. The probability of neither of them showing a six is $\frac{5^2}{6^2}$, which can be written as $\frac{5}{6} \cdot \frac{5}{6}$. Now P($b \neq 6$) = $\frac{5}{6}$, P($w \neq 6$) = $\frac{5}{6}$, and so P($b \neq 6$ and $w \neq 6$) = P($b \neq 6$) $\cdot$ P($w \neq 6$).

Again, the probability of both one event and the other occurring equals the product of the probabilities of the two events. This relation is illustrated by the diagram of Figure 4.3 in which the one shading represents $\{b \neq 6\}$, the other $\{w \neq 6\}$, and the double shaded 5×5 square represents $\{b \neq 6$ and $w \neq 6\} = \{b \neq 6\} \cap \{w \neq 6\}$. If we consider the length of each side of the big square to be one unit, then the length of the segment marked $b \neq 6$ is $\frac{5}{6}$, which is also the area of the corresponding vertical strip of $5 \cdot 6 = 30$ small squares. Thus, both this length and area have the same measure as the probability of $\{b \neq 6\}$. (The length can be thought of as representing the probability of $\{b \neq 6\}$ in the 6-point sample space for b alone and the area as representing $P(b \neq 6)$ in the 36-point sample space for b and w together.) Similarly $P(w \neq 6)$ too shows up as a vertical length of $\frac{5}{6}$ units and also as the area of the corresponding horizontal strip. $P(b \neq 6$ and $w \neq 6)$ shows up only as an area, namely, that of the corresponding 5×5 square.   ♦

From these examples, we abstract the following definition:

Definition 4.3.1. *Independence of Two Events*. *Two events A and B are said to be (statistically) independent[4] (of each other) if*

$$P(AB) = P(A) \cdot P(B). \tag{4.43}$$

Also, two collections $\mathcal{F}_1$ and $\mathcal{F}_2$ of events are independent if Equation 4.43 holds for all pairs of events $A \in \mathcal{F}_1$ and $B \in \mathcal{F}_2$, and two experiments $\mathcal{E}_1$ and $\mathcal{E}_2$ are independent if the corresponding collections $\mathcal{F}_1$ and $\mathcal{F}_2$ of events are independent.

The main use of this definition is in the assignment of probabilities to the joint occurrence of pairs of events that we know are independent in the everyday sense of the word. Using this definition, we make them statistically independent, too, as in the following example.

Example 4.3.3. Distribution of Voters.

Assume that the distribution of voters in a certain city is as described in the two tables below.

Party affiliation:	Republican	Democrat	Independent
% of all voters:	25	40	35

Age group:	Under 30	30 to 50	Over 50
% of all voters:	30	40	30

The probability of a randomly picked voter belonging to a given group is the decimal fraction corresponding to the group's percentage in the table, and the two tables each describe a probability distribution on the sample spaces

---

[4] Note that, terminology notwithstanding, it is the events with their probabilities that are here defined to be independent, not just the events themselves.

$S_1 = \{$Republican, Democrat, Independent$\}$ and $S_2 = \{$under 30, 30–50, over 50$\}$, respectively.

Assuming that party affiliation is independent of age, we can find each of the nine probabilities of a randomly picked voter belonging to a given possible classification according to party *and* age. These probabilities can be obtained according to Definition 4.3.1 by multiplying the probabilities (that were given as percentages) of the previous tables. The products are listed in the next table, describing a probability distribution on the sample space $S = S_1 \times S_2$.

Age\ Party	Republican	Democrat	Independent	Any affiliation
Under 30	.075	.12	.105	.30
30 to 50	.10	.16	.14	.40
Over 50	.075	.12	.105	.30
Any age	.25	.40	.35	1

The probabilities in this table are called the *joint probabilities* of party affiliation and age group, and the probabilities given in the first two tables are called the *marginal probabilities* of the two-way classification, because they are equal to the probabilities in the margins of the last table. For instance, $P(\text{Any age} \cap \text{Republican}) = 0.25$ in the nine-element sample space $S = S_1 \times S_2$, equals $P(\text{Republican}) = 0.25$ in the three-element sample space S_1. Notice that the marginal probabilities are the row and column sums of the joint probabilities of the nine elementary events, and all add up to 1, of course. ◆

The notion of independence can easily be extended to more than two events:

Definition 4.3.2. Independence of Several Events. *Let $A_1, A_2, \ldots$ be any events. We say that they are independent (of each other), if for all possible sets of two or more of them the probability of the intersection of the events in the set equals the product of the probabilities of the individual events in the set, that is,*

$$P(A_1 \cap A_2) = P(A_1)P(A_2), P(A_1 \cap A_3) = P(A_1)P(A_3), \ldots$$
$$P(A_1 \cap A_2 \cap A_3) = P(A_1)P(A_2)P(A_3), \ldots$$
$$\cdots$$

$$(4.44)$$

Also, collections $\mathcal{F}_1, \mathcal{F}_2, \ldots$ of events are independent if the Equations 4.44 hold for all events $A_1 \in \mathcal{F}_1, A_2 \in \mathcal{F}_2, \ldots$. Furthermore, experiments $\mathcal{E}_1, \mathcal{E}_2, \ldots$ are independent if the corresponding collections $\mathcal{F}_1, \mathcal{F}_2, \ldots$ of events are independent.

Note that it is not enough to require the product formula just for the intersections of all pairs of events or just for the intersection of all the events under consideration, but we have to require it for the intersections of all possible combinations. (See Exercises 4.3.3 and 4.3.4.)

A frequent misconception is to think that independence is a property of individual events. *No*, it is a relation among the members of a set of at least two events.

We can use this definition to derive the formula of the binomial distribution anew in a very general setting:

Example 4.3.4. Binomial Distribution.

Consider an experiment that consists of n identical subexperiments called trials. In each trial we have:

1. Two possible outcomes, which we call success and failure;
2. The trials are independent of each other,
3. The probability of success is the same number p in each trial, while the probability of failure is $q = 1 - p$.

Such trials are called Bernoulli trials.[5] For example, tossing a coin or throwing a die repeatedly or selecting a person from a given population with replacement and observing whether he or she has a certain trait are such trials. We ask for the probability $b(k; n, p)$ of obtaining exactly k successes in the n trials. Now, by the assumed independence, the probability of having k successes and $n - k$ failures in any fixed order is $p^k q^{n-k}$, and since the k successes and $n - k$ failures can be ordered in $\binom{n}{k}$ mutually exclusive ways:

$$b(k; n, p) = \binom{n}{k} p^k q^{n-k}. \tag{4.45}$$

Thus we have obtained the same binomial distribution as in Example 4.2.5, but in a more general setting.

The great importance of this distribution stems from the many possible applications of its scheme. Success and failure can mean head or tail in coin tossing, winning or losing in any game, curing a patient or not in a medical experiment, people answering yes or no to some question in a poll, people with life insurance surviving or dying, etc.   ◆

Example 4.3.5. De Méré's Paradox.

In the seventeenth century, a French nobleman, the Chevalier de Méré, posed the following question to the famous mathematician Blaise Pascal: If you throw a die four times, he said, gamblers know from experience that the probability of obtaining at least one six is a little more than $\frac{1}{2}$, and if you throw two dice 24 times, the probability of getting at least one double

---

[5] Named after one of the founders of the theory of probability, Jacob Bernoulli (1654–1705), the most prominent member of a Swiss family of at least six famous mathematicians.

six is a little less than $\frac{1}{2}$. How is it possible that you do not get the same probability in both cases, in view of the fact that P(double six for a pair of dice) $= \frac{1}{36} = \frac{1}{6} \cdot$ P(a six for a single die), but you compensate for the factor of $\frac{1}{6}$ by throwing not 4 but $6 \cdot 4 = 24$ times when using two dice?

Well, the facts do not lie, and so there must be a mistake in the argument. Indeed, there is one in the last step: If we multiply the number of throws by 6, the probability of getting at least one double six is not 6 times what it is in 4 throws or 24 times what it is in one throw.[6]

Applying de Méré's argument to throws of a single die, we can see at once that such multiplication must be wrong: If we throw one die six times, then his reasoning would give for the probability of getting at least one 6 $6 \cdot \frac{1}{6} = 1$, and if we throw it seven times, the probability of at least one 6 would be $7 \cdot \frac{1}{6} > 1$; clearly impossible. The source of the error lies in the inappropriate use of the additivity axiom, because the events $A_i =$ "the ith throw yields six," for $i = 1, 2, 3, 4$, are not mutually exclusive, and so P(at least one 6 in four throws of a single die) $= \mathrm{P}(A_1 \cup A_2 \cup A_3 \cup A_4)$ is not equal to $\mathrm{P}(A_1) + \mathrm{P}(A_2) + \mathrm{P}(A_3) + \mathrm{P}(A_4) = 4 \cdot \frac{1}{6}$.

Similarly, the events $B_i =$ "the ith throw yields a double six for a pair of dice," for $i = 1, 2, \ldots, 24$, are not mutually exclusive, and so P(at least one double six in 24 throws of a pair of dice) $= \mathrm{P}(B_1 \cup B_2 \cup \cdots \cup B_{24}) \neq 24 \cdot \frac{1}{36} = \frac{2}{3}$.

We could write correct formulas for $\mathrm{P}(A_1 \cup A_2 \cup A_3 \cup A_4)$ and $\mathrm{P}(B_1 \cup B_2 \cup \cdots \cup B_{24})$ using Theorem 4.2.1, but it is easier to compute the required probabilities by complementation: P(at least one 6 in four throws of a single die) $= 1 -$ P(no six in four throws of a single die) $= 1 - \left(\frac{5}{6}\right)^4 \approx .5177$. Similarly, P(at least one double six in 24 throws of a pair of dice) $= 1 - $ P(no double six in 24 throws of a pair of dice) $= 1 - \left(\frac{35}{36}\right)^{24} \approx .4914$.   ♦

In closing this section, let us mention that the marginal probabilities do not determine the joint probabilities without some assumption like independence, that is, it is possible to have different joint probabilities with the same marginals. For instance, the joint probability distribution in the following example has the same marginals as the one in Example 4.3.3

Example 4.3.6. Another Distribution of Voters.

Let the joint distribution of voters in a certain city be described by the table below.

Age\ Party	Republican	Democrat	Independent	Any affiliation
Under 30	.05	.095	.155	.30
30 to 50	.075	.21	.115	.40
Over 50	.125	.095	.08	.30
Any age	.25	.40	.35	1

---

[6] Note, however, that such a multiplication rule does hold for expected values. In this case, the expected number of double sixes in n throws is n times the expected number in one throw, as we shall see in Section 6.1.

It is easy to check that the various ages and party affiliations are not independent of each other. For instance, P(under 30)P(Republican) = .30 · .25 = .075, while P(under 30 and Republican) = .05.   ♦

Exercises

Exercise 4.3.1.

Three dice are thrown. Show that the events A = {the first die shows an even number} and B = {the sum of the numbers on the second and third dice is even} are independent.

Exercise 4.3.2.

If b and w stand for the results of a throw of two dice, show that the events $A = \{b + w < 8\}$ and $B = \{b = 3 \text{ or } 4\}$ are statistically independent (although it is difficult to see why they should be in the usual sense of the word).

Exercise 4.3.3.

Toss two dice. Let $A = \{b < 4\}, B = \{b = 3, 4, \text{ or } 5\}$ and $C = \{b + w = 9\}$. Show that these events are not independent pairwise, but $P(A \cap B \cap C) = P(A)P(B)P(C)$.

Exercise 4.3.4.

Toss two coins. Let $A = \{HH, HT\}, B = \{TH, HH\}$ and $C = \{HT, TH\}$. Show that these events are independent pairwise, but $P(A \cap B \cap C) \neq P(A)P(B)P(C)$.

Exercise 4.3.5.

Let A and B be independent events. Show that

a) A and $\overline{B}$ are also independent,
b) And so are $\overline{A}$ and $\overline{B}$.

Exercise 4.3.6.

a) Can two independent events with nonzero probabilities be mutually exclusive?
b) Can two mutually exclusive events with nonzero probabilities be independent?

(Prove your answers.)

Exercise 4.3.7.

A coin is tossed five times. Find the probabilities of obtaining exactly $0, 1, 2, 3, 4$, and 5 heads and plot them in a coordinate system.

Exercise 4.3.8.

A die is thrown six times. Find the probabilities of obtaining

a) Exactly 4 sixes,
b) Exactly 5 sixes,
c) Exactly 6 sixes,
d) At least 4 sixes,
e) At most 3 sixes.

Exercise 4.3.9.

An urn contains five red, five white, and five blue balls. We draw six balls independently, one after the other, with replacement. What is the probability of obtaining two of each color?

Exercise 4.3.10.

Let A, B, and C be independent events for which $A \cup B \cup C = S$. What are the possible values of P(A), P(B), and P(C)?

Exercise 4.3.11.

Let A, B, and C be pairwise independent events and A be independent of $B \cup C$. Prove that A, B, and C are totally independent.

Exercise 4.3.12.

Let A, B, and C be pairwise independent events and A be independent of $B\overline{C}$. Prove that A, B, and C are totally independent.

4.4 Conditional Probabilities

In this section, we discuss probabilities if certain events are known to have occurred. We start by considering two examples.

Example 4.4.1. Relative Frequencies in Repeated Tossings of Two Coins.

Suppose we toss two coins $n = 10$ times and observe the following outcomes: $HT, TT, HT, HH, TT, HH, HH, HT, TH, HT$.

If we denote the event that the first coin shows H by A and the event that the second coin shows H by B, then A occurs $n_A = 7$ times, B occurs $n_B = 4$ times, and $A \cap B$ occurs $n_{AB} = 3$ times. The relative frequencies of these events are $f_A = \frac{7}{10}$, $f_B = \frac{4}{10}$, and $f_{AB} = \frac{3}{10}$.

Let us now ask the question: what is the relative frequency of A among the outcomes in which B has occurred? Then we must relate the number n_{AB} of occurrences of A among these outcomes to the total number n_B of outcomes in which B has occurred. Thus, if we denote this relative frequency by $f_{A|B}$, then we have

$$f_{A|B} = \frac{n_{AB}}{n_B} = \frac{3}{4}. \tag{4.46}$$

We call $f_{A|B}$ the conditional relative frequency of A, given B (or, under the condition B). It is very simply related to the old "unconditional" relative frequencies:

$$f_{A|B} = \frac{n_{AB}/n}{n_B/n} = \frac{3/10}{4/10} = \frac{f_{AB}}{f_B}. \tag{4.47}$$

According to this example, we would want to define conditional probabilities in an analogous manner by $P(A|B) = \frac{P(AB)}{P(B)}$, for any events A and B with $P(B) \neq 0$. Indeed, this is what we shall do, but let us see another example first.   ◆

Example 4.4.2. Conditional Probabilities for Randomly Picked Points.

Assume that we pick a point at random from those shown in Figure 4.4. If $P(A)$, $P(B)$, and $P(AB)$ denote the probabilities of picking the point from A, B, and $A \cap B$, respectively, then $P(A) = \frac{5}{10}$, $P(B) = \frac{4}{10}$, $P(AB) = \frac{3}{10}$. If we restrict our attention to only those trials in which B has occurred, that is, if we know that the point has been picked from B, then obviously we want to define the conditional probability $P(A|B)$ of A given B as $\frac{3}{4}$, that is, as $P(A|B) = \frac{P(AB)}{P(B)}$ again.   ◆

These examples lead us to

Definition 4.4.1. *Conditional Probability. Let A and B be arbitrary events in a given probability space, with $P(B) \neq 0$. Then we define the conditional probability of A, given B, as*

$$P(A|B) = \frac{P(AB)}{P(B)}. \tag{4.48}$$

Notice that actually every probability may be regarded as a conditional probability, with the condition S, since

$$P(A|S) = \frac{P(AS)}{P(S)} = \frac{P(A)}{P(S)} = P(A), \tag{4.49}$$

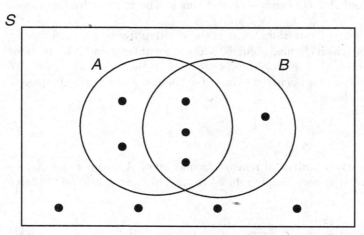

Fig. 4.4. Illustration of conditional probabilities by picking points at random

Conversely, every conditional probability $P(A|B)$ may be regarded as an unconditional probability in a new, reduced sample space, namely, in B, in place of S. (This fact is clearly true in sample spaces with equally likely outcomes, as in Example 4.4.2, but, in general, it needs to be proved from Definition 4.4.1. It will be the subject of Theorem 4.4.1 below.)

Let us see some further examples.

Example 4.4.3. Two Dice.

Two dice are thrown. What is the probability that the sum of the numbers that come up is two or three, given that at least one die shows a 1?

Let us call these events A and B, that is, let $A = \{(1,1), (1,2), (2,1)\}$ and $B = \{(1,1), (1,2), (1,3), (1,4), (1,5), (1,6), (2,1), (3,1), (4,1), (5,1), (6,1)\}$. Then $AB = A$, and so $P(AB) = \frac{3}{36}$, $P(B) = \frac{11}{36}$ and, by the definition of conditional probabilities, $P(A|B) = \frac{3/36}{11/36} = \frac{3}{11}$.

We could also have obtained this result directly, as the unconditional probability of the three-point event A in the eleven-point sample space B. ♦

Warning: We must be careful not to confuse the probability $P(AB)$ of A and B occurring jointly (or as we say, their joint probability) with the conditional probability $P(A|B)$. In the above example, for instance, it would be wrong to assume that the probability of the sum being two or three, if one die shows a 1, is $\frac{3}{36}$ since, under that condition, the 3 favorable cases must be related to a total of 11 cases, rather than to all 36.

Example 4.4.4. Modified Monty Hall Problem.

Assume the same setup as in Example 4.1.3, but assume that the host too does not know which door hides the car, but opens a door, other than the player's choice, at random and happens to reveal a goat. The question is again to switch or not to switch?

Denote the doors by what is behind them, that is, by "car," "g_1," and "g_2." Then the player and the host cannot both open the same door, and the remaining six possibilities for their choices are equally likely. Thus the joint probabilities for their choices are

Host\ Player	car	g_1	g_2
car	0	1/6	1/6
g_1	1/6	0	1/6
g_2	1/6	1/6	0

Let $C =$ "the player's choice hides the car" and $A =$ "the host finds a goat". Then

$$P(C|A) = \frac{P(AC)}{P(A)} = \frac{2/6}{4/6} = \frac{1}{2}, \tag{4.50}$$

that is, the probability of winning if we stay with the initial choice is $1/2$. Similarly, the probability of winning if we switch is $P\left(\overline{C}|A\right) = 1/2$. Thus, in this case, it does not matter which strategy we use.

Incidentally, the fact that the two versions show different results implies that in order to determine whether switching is advantageous or not, it is not enough to know that the host has revealed a goat; we must also know how he arrived at his choice. Further modifications of this choice will be examined in Exercise 4.5.15 and will show that switching is good in most cases and never bad. Thus one should always switch regardless of the host's strategy, and in the worst case, as above, it still leads to the same result as not switching. ◆

Example 4.4.5. Sex of Children in Randomly Selected Family.

From all families with three children, we select one family at random. What is the probability that the children are all boys, if we know that a) the first one is a boy, and b) at least one is a boy? (Assume that each child is a boy or a girl with probability $1/2$, independently of each other.)

The sample space is $S = \{bbb, bbg, bgb, bgg, gbb, gbg, ggb, ggg\}$ with 8 equally likely outcomes. The sample points are the possible types of families, with the children listed in the order of their births; for instance, bgg stands for a family in which the first child is a boy and the other two are girls.

The reduced sample space for Part a) is $\{bbb, bbg, bgb, bgg\}$, and so P(all are boys | the first one is a boy) = 1/4. Similarly, the reduced sample space for Part b) is $\{bbb, bbg, bgb, bgg, gbb, gbg, ggb\}$, and so P(all are boys | at least one is a boy) = 1/7.

It may seem paradoxical that the two answers are different. After all, if we know that one child is a boy, what does it matter whether it is the first child we know this about or about any one of the three? But in the first case, we know more: we know not just that one child is a boy but also that it is the first one who is a boy. Thus, in the first case, the reduced sample space is smaller than in the second case, and consequently the denominator of the conditional probability is smaller, while the numerator is the same.   ♦

Example 4.4.6. Sex of Sibling of Randomly Selected Child.

From all families with two children, we select one child at random. If the selected child is a boy, what is the probability that he comes from a family with two boys? (Assume that each child is a boy or a girl with probability 1/2, independently of each other.)

The main difference between this example and the preceding one is that we selected a family and here we select a child. Thus, here the sample points must be children, not families. We denote the child to be selected by $\mathbf{b}$ or $\mathbf{g}$, but we also want to indicate the type of family he or she comes from. So, denoting the other child by b or g, we write, for instance, $\mathbf{b}b$ for a boy with a younger brother, $g\mathbf{b}$ for a boy with an older sister, etc. Thus, we use the sample space $S = \{\mathbf{b}b, \mathbf{b}g, \mathbf{g}b, \mathbf{g}g, b\mathbf{b}, g\mathbf{b}, b\mathbf{g}, g\mathbf{g}\}$ with eight equally likely outcomes, which denote the eight different types of child that can be selected. The reduced sample space for which the selected child is a boy is $\{\mathbf{b}b, \mathbf{b}g, b\mathbf{b}, g\mathbf{b}\}$, and so P(both children of the family are boys | the selected child is a boy) = 2/4 = 1/2.

We may also solve this problem by ignoring the birth order. Then $S = \{\mathbf{b}b, \mathbf{b}g, \mathbf{g}b, \mathbf{g}g\}$, where $\mathbf{b}b$ stands for a boy with a brother, $\mathbf{b}g$ for a boy with a sister, etc. Now the reduced sample space is $\{\mathbf{b}b, \mathbf{b}g\}$. Hence P(both children of the family are boys | the selected child is a boy) = 1/2 again.   ♦

The definition of conditional probabilities is often used in the multiplicative form:

$$P(AB) = P(A|B)P(B) \tag{4.51}$$

for the assignment of probabilities to joint events, much as we used the definition of independence for that purpose. Let us show this use in some examples.

Example 4.4.7. Dealing two Aces.

Two cards are dealt without replacement from a regular deck of 52 cards. Find the probability of getting two aces.

Letting $A = \{$the second card is an Ace$\}$ and $B = \{$the first card is an ace$\}$, we have $P(B) = \frac{4}{52}$ and $P(A|B) = \frac{3}{51}$, because, when we get to dealing the second card, there are 3 aces and a total of 51 cards left. Hence $P($both cards are aces$) = P(AB) = P(A|B)P(B) = \frac{4}{52} \cdot \frac{3}{51}$.   ♦

In several of the preceding examples, we saw that conditional probabilities behave like unconditional probabilities on a reduced sample space. The following theorem shows that indeed they satisfy the three axioms of probabilities.[7]

Theorem 4.4.1. *For a Fixed Condition, Conditional Probabilities Satisfy the Axioms of Probabilities. Let B be an event with nonzero probability in a sample space S. The conditional probabilities under the condition B have the following properties:*

1. $P(A|B) \geq 0$ *for every event A,*
2. $P(S|B) = 1$,
3. $P(A_1 \cup A_2 \cup \cdots |B) = P(A_1|B) + P(A_2|B) + \cdots$ *for any finite or countably infinite number of mutually exclusive events $A_1, A_2, \ldots$.*

1. In the definition of $P(A|B)$ the numerator is nonnegative by Axiom 1, and the denominator is positive by assumption. Thus, the fraction is nonnegative.
2. Taking $A = S$ in the definition of $P(A|B)$, we get

$$P(S|B) = \frac{P(S \cap B)}{P(B)} = \frac{P(B)}{P(B)} = 1. \tag{4.52}$$

3.

$$P(A_1 \cup A_2 \cup \cdots |B) = \frac{P((A_1 \cup A_2 \cup \cdots)B)}{P(B)}$$
$$= \frac{P(A_1B \cup A_2B \cup \cdots)}{P(B)} = \frac{P(A_1B) + P(A_2B) + \cdots}{P(B)}$$
$$= P(A_1|B) + P(A_2|B) + \cdots \tag{4.53}$$

where the next to last equality followed from Axiom 3 and Definition 4.4.1. [8]   ∎

---

[7] This theorem does not quite make $P(A|B)$ for fixed B into a probability measure on B in place of S though, because in Definition 4.1.2 $P(A)$ was defined for events $A \subset S$, but in $P(A|B)$ we do not need to have $A \subset B$. See Corollary 4.4.1, however.

[8] Because of this theorem, a few authors use the notation $P_B(A)$ for $P(A|B)$ to emphasize the fact that P_B is a probability measure on S and in $P(A|B)$ we do not have a function of a conditional event $A|B$ but a function of A. In other words, $P(A|B) = $ (the probability of A) given B, and not the probability of $(A$ given $B)$. Conditional events have been defined but have not gained popularity.

Corollary 4.4.1. *If the events A and $A_1, A_2, \ldots$ are subsets of B, then for fixed B, the function $P(.|B)$ is a probability measure on the reduced sample space B in place of S.*

The definition of conditional probabilities leads to an important test for the independence of two events:

Theorem 4.4.2. A *Condition for Independence*. *Two events A and B, with $P(B) \neq 0$, are independent, if and only if*

$$P(A|B) = P(A). \tag{4.54}$$

Proof. By Definition 4.3.1 two events are independent, if and only if

$$P(AB) = P(A)P(B). \tag{4.55}$$

Substituting into the left side of this equation from Equation 4.51, we get equivalently, when $P(B) \neq 0$ (the conditional probability $P(A|B)$ is defined only if $P(B) \neq 0$),

$$P(A|B)P(B) = P(A)P(B) \tag{4.56}$$

or, by canceling $P(B)$,

$$P(A|B) = P(A). \tag{4.57}$$

∎

Note that the condition in Theorem 4.4.2 is asymmetric in A and B, but if $P(A) \neq 0$, then we could similarly prove that A and B are independent, if and only if

$$P(B|A) = P(B). \tag{4.58}$$

Exercises

Exercise 4.4.1.

Suppose the following sequence of tosses of two coins is observed: HH, TT, HT, TT, TH, HT, HT, HT, TH, TT, TH, HT, TT, TH, HH, TH, TT, HH, HT, TH.

Let $A = \{$the first coin shows $H\}$ and $B = \{$the second coin shows $T\}$.

a) Find the relative frequencies f_A, f_B, f_{AB}, $f_{A|B}$ and $f_{B|A}$.

b) Find the corresponding probabilities $P(A)$, $P(B)$, $P(AB)$, $P(A|B)$, $P(B|A)$. Assume that the coins are fair and the tosses independent.

Exercise 4.4.2.

Two dice are thrown, with b and w denoting their outcomes. (See Figure 2.4 on page 21.) Find $P(w \leq 3 \text{ and } b + w = 7)$, $P(w \leq 3 | b + w = 7)$ and $P(b + w = 7 | w \leq 3)$.

Exercise 4.4.3.

A card is drawn at random from a deck of 52 cards. What is the probability that it is a King or a 2, given that it is a face card (J, Q, K)?

Exercise 4.4.4.

In Example 4.3.6, voters of a certain district are classified according to age and party registration (e.g., the .05 in the under 30 and Republican category means that 5% of the total is under 30 and Republican, that is, $P(\{\text{under } 30\} \cap \{\text{Republican}\}) = .05$) for a randomly selected voter. Find the probabilities of a voter being

a) Republican,
b) Under 30,
c) Republican if under 30,
d) Under 30 if Republican,
e) Democrat,
f) Democrat if under 30,
g) Independent,
h) Independent if under 30.

Exercise 4.4.5.

In the previous problem, the sum of the answers to parts c, f, and h should be 1. Why?

Exercise 4.4.6.

Consider two events A and B with $P(A) = 8/10$ and $P(B) = 9/10$. Prove that $P(A|B) \geq 7/9$.

Exercise 4.4.7.

From a family of three children, a child is selected at random and is found to be a girl. What is the probability that she came from a family with two girls and one boy? (Assume that each child is a boy or a girl with probability $1/2$, independently of one another.)

Exercise 4.4.8.

Three dice were rolled. What is the probability that exactly one 6 came up if it is known that at least one 6 came up.

Exercise 4.4.9.

Two cards are drawn at random from a deck of 52 cards without replacement. What is the probability that they are both Kings, given that they are both face cards (J, Q, K)?

Exercise 4.4.10.

Prove that any two events A and B, with $P(B) \neq 0$ and $P(\overline{B}) \neq 0$, are independent of each other if and only if $P(A|B) = P(A|\overline{B})$.

Exercise 4.4.11.

Two cards are drawn at random from a deck of 52 cards without replacement. What is the probability that exactly one is a King, given that at most one is a King?

Exercise 4.4.12.

Two cards are drawn at random from a deck of 52 cards with replacement. What is the probability that exactly one is a King, given that at most one is a King?

Exercise 4.4.13.

A 13-card hand is dealt from a standard deck of 52 cards. What is the probability that

a) It contains no spades if it contains exactly five hearts,
b) It contains at least one spade if it contains exactly five hearts?

4.5 The Theorem of Total Probability and the Theorem of Bayes

In many applications, we need to combine the definition of conditional probabilities with the additivity property, as in the following examples.

Example 4.5.1. Picking Balls from Urns.

Suppose we have two urns, with the first one containing two white and six black balls and the second one containing two white and two black balls. We pick an urn at random and then pick a ball from the chosen urn at random. What is the probability of picking a white ball?

Let us denote the events that we choose urn 1 by U_1 and urn 2 by U_2 and that we pick a white ball by W and a black ball by B. We are given the probabilities $P(U_1) = P(U_2) = \frac{1}{2}$, since this is what it means that an urn is picked at random; and, given that urn 1 is chosen, the random choice of a ball gives us the conditional probability $P(W|U_1) = \frac{2}{8}$, and similarly $P(W|U_2) = \frac{2}{4}$. Then, by Formula 4.51,

$$P(W \cap U_1) = P(W|U_1)P(U_1) = \frac{2}{8} \cdot \frac{1}{2} = \frac{1}{8}, \tag{4.59}$$

and

$$P(W \cap U_2) = P(W|U_2)P(U_2) = \frac{2}{4} \cdot \frac{1}{2} = \frac{1}{4}, \tag{4.60}$$

Now obviously W is the union of the disjoint events $W \cap U_1$ and $W \cap U_2$, and so by the additivity of probabilities

$$P(W) = P(W \cap U_1) + P(W \cap U_2) = \frac{1}{8} + \frac{1}{4} = \frac{3}{8}. \tag{4.61}$$

Note that this result is not the same as that which we would get if we were to put all 12 balls into one urn and picked one at random from there. Then we would get $\frac{4}{12} = \frac{1}{3}$ for the probability of picking a white ball.

In problems such as this one, it is generally very helpful to draw a tree diagram, with the given conditional probabilities on the branches, as indicated in Figure 4.5.

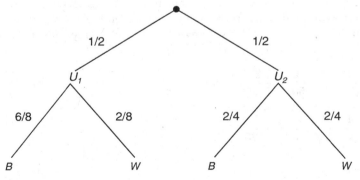

Fig. 4.5. Picking an urn and a ball from it

The unconditional probabilities of each path from top to bottom are obtained by multiplying the conditional probabilities written over it. For example, the probability of the path through U_1 and B is $P(U_1 \cap B) = \frac{1}{2} \cdot \frac{6}{8} = \frac{3}{8}$, and similarly, $P(U_2 \cap B) = \frac{1}{2} \cdot \frac{2}{4} = \frac{1}{4}$.

The probability of obtaining a given result in the end, regardless of the path, is the sum of the probabilities of all paths ending in that result. Thus $P(B) = \frac{3}{8} + \frac{1}{4} = \frac{5}{8}$.   ◆

The method just shown can be used in situations involving any number of alternatives and stages, whenever the conditional probabilities are known and we want to find the unconditional probabilities.

Example 4.5.2. Dealing Three Cards.

From a deck of 52 cards, three are drawn without replacement. What is the probability of the event E of getting two Aces and one King in any order?

If we denote the relevant outcomes by A, K, and O (for "other"), then we can illustrate the experiment by the tree in Figure 4.6.

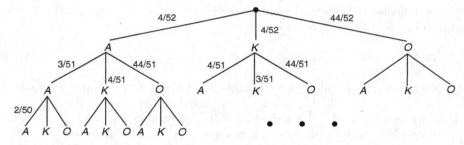

Fig. 4.6. Dealing three cards

The event E is the union of the three elementary events AAK, AKA, and KAA. The relevant conditional probabilities have been indicated on the corresponding paths. (The rest of the diagram is actually superfluous for answering this particular question.) Now

$$P(AAK) = \frac{4}{52} \cdot \frac{3}{51} \cdot \frac{4}{50} = \frac{2}{5525}, \tag{4.62}$$

$$P(AKA) = \frac{4}{52} \cdot \frac{4}{51} \cdot \frac{3}{50} = \frac{2}{5525}, \tag{4.63}$$

and

$$P(KAA) = \frac{4}{52} \cdot \frac{4}{51} \cdot \frac{3}{50} = \frac{2}{5525}. \tag{4.64}$$

Thus

$$P(E) = P(AAK) + P(AKA) + P(KAA) = \frac{6}{5525} \approx 0.11\%. \qquad (4.65)$$

Let us explain the reasons for these calculations: $P(\text{Ace first}) = \frac{4}{52}$, since there are 4 Aces and 52 cards at the beginning. $P(\text{Ace second} \mid \text{Ace first}) = \frac{3}{51}$, since after drawing an Ace first, we are left with 3 Aces and 51 cards. Then, from the definition of conditional probabilities,

$$P(\text{Ace first and Ace second}) = P(\text{Ace first})P(\text{Ace second}|\text{Ace first}) = \frac{4}{52} \cdot \frac{3}{51}.$$
$$(4.66)$$

After drawing two Aces, we have 4 Kings and 50 cards left, hence $P(\text{King}$ third $\mid$ Ace first and Ace second$) = \frac{4}{50}$. Then, again from the definition of conditional probabilities,

$$P(AAK) = P(\text{Ace first and Ace second and King third})$$
$$= P(\text{Ace first and Ace second})P(\text{King third}|\text{Ace first and Ace second})$$
$$= \frac{4}{52} \cdot \frac{3}{51} \cdot \frac{4}{50} = \frac{2}{5525}, \qquad (4.67)$$

which is the same as our previous value for $P(AAK)$. Now $P(AKA)$ and $P(KAA)$ can be obtained in a similar manner, and since these are the probabilities of mutually exclusive events whose union is E, we obtain $P(E)$ as their sum. ◆

The foregoing examples illustrate two general theorems:

Theorem 4.5.1. *Joint Probability of Three Events. For any three events A, B, and C with $P(BC) \neq 0$ we have*

$$P(ABC) = P(A|BC)P(B|C)P(C). \qquad (4.68)$$

We leave the proof to the reader.

Theorem 4.5.2. *The Theorem of Total Probability. If $B_1, B_2, \ldots, B_n$ are mutually exclusive events with nonzero probabilities, whose union is B, and A is any event, then*

$$P(AB) = P(A|B_1)P(B_1) + P(A|B_2)P(B_2) + \cdots + P(A|B_n)P(B_n). \quad (4.69)$$

Proof. Applying Equation 4.51 to each term on the right above, we get

$$P(A|B_1)P(B_1) + P(A|B_2)P(B_2) + \cdots + P(A|B_n)P(B_n)$$
$$= P(AB_1) + P(AB_2) + \cdots + P(AB_n) = P(AB). \qquad (4.70)$$

The last sum equals $P(AB)$, because

$$(AB_1) \cup (AB_2) \cup \ldots \cup (AB_n) = A(B_1 \cup B_2 \cup \ldots \cup B_n) = AB . \quad (4.71)$$

and the AB_i's are mutually exclusive because the B_i's are, that is,

$$(AB_i)(AB_j) = A(B_i B_j) = A\emptyset = \emptyset \quad\quad\quad (4.72)$$

for any pair B_i, B_j with $i \neq j$.   ∎

If $B = S$ or $B \supset A$, then $AB = A$ and the theorem reduces to the following special case:

Corollary 4.5.1. *If A is any event and $B_1, B_2, \ldots, B_n$ are mutually exclusive events with nonzero probabilities, whose union is S or contains A, then*

$$P(A) = P(A|B_1)P(B_1) + P(A|B_2)P(B_2) + \cdots + P(A|B_n)P(B_n). \quad (4.73)$$

Example 4.5.3. Second Card in a Deal.

From a well-shuffled deck of 52 cards, we deal out two cards. What is the probability that the second card is a spade?

We present two solutions.

First, letting S_1 denote the event that the first card is a spade and S_2 the event that the second one is a spade, Corollary 4.5.1 gives

$$P(S_2) = P(S_2|S_1)P(S_1) + P(S_2|\overline{S}_1)P(\overline{S}_1) = \frac{12}{51} \cdot \frac{13}{52} + \frac{13}{51} \cdot \frac{39}{52} = \frac{1}{4}. \quad (4.74)$$

On the other hand, we could have argued simply that the second card in the deck has just as much chance of being a spade as the first card, if we do not know whether the first card is a spade or not. Similarly, the probability that the nth card is a spade is also $1/4$ for any n from 1 to 52, since we may cut the deck just above the nth card, and start dealing from there.   ◆

Example 4.5.4. Suit of Cards Under Various Conditions.

From a deck of cards, two are dealt without replacement. Find the probabilities that

a) Both are clubs, given that the first one is a club,
b) Both are clubs, given that one is a club,
c) Both are clubs, given that one is the Ace of clubs,
d) One is the Ace of clubs, given that both are clubs.

a) Clearly, P(both are clubs | the first one is a club) = P(second card is a club | the first one is a club) = $\frac{12}{51} = \frac{4}{17}$.

b) In this case, the possible outcomes are $\{CC, C\overline{C}, \overline{C}C, \overline{C}\,\overline{C}\}$, with C denoting a club and $\overline{C}$ a non-club, and the first letter indicating the first card and the second letter the second card. The condition that one card is a club means that we know that one of the two cards is a club but the other can be anything or, in other words, that at least one of the two cards is a club. Thus,[9] P(one is a club) $= \mathrm{P}\left(CC, C\overline{C}, \overline{C}C\right) = \frac{1}{4}\cdot\frac{12}{51} + \frac{1}{4}\cdot\frac{39}{51} + \frac{3}{4}\cdot\frac{13}{51}$, and so

$$\mathrm{P}(\text{both are } C \mid \text{one is } C) = \mathrm{P}\left(CC | CC \cup C\overline{C} \cup \overline{C}C\right)$$

$$= \frac{\frac{1}{4}\cdot\frac{12}{51}}{\frac{1}{4}\cdot\frac{12}{51} + \frac{1}{4}\cdot\frac{39}{51} + \frac{3}{4}\cdot\frac{13}{51}} = \frac{2}{15}. \qquad (4.75)$$

Another way of computing this probability is by using a reduced sample space: There are $13 \cdot 51 + 39 \cdot 13$ ordered ways of dealing at least one club, because the first card can be a club 13 ways and then the second card can be any one of the remaining 51 cards or the first card can be other than a club in 39 ways but then the second card must be one of the 13 clubs. Also, there are $13 \cdot 12$ ways of dealing two clubs and $\{\text{both are } C\} \cap \{\text{one is } C\} = \{\text{both are } C\}$. Thus,

$$\mathrm{P}(\text{both are } C \mid \text{one is } C) = \frac{13 \cdot 12}{13 \cdot 51 + 39 \cdot 13} = \frac{2}{15}, \qquad (4.76)$$

the same as before.

It may seem surprising that the answers to Parts a) and b) are not the same. After all, why should it make a difference whether we know that the first card is a club or just that one of the cards is a club? The answer is that the conditions are different: in case a), we computed $\mathrm{P}\left(CC | CC \cup C\overline{C}\right) = \frac{\frac{1}{4}\cdot\frac{12}{51}}{\frac{1}{4}\cdot\frac{12}{51}+\frac{1}{4}\cdot\frac{39}{51}} = \frac{4}{17}$, whereas in case b), we computed $\mathrm{P}\left(CC | CC \cup C\overline{C} \cup \overline{C}C\right)$.

c) Again, at first glance, it may seem paradoxical that it makes a difference whether we know that one of the cards is the Ace of clubs or just any club, but, as we shall see, we are talking here of a different event under a different condition.

Computing with the reduced sample space, we have $1 \cdot 51 + 51 \cdot 1$ ordered ways of dealing the Ace of clubs, because the first card can be the Ace of clubs in just 1 way and then the second card can be any one of the remaining 51 cards, or the first card can be other than the Ace of clubs in 51 ways but then the second card must be the Ace of clubs. Similarly, there are $1\cdot12+12\cdot1$ ways of dealing two clubs, one of which is the Ace, and so

$$\mathrm{P}(\text{both are } C \mid \text{one is the } AC) = \frac{1 \cdot 12 + 12 \cdot 1}{1 \cdot 51 + 51 \cdot 1} = \frac{4}{17}. \qquad (4.77)$$

---

[9] We usually omit the braces or union signs around compound events when there are already parentheses there, and separate the components with commas. Thus we write $\mathrm{P}\left(CC, C\overline{C}, \overline{C}C\right)$ rather than $\mathrm{P}\left(\{CC, C\overline{C}, \overline{C}C\}\right)$ or $\mathrm{P}\left(CC \cup C\overline{C} \cup \overline{C}C\right)$.

d) In this case

$$\mathrm{P}(\text{one is the } AC \mid \text{both are } C) = \frac{\mathrm{P}(\text{one is the } AC \text{ and both are } C)}{\mathrm{P}(\text{both are } C)}$$

$$= \frac{\frac{1}{52} \cdot \frac{12}{51} + \frac{12}{52} \cdot \frac{1}{51}}{\frac{13}{52} \cdot \frac{12}{51}} = \frac{2}{13}. \qquad (4.78)$$

♦

Example 4.5.5. The Gambler's Ruin.

A gambler who has $m > 0$ dollars, bets 1 dollar each time on H in successive tosses of a coin, that is, he wins or loses 1 dollar each time, until he ends up with n dollars, for some $n > m$ or runs out of money. Find the probability of the gambler's ruin, assuming that his opponent cannot be ruined.

Let A_m denote the event that the gambler with initial capital m is ruined. Then, if he wins the first toss, he has $m + 1$ dollars, and the event of ruin in that case is denoted by A_{m+1}. That is, $\mathrm{P}(A_m | H) = \mathrm{P}(A_{m+1})$, where H denotes the outcome of the first toss. Similarly, $\mathrm{P}(A_m | T) = \mathrm{P}(A_{m-1})$.

On the other hand, by the Corollary,

$$\mathrm{P}(A_m) = \mathrm{P}(A_m | H)\mathrm{P}(H) + \mathrm{P}(A_m | T)\mathrm{P}(T) \text{ for } 0 < m < n, \qquad (4.79)$$

which can then also be written as

$$\mathrm{P}(A_m) = \mathrm{P}(A_{m+1}) \cdot \frac{1}{2} + \mathrm{P}(A_{m-1}) \cdot \frac{1}{2} \text{ for } 0 < m < n. \qquad (4.80)$$

If we regard $\mathrm{P}(A_m)$ as an unknown function $f(m)$, then this type of equation is called a difference equation and is known to have the general solution $f(m) = a + bm$, where a and b are arbitrary constants. (We shall deduce this fact in Section 6.3, but we need many other facts first.) For a particular solution, these constants can be determined by initial or boundary conditions. In the present case, obviously $\mathrm{P}(A_0) = 1$ and $\mathrm{P}(A_n) = 0$. Hence $a + b0 = 1$ and $a + bn = 0$, which give $a = 1$ and $b = -1/n$. Thus, the probability of the gambler's ruin is

$$\mathrm{P}(A_m) = 1 - \frac{m}{n}. \qquad (4.81)$$

This formula is indeed very reasonable. It shows, for instance, that if $n = 2m$, that is, that the gambler wants to double his money, then both the probability of ruin and the probability of success are $1/2$. Similarly, if the gambler wants to triple his money, that is, $n = 3m$, then the probability of ruin is $2/3$. Generally, the greedier he is, the larger the probability of ruin.♦

Example 4.5.6. Laplace's Rule of Succession.

The great eighteenth-century French mathematician Laplace used the following very interesting argument to estimate the chances of the sun's rising tomorrow.

Let sunrises be independent random events with an unknown probability p of occurrence. Let N be a large positive integer, $B_i = $ "p is i/N," for $i = 0, 1, 2, \ldots, N$, and $A = $ "the sun has risen every day for n days," where Laplace took n to be 1,826,213 days, which is 5,000 years. He assumed, since we have no advance knowledge of the value of p, that the possible values are equally likely and so $P(B_i) = \frac{1}{N+1}$ for each i. By the assumed independence,

$$P(A|B_i) = \left(\frac{i}{N}\right)^n. \tag{4.82}$$

Hence, by the Theorem of Total Probability,

$$P(A) = \sum_{i=0}^{N} \frac{1}{N+1} \left(\frac{i}{N}\right)^n. \tag{4.83}$$

Similarly, if $B = $ " the sun has risen for n days and will rise tomorrow," then

$$P(B) = \sum_{i=0}^{N} \frac{1}{N+1} \left(\frac{i}{N}\right)^{n+1}. \tag{4.84}$$

Consequently,

$$P(B|A) = \frac{P(AB)}{P(A)} = \frac{P(B)}{P(A)}. \tag{4.85}$$

For large values of N, the sums can be simplified by noting that $\sum_{i=0}^{N} \frac{1}{N} \left(\frac{i}{N}\right)^n$ is a Riemann sum for the integral $\int_0^1 x^n dx = \frac{1}{n+1}$. Therefore

$$P(A) \approx \frac{N}{N+1} \cdot \frac{1}{n+1}, \tag{4.86}$$

and

$$P(B) \approx \frac{N}{N+1} \cdot \frac{1}{n+2}. \tag{4.87}$$

Thus, the probability that the sun will rise tomorrow, if it has risen every day for n days is

$$P(B|A) \approx \frac{n+1}{n+2}. \tag{4.88}$$

For $n = 1,826,213$, this result is indeed very close to one. Unfortunately, however, the argument is on shaky grounds. First, it is difficult to see sunrises as random events. Second, why would sunrises on different days be independent of each other? Third, just because we don't know a probability, we cannot assume that it has a random value equally likely to be any number from zero to one. In the eighteenth century, however, probability theory was in its infancy, and its foundations were murky. Setting aside the application to the sun, we can easily build a model with urns and balls for which the probabilities above provide an accurate description.   ♦

The next theorem is a straightforward formula based on the definition of conditional probabilities and the theorem of total probability. It is important because it provides a scheme for many applications. Before discussing the general formula, however, we start with a simple example.

Example 4.5.7. Which Urn Did a Ball Come From?

We consider the same experiment as in Example 4.5.1 but ask a different question: We have two urns, with the first one containing two white and six black balls and the second one containing two white and two black balls. We pick an urn at random and then pick a ball from the chosen urn at random. We observe that the ball is white and ask: what is then the probability that it came from urn 1, that is, that in the first step we picked urn 1?

With the notation of Example 4.5.1, we are asking for the conditional probability $P(U_1|W)$. This probability can be computed as follows:

$$P(U_1|W) = \frac{P(WU_1)}{P(W)} = \frac{P(W|U_1)P(U_1)}{P(W|U_1)P(U_1) + P(W|U_2)P(U_2)}$$

$$= \frac{\frac{2}{8} \cdot \frac{1}{2}}{\frac{2}{8} \cdot \frac{1}{2} + \frac{2}{4} \cdot \frac{1}{2}} = \frac{1}{3}. \tag{4.89}$$

♦

The general scheme that this example illustrates is this: we have several possible outcomes of an experiment, like U_1 and U_2 of the first stage above, and we observe the occurrence of some other event like W. We ask then the question: What are the new probabilities of the original outcomes in light of this observation? The answer for the general case is given by

Theorem 4.5.3. Bayes' Theorem. *If A is any event with $P(A) \neq 0$ and $B_1, B_2, \ldots, B_n$ are mutually exclusive events with nonzero probabilities, whose union is S or contains A, then*

$$P(B_i|A) = \frac{P(A|B_i)P(B_i)}{P(A|B_1)P(B_1) + P(A|B_2)P(B_2) + \cdots + P(A|B_n)P(B_n)}$$

$$for \quad i = 1, 2, \ldots, n. \tag{4.90}$$

The proof is left to the reader.

Example 4.5.8. A Blood Test.

A blood test, when given to a person with a certain disease, shows the presence of the disease with a probability of .99 and fails to show it with a probability of .01. It also produces a false-positive result for healthy persons, with a probability of .02. We also know that .1% of the population has the disease. What is the probability that a person really has the disease if the test says so?

We use Bayes' theorem for a randomly selected person, with $B_1 =$ "the person has the disease," $B_2 =$ "the person does not have the disease," and $A =$ "the test gives a positive result." Then we are looking for $P(B_1|A)$, and we know that $P(A|B_1) = .99$, $P(B_1) = .001$, $P(A|B_2) = .02$, and $P(B_2) = .999$. Hence,

$$P(B_1|A) = \frac{.99 \cdot .001}{.99 \cdot .001 + .02 \cdot .999} = \frac{99}{99 + 1998} \approx .047. \tag{4.91}$$

Thus, the probability that a person really has the disease if the test says so turns out to be less than 5%. This number is unbelievably low. After all, the test is 99 or 98 percent accurate, so how can this be true? The explanation is this: The positive test result can arise in two ways. Either it is a *true* positive result, that is, the patient has the disease and the test shows it correctly, or it is a *false*-positive result, that is, the test has mistakenly diagnosed a healthy person as diseased. Now, because the disease is very rare (only one person in a thousand has it), the number of healthy persons is relatively large, and so the 2% of them who are falsely diagnosed as diseased still far outnumber, 1998 to 99, the correctly diagnosed, diseased people. Thus, the fraction of correct positive test results to all positive ones is small.

The moral of the example is that for a rare disease, we need a much more accurate test. The probability of a false-positive result must be of a lower order of magnitude than the fraction of people with the disease. On the other hand, the probability of a false-negative result does not have to be so low; it just depends on how many diseased persons we can afford to miss, regardless of the rarity of the disease.   ◆

Bayes' theorem is sometimes described as a formula for the probabilities of "causes." In the above example, for instance, B_1 and B_2 may be considered the two possible causes of the positive test result. The probabilities $P(B_1)$ and $P(B_2)$ are called the prior probabilities of these causes and $P(B_1|A)$ and $P(B_2|A)$ their posterior probabilities, because they represent the probabilities of B_1 and B_2 before and after consideration of the occurrence of A. The terminology of "causes" is, however, misleading in many applications where no causal relationship exists between A and the B_i.

Although Bayes' theorem is certainly true and quite useful, it has been controversial because of philosophical problems with the assignment of prior probabilities in some applications.

Exercises

Exercise 4.5.1.

In an urn, there are one white and three black balls and in a second urn three white and two black balls. One of the urns is chosen at random, and then a ball is picked from it at random:

a) Illustrate the possibilities by a tree diagram,
b) Find the branch probabilities,
c) Find the probability of picking a white ball.

Exercise 4.5.2.

Given two urns with balls as in the previous problem, we choose an urn at random, and then we pick two balls from it without replacement.

a) Illustrate the possibilities with a tree diagram,
b) Find the branch probabilities,
c) Find the probability of picking a white and a black ball (in any order).

Exercise 4.5.3.

From a deck of cards, two are drawn without replacement. Find the probabilities that

a) Both are Aces, given that one is an Ace,
b) Both are Aces, given that one is a red Ace,
c) Both are Aces, given that one is the Ace of spades,
d) One is the Ace of spades, given that both are Aces.

Exercise 4.5.4.

Modify the Gambler's ruin problem as follows: Suppose there are two players, Alice and Bob, who bet on successive flips of a coin until one of them wins all the money of the other. Alice has m dollars and bets one dollar each time on H, while Bob has n dollars and bets one dollar each time on T. In each play, the winner takes the dollar of the loser. Find the probability of ruin for each player.

Exercise 4.5.5.

Modify the Gambler's ruin problem by changing the probability of winning from $1/2$ to p in each trial. (Hint: Modify Equation 4.80 and try to find constants λ such that $P(A_m) = \lambda^m$ for $0 < m < n$. The general solution should be of the form $P(A_m) = a\lambda_1^m + b\lambda_2^m$, and the constants a and b are to be determined from the boundary conditions.)

Exercise 4.5.6.

In a mythical kingdom, a prisoner is given two urns and 50 black and 50 white marbles. The king says that the prisoner must place all the marbles in the urns with neither urn remaining empty and he will return later and pick an urn and then a marble from it at random. If the marble is white, the prisoner will be released, but if it is black, he will remain in jail. How should the prisoner distribute the marbles? Prove that your answer indeed maximizes the prisoner's chances of going free.

Exercise 4.5.7.

In an urn, there are one white and three black balls and in a second urn three white and two black balls as in Exercise 4.5.1. One of the urns is chosen at random, and then a ball is picked from it at random and turns out to be white. What is then the probability that it came from urn 1?

Exercise 4.5.8.

Given two urns with balls as in the previous problem, we choose an urn at random and then we pick two balls from it without replacement. (See also Exercise 4.5.2.) What is the probability that the two balls came from urn 1 if they have different colors?

Exercise 4.5.9.

From all families with two children, one is selected at random, and then a child is selected from it at random and is found to be a girl. What is the probability that she came from a family with two girls? (Assume that each child is a boy or a girl with probability $1/2$, independently of one another.) Use Bayes' theorem.

Exercise 4.5.10.

From all families with three children, one is selected at random, and then a child is selected from it at random and is found to be a girl. What is the probability that she came from a family with two girls and one boy? (Assume that each child is a boy or a girl with probability 1/2, independently of one another.) Use Bayes' theorem.

Exercise 4.5.11.

Given two urns with balls as in Exercise 4.5.1, we choose a ball from each urn. If one ball is white and the other black, what is the probability that the white ball came from urn 1?

Exercise 4.5.12.

On a multiple-choice question with five choices, a certain student either knows the answer and then marks the correct choice or does not know the answer and then marks one of the choices at random. What is the probability that he knew the answer if he marked the correct choice? Assume that the prior probability that he knew the answer is $\frac{3}{4}$.

Exercise 4.5.13.

Keith Devlin attributes this problem to Amos Tversky[10]: Imagine you are a member of a jury judging a hit-and-run case. A taxi hit a pedestrian one night and fled the scene. The entire case against the taxi company rests on the evidence of one witness, an elderly man, who saw the accident from his window some distance away. He says that he saw the pedestrian struck by a blue taxi. In trying to establish her case, the lawyer for the injured pedestrian establishes the following facts:

1. There are only two taxi companies in town, "Blue Cabs" and "Black Cabs." On the night in question, 85% of all taxies on the road were black and 15% were blue.
2. The witness has undergone an extensive vision test under conditions similar to those on the night in question and has demonstrated that he can successfully distinguish a blue taxi from a black taxi 80% of the time.

If you were on the jury, how would you decide?

---

[10] Tversky's Legacy Revisited, by Keith Devlin,
www.maa.org/devlin/devlin_july.html, 1996.

Exercise 4.5.14.

This is Bertrand's box problem, published in 1889.

1. There are three boxes: a box containing two gold coins, a box containing two silver coins, and a box containing one gold coin and one silver coin. A box is picked at random, and then a coin is picked from it at random and is found to be a gold coin. What is the probability that the other coin in the selected box is also gold?
2. This problem is often presented as a paradox with the following wrong solution: If we obtain a gold coin, then it had to come from box 1 or from box 3. Since those two boxes are equally likely to be picked, the probability that the other coin in the selected box is also gold is 1/2. What is wrong here?

Exercise 4.5.15.

What is the probability that the car is behind door No. 2 in our description of the Monty Hall problem, that is, given that the player picks door No. 1 and the host opens door No.3 (Example 4.1.3), if we modify it so that the host knows where the car is and

1. If door No. 1, which the player has picked, hides the car, then the host will always open door No. 3?
2. If door No. 1, which the player has picked, hides the car, then the host will open door No. 3 with probability p and door No. 2 with probability $1 - p$? Show that P(car is behind 2|3 is opened) $> 1/2$ holds.for any $p < 1$.

5. Random Variables

5.1 Probability Functions and Distribution Functions

In many applications, the outcomes of a probabilistic experiment are numbers
or have some numbers associated with them, and we can use these numbers
to obtain important information beyond what we have seen so far. We can,
for instance, describe in various ways how large or small these numbers are
likely to be and compute likely averages and measures of spread. For example,
in three tosses of a coin, the number of heads obtained can range from 0 to
3, and there is one of these numbers associated with each possible outcome.
Informally, the quantity "number of heads" is called a random variable and
the numbers 0 to 3 its possible values. In general, such an association of
numbers with each member of a set is called a function. For most functions
whose domain is a sample space, we have a new name:

Definition 5.1.1. _Random Variable._ _A random variable (abbreviated r.v.)_
is a real-valued function on a sample space.

Random variables are usually denoted by capital letters from the end of
the alphabet, such as X, Y, Z, and sets like $\{s : X(s) = x\}$, $\{s : X(s) \leq x\}$,
and $\{s : X(s) \in I\}$, for any number x and any interval I, are events[1] in S.
They are usually abbreviated as $\{X = x\}$, $\{X \leq x\}$, and $\{X \in I\}$ and have
probabilities associated with them. The assignment of probabilities to all such
events, for a given random variable X, is called the _probability distribution_
_of_ X. Furthermore, in the notation for such probabilities, it is customary
to drop the braces, that is, to write $P(X = x)$, for instance, rather than
$P(\{X = x\})$.

Hence, the preceding example can be formalized thus:

---

[1] Actually, in infinite sample spaces, there exist complicated functions for which
not all such sets are events, and so we define a r.v. as not just any real-valued
function X, but a so-called measurable function, that is, one for which all such
sets _are_ events. We shall ignore this issue; it is explored in more advanced books.

© Springer International Publishing Switzerland 2016
G. Schay, _Introduction to Probability with Statistical Applications_,
DOI 10.1007/978-3-319-30620-9_5

Example 5.1.1. Three Tosses of a Coin.

Let $S = \{HHH, HHT, HTH, HTT, THH, THT, TTH, TTT\}$ describe three tosses of a coin, and let X denote the number of heads obtained. Then the values of X, for each outcome s in S, are given in the following table:

$s:$	HHH	HHT	HTH	HTT	THH	THT	TTH	TTT
$X(s):$	3	2	2	1	2	1	1	0

Thus, in the case of three independent tosses of a fair coin, $P(X = 0) = 1/8$, $P(X = 1) = 3/8$, $P(X = 2) = 3/8$, and $P(X = 3) = 1/8$. ♦

The following functions are generally used to describe the probability distribution of a random variable:

Definition 5.1.2. Probability Function. *For any probability space and any random variable X on it, the function $f(x) = P(X = x)$, defined for all possible values[2] x of X, is called the probability function (abbreviated p.f.) of X.*

Definition 5.1.3. Distribution Function. *For any probability space and any random variable X on it, the function $F(x) = P(X \leq x)$, defined for all real numbers x, is called the distribution function (abbreviated d.f.) of X.*

Example 5.1.2. Three Tosses of a Coin, Continued.

Let X be the number of heads obtained in three independent tosses of a fair coin, as in the previous example. Then the p.f. of X is given by

$$f(x) = \begin{cases} 1/8 \text{ if } & x = 0 \\ 3/8 \text{ if } & x = 1 \\ 3/8 \text{ if } & x = 2 \\ 1/8 \text{ if } & x = 3 \end{cases} \tag{5.1}$$

and the d.f. of X is given by

$$F(x) = \begin{cases} 0 \text{ if } x < 0 \\ 1/8 \text{ if } 0 \leq x < 1 \\ 4/8 \text{ if } 1 \leq x < 2 \\ 7/8 \text{ if } 2 \leq x < 3 \\ 1 \text{ if } x \geq 3 \end{cases} \tag{5.2}$$

The graphs of these functions are shown in Figures 5.1 and 5.2 below.

It is also customary to picture the probability function by a histogram, which is a bar chart with the probabilities represented by areas. For the X above, this is shown in Figure 5.3. (In this case, the bars all have width one, and so their heights and areas are equal.) ♦

---

[2] Sometimes $f(x)$ is considered to be a function on all of $\mathbb{R}$, with $f(x) = 0$ if x is not a possible value of X. This is a minor distinction, and it should be clear from the context which definition is meant.

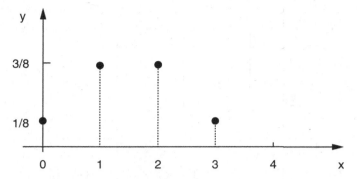

Fig. 5.1. Graph of the p.f. f of a binomial random variable with parameters $n = 3$ and $p = 1/2$

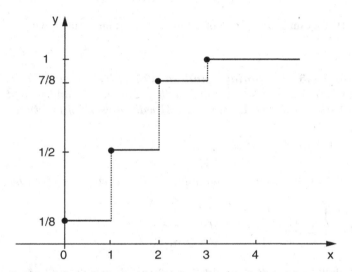

Fig. 5.2. Graph of the d.f. F of a binomial random variable with parameters $n = 3$ and $p = 1/2$

Certain frequently occurring random variables and their distributions have special names. Two of these are generalizations of the number of heads in the above example. The first one is for a single toss, but with a not necessarily fair coin, and the second one for an arbitrary number of tosses.

Definition 5.1.4. Bernoulli Random Variables. *A random variable X is called a Bernoulli random variable with parameter p, if it has two possible values, 0 and 1, with $P(X = 1) = p$ and $P(X = 0) = 1 - p = q$, where p is any number from the interval $[0,1]$. An experiment whose outcome is a Bernoulli random variable is called a Bernoulli trial.*

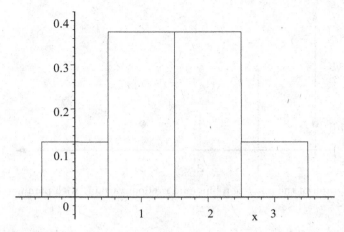

Fig. 5.3. Histogram of the p.f. f of a binomial random variable with parameters $n = 3$ and $p = 1/2$

Definition 5.1.5. *Binomial Random Variables. A random variable X is called a binomial random variable with parameters n and p, if it has the binomial distribution (see Example 4.3.4) with probability function*

$$f(x) = \binom{n}{x} p^x q^{n-x} \quad if \ \ x = 0, 1, 2, \ldots, n. \tag{5.3}$$

The distribution function of a binomial random variable is given by

$$F(x) = \begin{cases} 0 & if \ x < 0 \\ \sum_{k=0}^{\lfloor x \rfloor} \binom{n}{k} p^k q^{n-k} & if \ 0 \leq x < n \\ 1 & if \ x \geq n. \end{cases} \tag{5.4}$$

Here $\lfloor x \rfloor$ denotes the floor or greatest integer function, that is, $\lfloor x \rfloor =$ the greatest integer $\leq x$.

Example 5.1.3. Sum of Two Dice.

Let us consider again the tossing of two dice, with 36 equiprobable elementary events, and let X be the sum of the points obtained. Then $f(x)$ and $F(x)$ are given by the following tables. (Count the appropriate squares in Figure 2.4 on p. 21.)

x :	2	3	4	5	6	7	8	9	10	11	12
$f(x)$:	1/36	2/36	3/36	4/36	5/36	6/36	5/36	4/36	3/36	2/36	1/36

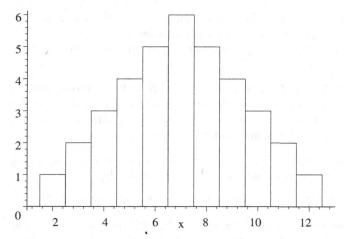

Fig. 5.4. Histogram of the d.f. of the sum thrown with two dice. The y-scale shows multiples of $1/36$

$x \in$	$(-\infty, 2)$	$[2, 3)$	$[3, 4)$	$[4, 5)$	$[5, 6)$	$[6, 7)$
$F(x):$	0	1/36	3/36	6/36	10/36	15/36

$[7, 8)$	$[8, 9)$	$[9, 10)$	$[10, 11)$	$[11, 12)$	$[12, \infty)$
21/36	26/36	30/36	33/36	35/36	1

The histogram of $f(x)$ and the graph of $F(x)$ are given by Figures 5.4 and 5.5. ♦

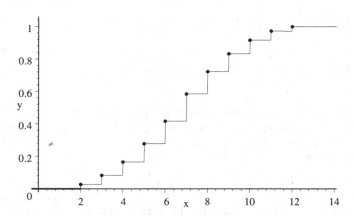

Fig. 5.5. Graph of the d.f. of the sum thrown with two dice

A random variable is said to be *discrete* if it has only a finite or a countably infinite number of possible values. The random variables we have seen so far are discrete. In the next section, we shall discuss the most important class of non-discrete random variables: continuous ones.

Another important type of discrete variable is named in the following definition:

Definition 5.1.6. *Discrete Uniform Random Variables.* *A random variable X and its distribution are called discrete uniform if X has a finite number of possible values, say $x_1, x_2, \ldots, x_n$, for any positive integer n, and $P(X = x_i) = \frac{1}{n}$ for all $i = 1, 2, \ldots, n$.*

Random variables with a countably infinite number of possible values occur in many applications, as in the next example.

Example 5.1.4. *Throwing a Die Until a Six Comes Up.*

Suppose we throw a fair die repeatedly, with the throws being independent of each other, until a six comes up. Let X be the number of throws. Clearly, X can take on any positive integer value, for it is possible (though unlikely) that we do not get a six in 100 throws, or 1000 throws, or in any large number of throws.

The probability function of X can be computed easily as follows:
$f(1) = \mathrm{P}(X = 1) = \mathrm{P}(\text{six on the first throw}) = \frac{1}{6}$,
$f(2) = \mathrm{P}(X = 2) = \mathrm{P}(\text{non} - \text{six on the first throw and six on the second})$
$= \frac{5}{6} \cdot \frac{1}{6}$,
$f(3) = \mathrm{P}(X = 3) = \mathrm{P}$ (non-six on the first two throws and six on the third) $= \left(\frac{5}{6}\right)^2 \cdot \frac{1}{6}$, and so on.
Thus

$$f(k) = \mathrm{P}(X = k) = \left(\frac{5}{6}\right)^{k-1} \cdot \frac{1}{6} \quad \text{for} \quad k = 1, 2, \ldots . \tag{5.5}$$

♦

The above example is a special case of another named family of random variables:

Definition 5.1.7. *Geometric Random Variables.* *Suppose we perform independent Bernoulli trials with parameter p, with $0 < p < 1$, until we obtain a success. The number X of trials is called a geometric random variable with parameter p. It has the probability function*

$$f(k) = P(X = k) = pq^{k-1} \quad for \quad k = 1, 2, \ldots . \tag{5.6}$$

The name "geometric" comes from the fact that the $f(k)$ values are the terms of a geometric series. Using the formula for the sum of a geometric series, we can confirm that they form a probability distribution:

$$\sum_{k=1}^{\infty} f(k) = \sum_{k=1}^{\infty} pq^{k-1} = \frac{p}{1-q} = 1. \tag{5.7}$$

From the preceding examples, we can glean some general observations about the probability and distribution functions of discrete random variables.

If $x_1, x_2, \ldots$ are the possible values of a discrete random variable X, then $f(x_i) \geq 0$ for all these values and $f(x) = 0$ otherwise. Furthermore, $\sum f(x_i) = 1$, because this sum equals the probability that X takes on any of its possible values, which is certain. Hence the total area of all the bars in the histogram of $f(x)$ is 1. Also, we can easily read off the histogram the probability of X falling in any given interval I, as the total area of those bars that cover the x_i values in I. For instance, for the X of Example 5.1.3, $P(3 < X \leq 6) = P(X = 4) + P(X = 5) + P(X = 6) = \frac{3}{36} + \frac{4}{36} + \frac{5}{36} = \frac{1}{3}$, which is the total area of the bars over 4, 5, and 6.

The above observations, when applied to infinite intervals of the type $(-\infty, x]$, lead to the equation $F(x) = P(X \in (-\infty, x]) = \sum_{x_i \leq x} P(X = x_i) = $ sum of the areas of the bars over each $x_i \leq x$ and to the following properties of the distribution function:

Theorem 5.1.1. *Properties of Distribution Functions. The distribution function F of any random variable X has the following properties:*

1. $F(-\infty) = \lim_{x \to -\infty} F(x) = 0$, *since as $x \to -\infty$, the interval $(-\infty, x]$* $\to \emptyset$.
2. $F(\infty) = \lim_{x \to \infty} F(x) = 1$, *since as $x \to \infty$, the interval $(-\infty, x] \to \mathbb{R}$.*
3. *F is a nondecreasing function, since if $x < y$, then*

$$F(y) = P(X \in (-\infty, y]) = P(X \in (-\infty, x]) + P(X \in (x, y])$$
$$= F(x) + P(X \in (x, y]), \tag{5.8}$$

and so, $F(y)$ being the sum of $F(x)$ and a nonnegative term, we have $F(y) \geq F(x)$.
4. *F is continuous from the right at every x.*

These four properties of F hold not just for discrete random variables but for all types. Their proofs are outlined in Exercise 5.1.13 and those following it. Also, in more advanced courses, it is proved that any function with these four properties is the distribution function of some random variable.

While the distribution function can be used for any random variable, the probability function is useful only for discrete ones. To describe continuous random variables, we need another function, the so-called density function, instead, as will be seen in the next section.

The next theorem shows that the distribution function of a random variable X completely determines the distribution of X, that is, the probabilities $P\{X \in I\}$ for all intervals I.

Theorem 5.1.2. *Probabilities of a Random Variable Falling in Various Intervals. For any random variable X and any real numbers x and y,*

1. $P(X \in (x,y]) = F(y) - F(x)$,
2. $P(X \in (x,y)) = \lim_{t \to y^-} F(t) - F(x)$,
3. $P(X \in [x,y]) = F(y) - \lim_{t \to x^-} F(t)$,
4. $P(X \in [x,y)) = \lim_{t \to y^-} F(t) - \lim_{t \to x^-} F(t)$.

For discrete random variables, the probability function and the distribution function determine each other: Let x_i, for $i = 1, 2, \ldots$, denote the possible values of X. Then clearly, for any x,

$$F(x) = \sum_{x_i \leq x} f(x_i) \tag{5.9}$$

and

$$f(x) = F(x) - \lim_{t \to x^-} F(t). \tag{5.10}$$

The first of these equations shows that $F(x)$ is constant between successive x_i values, and the latter equation shows that $f(x_i)$ equals the value of the jump of F at $x = x_i$.

Exercises

Exercise 5.1.1.

Let X be the number of hearts in a randomly dealt poker hand of five cards. Draw a histogram for its probability function and a graph for its distribution function.

Exercise 5.1.2.

Let X be the number of heads obtained in five independent tosses of a fair coin. Draw a histogram for its probability function and a graph for its distribution function.

Exercise 5.1.3.

Let X be the number of heads minus the number of tails obtained in four independent tosses of a fair coin. Draw a histogram for its probability function and a graph for its distribution function.

Exercise 5.1.4.

Let X be the absolute value of the difference between the number of heads and the number of tails obtained in four independent tosses of a fair coin. Draw a histogram for its probability function and a graph for its distribution function.

Exercise 5.1.5.

Let X be the larger of the number of heads and the number of tails obtained in five independent tosses of a fair coin. Draw a histogram for its probability function and a graph for its distribution function.

Exercise 5.1.6.

Let X be the number of heads minus the number of tails obtained in n independent tosses of a fair coin. Find a formula for its probability function and one for its distribution function.

Exercise 5.1.7.

Suppose we perform independent Bernoulli trials with parameter p, until we obtain two consecutive successes or two consecutive failures. Draw a tree diagram and find the probability function of the number of trials.

Exercise 5.1.8.

Suppose two players, A and B, play a game consisting of independent trials, each of which can result in a win for A or for B or in a draw D, until one player wins a trial. In each trial, $P(A \text{ wins}) = p_1$, $P(B \text{ wins}) = p_2$, and $P(\text{draw}) = q = 1 - (p_1 + p_2)$. Let $X = n$ if A wins the game in the nth trial, and $X = 0$ if A does not win the game ever. Draw a tree diagram and find the probability function of X. Find also the probability that A wins (in any number of trials) and the probability that B wins. Show also that the probability of an endless sequence of draws is 0.

Exercise 5.1.9.

Let X be the number obtained in a single roll of a fair die. Draw a histogram for its probability function and a graph for its distribution function.

Exercise 5.1.10.

We roll two fair dice, a blue and a red one, independently of each other. Let X be the number obtained on the blue die minus the number obtained on the red die. Draw a histogram for its probability function and a graph for its distribution function.

Exercise 5.1.11.

We roll two fair dice independently of each other. Let X be the absolute value of the difference of the numbers obtained on them. Draw a histogram for its probability function and a graph for its distribution function.

Exercise 5.1.12.

Let the distribution function of a random variable X be given by

$$F(x) = \begin{cases} 0 & \text{if } x < -2 \\ 1/4 & \text{if } -2 \leq x < 2 \\ 7/8 & \text{if } 2 \leq x < 3 \\ 1 & \text{if } x \geq 3 \end{cases} \qquad (5.11)$$

Find the probability function of X and graph both F and f.

Exercise 5.1.13.

Let $A_1, A_2, \ldots$ be a nondecreasing sequence of events on a sample space S, that is, let $A_n \subset A_{n+1}$ for $n = 1, 2, \ldots$, and let $A = \cup_{k=1}^{\infty} A_k$. Prove that $P(A) = \lim_{n \to \infty} P(A_n)$. *Hint:* Write A as the disjoint union $A_1 \cup [\cup_{k=2}^{\infty} (A_k - A_{k-1})]$, and apply the axiom of countable additivity.

Exercise 5.1.14.

Let $A_1, A_2, \ldots$ be a nonincreasing sequence of events on a sample space S, that is, let $A_n \supset A_{n+1}$ for $n = 1, 2, \ldots$, and let $A = \cap_{k=1}^{\infty} A_k$. Prove that $P(A) = \lim_{n \to \infty} P(A_n)$. *Hint:* Apply DeMorgan's laws to the result of the preceding exercise.

Exercise 5.1.15.

Prove that for the distribution function of any random variable, $\lim_{x \to -\infty} F(x) = 0$. *Hint:* Use the result of the preceding exercise and the theorem from real analysis that if $\lim_{n \to \infty} F(x_n) = L$ for every sequence $\langle x_n \rangle$ decreasing to $-\infty$, then $\lim_{x \to -\infty} F(x) = L$.

Exercise 5.1.16.

Prove that for the distribution function of any random variable, $\lim_{x \to \infty} F(x) = 1$. *Hint:* Use the result of Exercise 5.1.13 and the theorem from real analysis that if $\lim_{n \to \infty} F(x_n) = L$ for every sequence $\langle x_n \rangle$ increasing to ∞, then $\lim_{x \to \infty} F(x) = L$.

Exercise 5.1.17.

Prove that the distribution function F of any random variable is continuous from the right at every x. *Hint:* Use a modified version of the hints of the preceding exercises.

5.2 Continuous Random Variables

In this section, we consider random variables X whose possible values constitute a finite or infinite interval and whose distribution function is not a step function, but a continuous function. Such random variables are called *continuous* random variables.

The continuity of F implies that in Equation 5.10 $\lim_{t \to x^-} F(t) = \lim_{t \to x} F(t) = F(x)$, for every x, and so $f(x) = 0$, for every x. Thus, the probability function does not describe the distribution of such random variables because, in this case, the probability of X taking on any single value is zero. The latter statement can also be seen directly in the case of choosing a number at random from an interval, say from $[0, 1]$: If the probability of every value x were some positive c, then the total probability for obtaining any $x \in [0, 1]$ would be $\infty \cdot c = \infty$, in contradiction to the axiom requiring the total to be 1. On the other hand, we have no problem with $f(x) = 0$, for every x, since $\infty \cdot 0$ is indeterminate.

However, even if the probability of X taking on any single value is zero, the probability of X taking on any value in an *interval* need not be zero. Now, for a discrete random variable, the histogram of $f(x)$ readily displayed the probabilities of X falling in an interval I as the sum of the areas of the rectangles over I. Hence, a very natural generalization of such histograms suggests itself for continuous random variables: Just consider a continuous curve instead of the jagged top of the rectangles, and let the probability of X falling in I be the area under the curve over I. Thus we make the following formal definition:

Definition 5.2.1. *Probability Density.* *Let X be a continuous random variable with a given distribution function F. If there exists a nonnegative function[3] f that is integrable over $\mathbb{R}$ and for which*

$$\int_{-\infty}^{x} f(t)dt = F(x), \text{for all } x, \tag{5.12}$$

then f is called a probability density function[4] (or briefly, density or p.d.f.) of X, and X is called absolutely continuous.

Thus, if X has a density function, then

---

[3] Note that we are using the same letter f for this function as for the p.f. of a discrete r.v. This notation cannot lead to confusion though, since here we are dealing with continuous random variables rather than discrete ones. On the other hand, using the same letter for both functions will enable us to combine the two cases in some formulas later.

[4] The function f is not unique, because the integral remains unchanged if we change the integrand in a countable number of points. Usually, however, there is a version of f that is continuous wherever possible, and we shall call this version *the* density function of X, ignoring the possible ambiguity at points of discontinuity.

$$P\left(X \in [x, y]\right) = F(y) - F(x) = \int_x^y f(t)dt, \tag{5.13}$$

and the probability remains the same whether we include or exclude one or both endpoints x and y of the interval. Also, if we set $x = \infty$ in Equation 5.12, we see that every p.d.f. must satisfy

$$\int_{-\infty}^{\infty} f(t)dt = 1. \tag{5.14}$$

In fact, any nonnegative piecewise continuous function f satisfying Equation 5.14 is a suitable density function and can be used to obtain the distribution function of a continuous random variable via Equation 5.12.

While the density function is not a probability, it is often used with differential notation to write the probability of X falling in an infinitesimal interval as[5]

$$P\left(X \in [x, x + dx]\right) = \int_x^{x+dx} f(t)dt \sim f(x)dx. \tag{5.15}$$

By the fundamental theorem of calculus, the definition of the density function shows that, wherever f is continuous, F is differentiable and

$$F'(x) = f(x). \tag{5.16}$$

There exist, however, continuous random variables whose F is everywhere continuous but not differentiable and which therefore do not have a density function. Such random variables occur only very rarely in applications, and we do not discuss them in this book. In fact, we shall use the term continuous random variable—as most introductory books do—to denote random variables that possess a density function, instead of the precise term "absolutely continuous."

Let us turn now to examples of continuous random variables.

Example 5.2.1. Uniform Random Variable.

Consider a finite interval $[a, b]$, with $a < b$, and pick a point[6] X at random from it, that is, let the possible values of X be the numbers of $[a, b]$, and let X fall in each subinterval $[c, d]$ of $[a, b]$ with a probability that is proportional to the length of $[c, d]$ but does not depend on the location of $[c, d]$ within $[a, b]$. This distribution is achieved by the density function[7]

---

[5] The symbol $\sim$ means that the ratio of the expressions on either side of it tends to 1 as dx tends to 0 or, equivalently, that the limits of each side divided by dx are equal.

[6] We frequently use the words "point" and "number" interchangeably, ignoring the distinction between a number and its representation on the number line, just as the word "interval" is commonly used for both numbers and points.

[7] f is not unique: its values can be changed at a countable number of points, such as a and b, for instance, without affecting the probabilities, which are integrals of f.

$$f(x) = \begin{cases} \dfrac{1}{b-a} & \text{if} \quad a < x < b \\ 0 & \text{if } x \leq a \text{ or } x \geq b \end{cases}.$$ (5.17)

See Figure 5.6. Then, for $a \leq c \leq d \leq b$,

$$P(X \in [c,d]) = \int_c^d f(t)dt = \frac{d-c}{b-a},$$ (5.18)

which is indeed proportional to the length $d - c$ and does not depend on c and d in any other way.

The corresponding distribution function is given by

$$F(x) = \begin{cases} 0 & \text{if} \quad x < a \\ \dfrac{x-a}{b-a} & \text{if } a \leq x < b \\ 1 & \text{if} \quad x \geq b \end{cases}.$$ (5.19)

See Figure 5.7.   ◆

Definition 5.2.2. *Uniform Random Variable.* *A random variable X with the above density is called uniform over* $[a,b]$ *or uniformly distributed over* $[a,b]$. *Its distribution is called the uniform distribution over* $[a,b]$ *and its density and distribution functions the uniform density and distribution functions over* $[a,b]$.

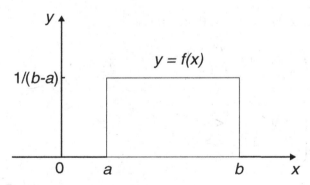

Fig. 5.6. The uniform density function over $[a,b]$

Often we know only the general shape of the density function, and we need to find the value of an unknown constant in its equation. Such constants can be determined by the requirement that f must satisfy $\int_{-\infty}^{\infty} f(t)dt = 1$, because the integral here equals the probability that X takes on any value whatsoever. The next two examples are of this type.

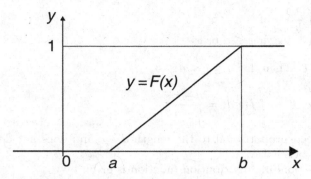

Fig. 5.7. The uniform distribution function over $[a, b]$

Example 5.2.2. Normalizing a p.d.f.

Let X be a random variable with p.d.f.

$$f(x) = \begin{cases} Cx^2 & \text{if } x \in [-1, 1] \\ 0 & \text{if } x \notin [-1, 1]. \end{cases} \tag{5.20}$$

Find the constant C and the distribution function of X.

Then,

$$1 = \int_{-\infty}^{\infty} f(t)dt = \int_{-1}^{1} Cx^2 dx = C\frac{x^3}{3}\big|_{-1}^{1} = \frac{2}{3}C. \tag{5.21}$$

Hence, $C = 3/2$. For $x \in [-1, 1]$, the d.f. is

$$F(x) = \int_{-\infty}^{x} f(t)dt = \int_{-1}^{x} \frac{3}{2}t^2 dt = \frac{1}{2}x^3 + \frac{1}{2}. \tag{5.22}$$

Thus,

$$F(x) = \begin{cases} 0 & \text{if } x < -1 \\ \frac{1}{2}x^3 + \frac{1}{2} & \text{if } -1 \leq x < 1 \\ 1 & \text{if } x \geq 1. \end{cases} \tag{5.23}$$

◆

Example 5.2.3. Exponential Waiting Time.

Assume that the time T in minutes you have to wait on a certain summer night to see a shooting star has a probability density of the form

$$f(t) = \begin{cases} 0 & \text{if } t \leq 0 \\ Ce^{-t/10} & \text{if } t > 0 \end{cases}. \tag{5.24}$$

Find the value of C and the distribution function of T and compute the probability that you have to wait more than 10 minutes.

Now,

$$1 = \int_{-\infty}^{\infty} f(t)dt = \int_{0}^{\infty} Ce^{-t/10}dt = -10Ce^{-t/10}|_{0}^{\infty} = 10C, \qquad (5.25)$$

and so $C = 1/10$. Thus

$$f(t) = \begin{cases} 0 & \text{if } t \leq 0 \\ \frac{1}{10}e^{-t/10} & \text{if } t > 0 \end{cases} \qquad (5.26)$$

and, for $t > 0$,

$$F(t) = \ \mathrm{P}(T \leq t) = \int_{0}^{t} \frac{1}{10}e^{-u/10}du = 1 - e^{-t/10}. \qquad (5.27)$$

Consequently,

$$\mathrm{P}(T > 10) = 1 - F(10) = e^{-1} \simeq 0.368. \qquad (5.28)$$

◆

The distribution of the example above is typical of many waiting time distributions occurring in real life, at least approximately. For instance, the time between the decay of atoms in a radioactive sample, the time one has to wait for the phone to ring in an office, and the time between customers showing up at some store are of this type; just the constants differ. (The reasons for the prevalence of this distribution will be discussed later under the heading "Poisson process.")

Definition 5.2.3. *Exponential Random Variable.* *A random variable T is called exponential with parameter $\lambda > 0$ if it has density*

$$f(t) = \begin{cases} 0 & \text{if } t < 0 \\ \lambda e^{-\lambda t} & \text{if } t \geq 0 \end{cases} \qquad (5.29)$$

and distribution function

$$F(t) = \begin{cases} 0 & \text{if } t < 0 \\ 1 - e^{-\lambda t} & \text{if } t \geq 0. \end{cases} \qquad (5.30)$$

There exist random variables that are neither discrete nor continuous; they are said to be of *mixed type*. Here is an example:

Example 5.2.4. *A Mixed Random Variable.*

Suppose we toss a fair coin, and if it comes up H, then $X = 1$, and if it comes up T, then X is determined by spinning a pointer and noting its final position on a scale from 0 to 2, that is, X is then uniformly distributed over the interval $[0, 2]$.

Let

$$F_1(x) = P(X \le x | H) = \begin{cases} 0 \text{ if } x < 1 \\ 1 \text{ if } x \ge 1 \end{cases} \tag{5.31}$$

and

$$F_2(x) = P(X \le x | T) = \begin{cases} 0 & \text{if } x < 0 \\ \frac{1}{2}x & \text{if } 0 \le x < 2 \\ 1 & \text{if } 2 \le x. \end{cases} \tag{5.32}$$

Then, according to the theorem of total probability, the distribution function F is given by

$$F(x) = \frac{1}{2}F_1(x) + \frac{1}{2}F_2(x) = \begin{cases} 0 & \text{if } x < 0 \\ \frac{1}{4}x & \text{if } 0 \le x < 1 \\ \frac{1}{4}x + \frac{1}{2} & \text{if } 1 \le x < 2 \\ 1 & \text{if } 2 \le x \end{cases} \tag{5.33}$$

and its graph is given by Figure 5.8.

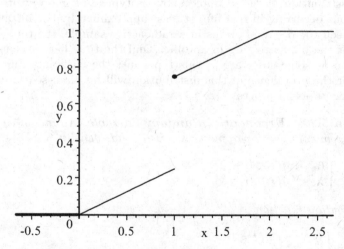

Fig. 5.8. A mixed-type distribution function

Note that

$$F'(x) = f(x) = \begin{cases} 0 \text{ if } x < 0 \\ \frac{1}{4} \text{ if } 0 < x < 1 \\ \frac{1}{4} \text{ if } 1 < x < 2 \\ 0 \text{ if } 2 < x \end{cases} \tag{5.34}$$

exists everywhere except at $x = 0$, 1, and 2, but because of the jump of F at 1, it is not a true density function. Indeed,

$$F(x) = \begin{cases} \int_{-\infty}^{x} f(t)dt & \text{if } x < 1 \\ \int_{-\infty}^{x} f(t)dt + \frac{1}{2} & \text{if } 1 \leq x, \end{cases} \tag{5.35}$$

and so $F(x) \neq \int_{-\infty}^{x} f(t)dt$ for all x, as required by the definition of density functions.   ◆

Exercises

Exercise 5.2.1.

A continuous random variable X has a density of the form

$$f(x) = \begin{cases} Cx & \text{if } 0 \leq x \leq 4 \\ 0 & \text{if } x < 0 \text{ or } x > 4 \end{cases}. \tag{5.36}$$

1. Find C.
2. Sketch the density function of X.
3. Find the distribution function of X and sketch its graph.
4. Find the probability $P(X < 1)$.
5. Find the probability $P(2 < X)$.

Exercise 5.2.2.

A continuous random variable X has a density of the form $f(x) = Ce^{-|x|}$, defined on all of $\mathbb{R}$:

1. Find C.
2. Sketch the density function of X.
3. Find the distribution function of X and sketch its graph.
4. Find the probability $P(-2 < X < 1)$.
5. Find the probability $P(2 < |X|)$.

Exercise 5.2.3.

A continuous random variable X has a density of the form

$$f(x) = \begin{cases} \frac{C}{x^2} & \text{if } x \geq 1 \\ 0 & \text{if } x < 1 \end{cases}. \tag{5.37}$$

1. Find C.
2. Sketch the density function of X.
3. Find the distribution function of X and sketch its graph.
4. Find the probability $P(X < 2)$.
5. Find the probability $P(2 < |X|)$.

Exercise 5.2.4.

A continuous random variable X has a density of the form

$$f(x) = \begin{cases} \frac{C}{x^2} & \text{if } |x| \geq 1 \\ 0 & \text{if } |x| < 1 \end{cases}. \tag{5.38}$$

1. Find C.
2. Sketch the density function of X.
3. Find the distribution function of X and sketch its graph.
4. Find the probability $P(X < 2)$.
5. Find the probability $P(2 < |X|)$.

Exercise 5.2.5.

Let X be a mixed random variable with distribution function

$$F(x) = \begin{cases} 0 & \text{if } x < 0 \\ \frac{1}{6}x & \text{if } 0 \leq x < 1 \\ \frac{1}{3} & \text{if } 1 \leq x < 2 \\ 1 & \text{if } 2 \leq x \end{cases}. \tag{5.39}$$

1. Devise an experiment whose outcome is this X.
2. Find the probability $P(X < 1/2)$.
3. Find the probability $P(X < 3/2)$.
4. Find the probability $P(1/2 < X < 2)$.
5. Find the probability $P(X = 1)$.
6. Find the probability $P(X > 1)$.
7. Find the probability $P(X = 2)$.

Exercise 5.2.6.

Let X be a mixed random variable with distribution function

$$F(x) = \begin{cases} 0 & \text{if } x < 0 \\ \frac{1}{3}x + \frac{1}{6} & \text{if } 0 \leq x < 1 \\ \frac{2}{3} & \text{if } 1 \leq x < 2 \\ 1 & \text{if } 2 \leq x \end{cases}. \tag{5.40}$$

1. Devise an experiment whose outcome is this X.
2. Find the probability $P(X < 1/2)$.
3. Find the probability $P(X < 3/2)$.
4. Find the probability $P(1/2 < X < 2)$.
5. Find the probability $P(X = 1)$.
6. Find the probability $P(X > 1)$.
7. Find the probability $P(X = 3/2)$.

Exercise 5.2.7.

Let X be a mixed random variable with distribution function F given by the graph in Figure 5.9:

1. Find a formula for $F(x)$.
2. Find the probability $P(X < 1/2)$.
3. Find the probability $P(X < 3/2)$.
4. Find the probability $P(1/2 < X < 2)$.
5. Find the probability $P(X = 1)$.
6. Find the probability $P(X > 1)$.
7. Find the probability $P(X = 2)$.

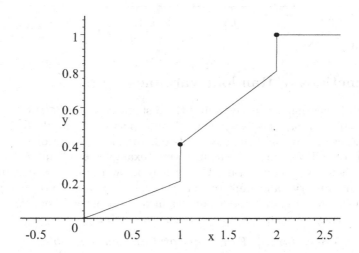

Fig. 5.9.

Exercise 5.2.8.

Let X be a mixed random variable with distribution function F given by the graph in Figure 5.10:

1. Find a formula for $F(x)$.
2. Find the probability $P(X < 1/2)$.
3. Find the probability $P(X < 3/2)$.
4. Find the probability $P(1/2 < X < 2)$.
5. Find the probability $P(X = 1)$.
6. Find the probability $P(X > 1)$.
7. Find the probability $P(X = 2)$.

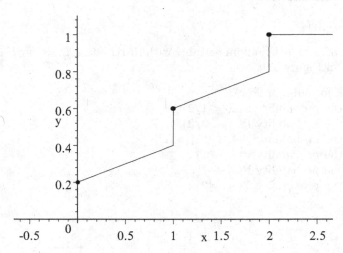

Fig. 5.10.

5.3 Functions of Random Variables

In many applications we need to find the distribution of a function of a random variable. For instance, we may know from measurements the distribution of the radius of stars, and we may want to know the distribution of their volumes. (Probabilities come in—as in several examples of Chapter 4—from a random choice of a single star.) Or we may know the income distributions in different countries and want to change scales to be able to compare them. We shall encounter many more examples in the rest of the book. We start off with the change of scale example in a general setting.

Example 5.3.1. Linear Functions of Random Variables.

Let X be a random variable with a known distribution function F_X and define a new random variable as $Y = aX + b$, where $a \neq 0$ and b are given constants.

If X is discrete, then we can obtain the probability function f_Y of Y very easily by solving the equation in its definition:

$$f_Y(y) = P(Y = y) = P(aX + b = y) = P\left(X = \frac{y - b}{a}\right) = f_X\left(\frac{y - b}{a}\right).$$

$$(5.41)$$

Equivalently, if x is a possible value of X, that is, $f_X(x) \neq 0$, then $f_Y(y) = f_X(x)$ for $y = ax + b$, which is the corresponding possible value of Y.

If X is continuous, then we cannot imitate the above procedure, because the density function is not a probability. We can, however, obtain the distribution function F_Y of Y similarly, by solving the inequality in its definition: For $a > 0$,

$$F_Y(y) = P(Y \leq y) = P(aX + b \leq y) = P\left(X \leq \frac{y-b}{a}\right) = F_X\left(\frac{y-b}{a}\right),$$

$$(5.42)$$

and for $a < 0$,

$$F_Y(y) = P(Y \leq y) = P(aX + b \leq y) = P\left(X \geq \frac{y-b}{a}\right) = 1 - F_X\left(\frac{y-b}{a}\right).$$

$$(5.43)$$

If X is continuous with density f_X, then F_X is differentiable and $f_X = F_X'$. As Equations 5.42 and 5.43 show, then F_Y is also differentiable. Hence Y too is continuous, with density function

$$f_Y(y) = F_Y'(y) = \pm\frac{d}{dy}F_X\left(\frac{y-b}{a}\right) = \frac{1}{|a|}F_X'\left(\frac{y-b}{a}\right) = \frac{1}{|a|}f_X\left(\frac{y-b}{a}\right).$$

$$(5.44)$$

◆

Example 5.3.2. Shifting and Stretching a Discrete Uniform Variable.

Let X denote the number obtained in the roll of a die and let $Y = 2X + 10$. Then the p.f. of X is

$$f_X(x) = \begin{cases} 1/6 \text{ if } x = 1, 2, \ldots, 6 \\ 0 \quad \text{otherwise.} \end{cases} \qquad (5.45)$$

Thus, using Equation 5.41 with this f_X and with $a = 2$ and $b = 10$, we get the p.f. of Y as

$$f_Y(y) = f_X\left(\frac{y-10}{2}\right) = \begin{cases} 1/6 \text{ if } y = 12, 14, \ldots, 22 \\ 0 \quad \text{otherwise.} \end{cases} \qquad (5.46)$$

We can obtain the same result more simply, by tabulating the possible x and $y = 2x + 10$ values and the corresponding probabilities:

x	1	2	3	4	5	6
y	12	14	16	18	20	22
$f_X(x) = f_Y(y)$	1/6	1/6	1/6	1/6	1/6	1/6

◆

Example 5.3.3. Shifting and Stretching a Uniform Variable.

Let X be uniform on the interval $[-1, 1]$ and let $Y = 2X + 10$. Then the p.d.f. of X is

$$f_X(x) = \begin{cases} 1/2 & \text{if } x \in [-1, 1] \\ 0 & \text{otherwise.} \end{cases} \tag{5.47}$$

If $X = -1$, then $Y = 2(-1) + 10 = 8$, and if $X = 1$, then $Y = 2 \cdot 1 + 10 = 12$. Thus, the interval $[-1, 1]$ gets changed into $[8, 12]$, and so Equation 5.44, with the present f_X and with $a = 2$ and $b = 10$, yields

$$f_Y(y) = \frac{1}{2} f_X \left(\frac{y - 10}{2} \right) = \begin{cases} 1/4 & \text{if } y \in [8, 12] \\ 0 & \text{otherwise.} \end{cases} \tag{5.48}$$

Notice that here the p.d.f. got shifted and stretched in much the same way as the p.f. in the preceding example, but there the values of the p.f. remained $\frac{1}{6}$, while here the values of the p.d.f. have become halved. The reason for this difference is clear: In the discrete case, the number of possible values has not changed (both X and Y had six), but in the continuous case, the p.d.f. got stretched by a factor of 2 (from width 2 to width 4) and so, to compensate for that, in order to have a total area of 1, we had to halve the density.

The foregoing examples can easily be generalized to the case in which $Y = g(X)$, for any invertible function g. Rather than summarizing the results in a theorem, we just give prescriptions for the procedures and illustrate them with examples:

1. For discrete X tabulate the possible values x of X together with $y = g(x)$ and $f_X(x) = f_Y(y)$.
2. For continuous X, solve the inequality in $F_Y(y) = P(Y \leq y) = P(g(X) \leq y)$ to obtain $F_Y(y)$ in terms of $F_X(g^{-1}(y))$. (We must be careful to reverse the inequality when solving for X if g is decreasing.) To obtain the p.d.f. $f_Y(y)$, differentiate $F_Y(y)$.

Example 5.3.4. Squaring a Binomial.

Let X be binomial with parameters $n = 3$ and $p = \frac{1}{2}$ and let $Y = X^2$. Then we can obtain f_Y by tabulating the possible X and $Y = X^2$ values and the corresponding probabilities:

x	0	1	2	3
y	0	1	4	9
$f_X(x) = f_Y(y)$	1/8	3/8	3/8	1/8

♦

Example 5.3.5. Squaring a Positive Uniform Random Variable.

Let X be uniform on the interval $[1, 3]$ and let $Y = X^2$. Then the p.d.f. of X is

$$f_X(x) = \begin{cases} 1/2 & \text{if } x \in [1,3] \\ 0 & \text{otherwise.} \end{cases} \tag{5.49}$$

Now, $g(X) = X^2$ is one-to-one for the possible values of X, which are positive, and so, for $y \geq 0$,

$$F_Y(y) = \mathrm{P}(Y \leq y) = \mathrm{P}(X^2 \leq y) = \mathrm{P}\left(X \leq \sqrt{y}\right) = F_X\left(\sqrt{y}\right). \tag{5.50}$$

Hence, by the chain rule,

$$f_Y(y) = \frac{d}{dy}F_X\left(\sqrt{y}\right) = f_X\left(\sqrt{y}\right)\frac{d\sqrt{y}}{dy} = \begin{cases} \frac{1}{2}\frac{1}{2\sqrt{y}} & \text{if } y \in [1,9] \\ 0 & \text{otherwise.} \end{cases} \tag{5.51}$$

We can check that this f_Y is indeed a density function:

$$\int_1^9 \frac{1}{4\sqrt{y}}dy = \left.\frac{\sqrt{y}}{2}\right|_1^9 = \frac{1}{2}\left(\sqrt{9} - \sqrt{1}\right) = 1. \tag{5.52}$$

◆

Example 5.3.6. Random Number Generation.

An important application of the procedures for changing of random variables described above is to the computer simulation of physical systems with random inputs. Most mathematical and statistical software packages produce so-called random numbers (or more precisely, pseudorandom numbers) that are uniformly distributed on the interval $[0,1]$. (Though such numbers are generated by deterministic algorithms, they are for most practical purposes a good substitute for samples of independent, uniform random variables on the interval $[0,1]$.) Often, however, we need random numbers with a different distribution and want to transform the uniform random numbers to new numbers that have the desired distribution.

Suppose we need random numbers that have the continuous distribution function F and that F is strictly increasing where it is not 0 or 1. (The restrictions on F can be removed, but we do not want to get into this.) Then F has a strictly increasing inverse F^{-1} over $[0,1]$, which we can use as the function g in Part 2 of the general procedure given above. Thus, letting $Y = F^{-1}(X)$, with X being uniform on $[0,1]$, we have

$$F_Y(y) = \mathrm{P}(Y \leq y) = \mathrm{P}(F^{-1}(X) \leq y) = \mathrm{P}(X \leq F(y)) = F(y), \tag{5.53}$$

where the last step follows from the fact that $\mathrm{P}(X \leq x) = x$ on $[0,1]$ for an X that is uniform on $[0,1]$, with the substitution $x = F(y)$. (See Equation 5.19.)

Thus, if $x_1, x_2, \ldots$ are random numbers uniform on $[0,1]$, produced by the generator, then the numbers $y_1 = F^{-1}(x_1)$, $y_2 = F^{-1}(x_2), \ldots$ are random numbers with the distribution function F.   ◆

If g is not one-to-one, we can still follow the procedures of the examples above, but, for some y, we have more than one solution of the equation $y = g(x)$ or of the corresponding inequality, and we must consider all of those solutions, as in the following examples.

Example 5.3.7. The X^2 Function.

Let X be a random variable with a known distribution function F_X and define a new random variable as $Y = X^2$.

If X is discrete, then we can obtain the probability function f_Y of Y as

$$f_Y(y) = P(X^2 = y) = \begin{cases} P(X = \pm\sqrt{y}) = f_X(\sqrt{y}) + f_X(-\sqrt{y}) & \text{if } y > 0 \\ P(X = 0) = f_X(0) & \text{if } y = 0 \\ 0 & \text{if } y < 0. \end{cases}$$
(5.54)

For continuous X, the distribution function F_Y of Y is given by

$$F_Y(y) = P(X^2 \leq y) = \begin{cases} P(-\sqrt{y} \leq X \leq \sqrt{y}) = F_X(\sqrt{y}) - F_X(-\sqrt{y}) & \text{if } y > 0 \\ 0 & \text{if } y \leq 0, \end{cases}$$
(5.55)

and for discrete X, we have

$$F_Y(y) = \begin{cases} P(-\sqrt{y} \leq X \leq \sqrt{y}) = F_X(\sqrt{y}) - F_X(-\sqrt{y}) + f_X(-\sqrt{y}) & \text{if } y \geq 0 \\ 0 & \text{if } y < 0. \end{cases}$$
(5.56)

If X is continuous and has density function f_X, then differentiating Equation 5.55 we get

$$f_Y(y) = F_Y'(y) = \begin{cases} \frac{1}{2\sqrt{y}} \left[f_X(\sqrt{y}) + f_X(-\sqrt{y}) \right] & \text{if } y > 0 \\ 0 & \text{if } y \leq 0. \end{cases}$$
(5.57)

◆

Example 5.3.8. Distribution of $(X-2)^2$ for a Binomial.

Let X be binomial with parameters $n = 3$ and $p = \frac{1}{2}$, and let $Y = (X-2)^2$. Rather than developing a formula like Equation 5.54, the best way to proceed is to tabulate the possible values of X and Y and the corresponding probabilities, as in Example 5.3.4:

x	0	1	2	3
y	4	1	0	1
$f_X(x)$	1/8	3/8	3/8	1/8

Now, $Y = 1$ occurs when $X = 1$ or 3. Since these cases are mutually exclusive, $P(Y = 1) = P(X = 1) + P(X = 3) = 3/8 + 1/8 = 1/2$. Hence, the table of f_Y is

y	0	1	4
$f_Y(y)$	3/8	1/2	1/8

◆

Example 5.3.9. Distribution of X^2 for a Uniform X.

Let X be uniform on the interval $[-1,1]$ and let $Y = X^2$. Then, by Formula 5.19,

$$F_X(x) = \begin{cases} 0 & \text{if } x < -1 \\ \dfrac{x+1}{2} & \text{if } -1 \leq x < 1 \\ 1 & \text{if } x \geq 1. \end{cases} \tag{5.58}$$

Substituting this F_X into Equation 5.55 and observing that $\dfrac{\sqrt{y}+1}{2} - \dfrac{-\sqrt{y}+1}{2} = \sqrt{y}$, we get

$$F_Y(y) = \begin{cases} 0 & \text{if } y < 0 \\ \sqrt{y} & \text{if } 0 \leq y < 1 \\ 1 & \text{if } y \geq 1. \end{cases} \tag{5.59}$$

We can obtain the density of Y by differentiating F_Y, as

$$f_Y(y) = \begin{cases} \frac{1}{2\sqrt{y}} & \text{if } 0 < y < 1 \\ 0 & \text{otherwise.} \end{cases} \tag{5.60}$$

♦

Example 5.3.10. Distribution of X^2 for a Nonuniform X.

Let X be a random variable with p.d.f. $f(x) = \frac{3x^2}{2}$ on the interval $[-1,1]$ and 0 elsewhere and $Y = X^2$. Find the distribution function and the density function of Y.

Solution: If $0 < y < 1$, then

$$F_Y(y) = P(X^2 \leq y) = P\left(-\sqrt{y} \leq X \leq \sqrt{y}\right) \tag{5.61}$$

$$= \int_{-\sqrt{y}}^{\sqrt{y}} \frac{3x^2}{2} dx = \left[\frac{x^3}{2}\right]_{-\sqrt{y}}^{\sqrt{y}} = y^{3/2}.$$

Thus

$$F_Y(y) = \begin{cases} 0 & \text{if } y \leq 0 \\ y^{3/2} & \text{if } 0 < y < 1 \\ 1 & \text{if } y \geq 1. \end{cases} \tag{5.62}$$

Hence

$$f_Y(y) = F_Y'(y) = \begin{cases} \frac{3\sqrt{y}}{2} & \text{if } 0 < y < 1 \\ 0 & \text{if } y \leq 0 \text{ or } y \geq 1. \end{cases} \tag{5.63}$$

♦

Example 5.3.11. Coordinates of a Uniform Random Variable on a Circle.

Suppose that a point is moving around a circle of radius r centered at the origin of the xy coordinate system with constant speed, and we observe it at a random instant. What is the distribution of each of the point's coordinates at that time?

Since the point is observed at a random instant, its position is uniformly distributed on the circle. Thus its polar angle Θ is a uniform random variable on the interval $[0, 2\pi]$, with constant density $f_\Theta(\theta) = \frac{1}{2\pi}$ there and 0 elsewhere. We want to find the distributions of $X = r\cos\Theta$ and $Y = r\sin\Theta$.

Now, for a given $x = r\cos\theta$, there are two solutions modulo 2π: $\theta_1 = \arccos\frac{x}{r}$ and $\theta_2 = 2\pi - \arccos\frac{x}{r}$. So if $X \le x$, then Θ falls in the angle on the left between these two values. Thus

$$F_X(x) = \mathrm{P}(X \le x) = \begin{cases} 0 & \text{if } \quad x < -r \\ \frac{\theta_2 - \theta_1}{2\pi} = 1 - \frac{1}{\pi}\arccos\frac{x}{r} & \text{if } -r \le x < r \\ 1 & \text{if } \quad r \le x. \end{cases} \quad (5.64)$$

Hence

$$f_X(x) = F_X'(x) = \begin{cases} \frac{1}{\pi\sqrt{r^2 - x^2}} & \text{if } -r < x < r \\ 0 & \text{otherwise.} \end{cases} \quad (5.65)$$

The density of X can also be obtained directly from Figure 5.11 by using Equation 5.15. For $x > 0$ and $dx > 0$, the variable X falls into the interval $[x, x + dx]$ if and only if Θ falls into either of the intervals of size $d\theta$ at θ_1 and θ_2. (For negative x or dx, we need obvious modifications.) Thus, $f_X(x)dx = 2 \cdot \frac{1}{2\pi}d\theta$, and so $f_X(x) = \frac{1}{\pi} \cdot \frac{d\theta}{dx} = \frac{1}{\pi} \cdot 1 / \frac{dx}{d\theta} = \frac{1}{\pi\sqrt{r^2 - x^2}}$ as before.

We leave the analogous computation for the distribution of the y-coordinate as an exercise.   ◆

Exercises

Exercise 5.3.1.

Let X be a discrete uniform random variable with possible values $-5, -4, \ldots, 4, 5$. Find the probability function and the distribution function of $Y = X^2 - 3X$.

Exercise 5.3.2.

Let X be a binomial random variable with parameters $p = \frac{1}{2}$ and $n = 6$. Find the probability function and the distribution function of $Y = X^2 - 2X$.

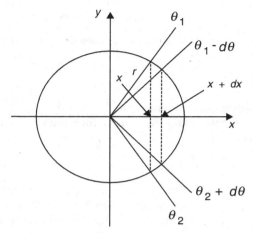

Fig. 5.11. Density of the x-coordinate of a random point on a circle

Exercise 5.3.3.

Let X be a Bernoulli random variable with $p = \frac{1}{2}$ and $Y = \arctan X$. Find the probability function and the distribution function of Y.

Exercise 5.3.4.

Let X be a discrete random variable with probability function f_X. Find formulas for the probability function and the distribution function of $Y = (X - a)^2$, where a is an arbitrary constant.

Exercise 5.3.5.

Let X be a random variable uniformly distributed on the interval $(0,1)$ and $Y = \ln X$. Find the distribution function and the density function of Y.

Exercise 5.3.6.

Let X be a random variable uniformly distributed on the interval $[-1,1]$ and $Y = |X|$. Find the distribution function and the density function of Y.

Exercise 5.3.7.

Let X be a continuous random variable with density function f_X. Find formulas for the distribution function and the density function of $Y = |X|$.

Exercise 5.3.8.

Assume that the distribution of the radius R of stars has a density function f_R. Find formulas for the density and the distribution function of their volume $V = \frac{4}{3}R^3\pi$.

Exercise 5.3.9.

Find the distribution function and the density function of Y in Example 5.3.11.

Exercise 5.3.10.

Let X be a continuous random variable with density f_X. Find formulas for the distribution function and the density function of $Y = (X - a)^2$, where a is an arbitrary constant.

Exercise 5.3.11.

Let X be a continuous random variable with a continuous distribution function F that is strictly increasing where it is not 0 or 1. Show that the random variable $Y = F(X)$ is uniformly distributed on the interval $[0,1]$.

Exercise 5.3.12.

Let X be a random variable uniformly distributed on the interval $[-2,2]$ and $Y = (X - 1)^2$:

a) Find the density function and the distribution function of X.
b) Find the distribution function and the density function of Y.

5.4 Joint Distributions

In many applications, we need to consider two or more random variables simultaneously. For instance, the two-way classification of voters in Example 4.3.3 can be regarded to involve two random variables, if we assign numbers to the various age groups and party affiliations.

In general, we want to consider joint probabilities of events defined by two or more random variables on the same sample space. The probabilities of all such events constitute *the joint distribution* or the *bivariate* (for two variables) or *multivariate* (for two or more variables) *distribution* of the given random variables and can be described by their joint p.f., d.f., or p.d.f., much as for single random variables.

Definition 5.4.1. *Joint Probability Function.*

Let X and Y be two discrete random variables on the same sample space. The function of two variables defined by[8] $f(x,y) = \mathrm{P}(X = x, Y = y)$, *for all possible values*[9] x *of X and y of Y, is called the joint or bivariate probability function of X and Y or of the pair (X,Y).*

---

[8] $\mathrm{P}(X = x, Y = y)$ stands for $\mathrm{P}(X = x \text{ and } Y = y) = \mathrm{P}(\{X = x\} \cap \{Y = y\})$.
[9] Sometimes $f(x,y)$ is defined for all real numbers x, y, with $f(x,y) = 0$ if $\mathrm{P}(X = x) = 0$ or $\mathrm{P}(Y = y) = 0$.

Similarly, for a set of n random variables on the same sample space, with n a positive integer greater than 2, we define the joint or multivariate probability function of $(X_1, X_2, \ldots, X_n)$ as the function given by $f(x_1, x_2, \ldots, x_n) = P(X_1 = x_1, X_2 = x_2, \ldots, X_n = x_n)$, for all possible values x_i of each X_i or for all $(x_1, x_2, \ldots, x_n) \in \mathbb{R}^n$.

If for two random variables we sum $f(x, y)$ over all possible values y of Y, then we get the $(marginal)$[10] probability function f_X (or f_1) of X. Indeed,

$$\sum_y f(x, y) = \sum_y P(X = x, Y = y) = P(\{X = x\} \cap (\cup_y \{Y = y\}))$$

$$= P(\{X = x\} \cap S) = P(X = x) = f_X(x). \tag{5.66}$$

Similarly, if we sum $f(x, y)$ over all possible values x of X, then we get the probability function f_Y (or f_2) of Y, and if we sum $f(x, y)$ over all possible values x of X and y of Y both, in either order then, of course, we get 1.

For n random variables, if we sum $f(x_1, x_2, \ldots, x_n)$ over all possible values x_i of any X_i, then we get the joint (marginal) probability function of the $n-1$ random variables X_j with $j \neq i$, and if we sum over all possible values of any k of them, then we get the joint (marginal) probability function of the remaining $n - k$ random variables.

Definition 5.4.2. *Joint Distribution Function.*

Let X and Y be two arbitrary random variables on the same sample space. The function of two variables defined by $F(x, y) = P(X \leq x, Y \leq y)$, for all real x and y, is called the joint or bivariate distribution function of X and Y or of the pair (X, Y).

The functions[11] $F_X(x) = F(x, \infty)$ and $F_Y(y) = F(\infty, y)$ are called the (marginal) distribution functions of X and Y.

Similarly, for a set of n random variables on the same sample space, with n a positive integer greater than 2, we define the joint or multivariate distribution function of $(X_1, X_2, \ldots, X_n)$ as the function given by $F(x_1, x_2, \ldots, x_n) = P(X_1 \leq x_1, X_2 \leq x_2, \ldots, X_n \leq x_n)$, for all real numbers $x_1, x_2, \ldots, x_n$.

If we substitute ∞ for any of the arguments of $F(x_1, x_2, \ldots, x_n)$, we get the marginal d.f.'s of the random variables that correspond to the remaining arguments.

For joint distributions, we have the following obvious theorem:

Theorem 5.4.1. *Joint Distribution of Two Functions of Two Discrete Random Variables. If X and Y are two discrete random variables with joint probability function $f_{X,Y}(x, y)$ and $U = g(X, Y)$ and $V = h(X, Y)$ any two functions, then the joint probability function of U and V is given by*

---

[10] The adjective "marginal" is really unnecessary; we just use it occasionally to emphasize the relation to the joint distribution.

[11] $F(x, \infty)$ is a shorthand for $\lim_{y \to \infty} F(x, y)$, etc.

$$f_{U,V}(u,v) = \sum_{(x,y):g(x,y)=u,\ h(x,y)=v} f_{X,Y}(x,y). \tag{5.67}$$

Example 5.4.1. Sum and Absolute Difference of Two Dice.

Roll two fair dice as in 2.3.3, and let X and Y denote the numbers obtained with them. Find the joint probability function of $U = X+Y$ and $V = |X-Y|$.

First, we construct a table of the values of U and V, for all possible outcomes x and y (Table 5.1):

$y\backslash x$	1	2	3	4	5	6
1	2,0	3,1	4,2	5,3	6,4	7,5
2	3,1	4,0	5,1	6,2	7,3	8,4
3	4,2	5,1	6,0	7,1	8,2	9,3
4	5,3	6,2	7,1	8,0	9,1	10,2
5	6,4	7,3	8,2	9,1	10,0	11,1
6	7,5	8,4	9,3	10,2	11,1	12,0

Table 5.1. The values of $U = X + Y$ and $V = X - Y$ for the numbers X and Y showing on two dice

By assumption, each pair of x and y values has probability $1/36$, and so each pair (u, v) of U and V values has as its probability $1/36$ times the number of boxes in which it appears. Hence, for instance, $f_{U,V}(3,1) = P(U = 3, V = 1) = P(X = 1, Y = 2) + P(X = 2, Y = 1) = \frac{2}{36}$. Thus, the joint probability function $f_{U,V}(u,v)$ of U and V is given by the table below (Table 5.2), with the marginal probability function $f_U(u)$ shown as the row sums on the right margin and the marginal probability function $f_V(v)$ shown as the column sums on the bottom margin:

$u\backslash v$	0	1	2	3	4	5	$f_U(u)$
2	1/36	0	0	0	0	0	1/36
3	0	2/36	0	0	0	0	2/36
4	1/36	0	2/36	0	0	0	3/36
5	0	2/36	0	2/36	0	0	4/36
6	1/36	0	2/36	0	2/36	0	5/36
7	0	2/36	0	2/36	0	2/36	6/36
8	1/36	0	2/36	0	2/36	0	5/36
9	0	2/36	0	2/36	0	0	4/36
10	1/36	0	2/36	0	0	0	3/36
11	0	2/36	0	0	0	0	2/36
12	1/36	0	0	0	0	0	1/36
$f_V(v)$	6/36	10/36	8/36	6/36	4/36	2/36	1

Table 5.2. The joint and marginal probability functions of $U = X + Y$ and $V = X - Y$ for the numbers X and Y showing on two dice

Example 5.4.2. Maximum and Minimum of Three Integers.

Choose three numbers X_1, X_2, X_3 without replacement and with equal probabilities from the set $\{1, 2, 3, 4\}$, and let $X = \max\{X_1, X_2, X_3\}$ and $Y = \min\{X_1, X_2, X_3\}$. Find the joint probability function of X and Y.

First, we list the set of all 24 possible outcomes (Table 5.3), together with the values of X and Y:

X_1	1	1	1	1	1	1	2	2	2	2	2	2	3	3	3	3	3	3	4	4	4	4	4	4
X_2	2	2	3	3	4	4	1	1	3	3	4	4	1	1	2	2	4	4	1	1	2	2	3	3
X_3	3	4	2	4	2	3	3	4	1	4	1	3	2	4	1	4	1	2	2	3	1	3	1	2
X	3	4	3	4	4	4	3	4	3	4	4	4	3	4	3	4	4	4	4	4	4	4	4	4
Y	1	1	1	1	1	1	1	1	1	1	2	1	2	1	1	1	1	2	1	2	1	1	1	2

Table 5.3. The values of $X = \max(X_1, X_2, X_3)$ and $Y = \min(X_1, X_2, X_3)$

Now, each possible outcome has probability $1/24$, and so we just have to count the number of times each pair of X, Y values occurs and multiply it by $1/24$ to get the probability function $f(x, y)$ of (X, Y). This p.f. is given in the following table (Table 5.4), together with the marginal probabilities $f_Y(y)$ on the right and $f_X(x)$ at the bottom:

$y \backslash x$	3	4	Any x
1	1/4	1/2	3/4
2	0	1/4	1/4
Any y	1/4	3/4	1

Table 5.4. The joint p.f. and marginals of $X = \max(X_1, X_2, X_3)$ and $Y = \min(X_1, X_2, X_3)$

$\blacklozenge$

Example 5.4.3. Multinomial Distribution.

Suppose we have k types of objects and we perform n independent trials of choosing one of these objects, with probabilities $p_1, p_2, \ldots, p_k$ for the different types in each of the trials, where $p_1 + p_2 + \cdots + p_k = 1$. Let $N_1, N_2, \ldots, N_k$ denote the numbers of objects obtained in each category. Then clearly, the joint probability function of $N_1, N_2, \ldots, N_k$ is given by

$$f(n_1, n_2, \ldots, n_k) = P(N_1 = n_1, N_2 = n_2, \ldots, N_k = n_k)$$

$$= \binom{n}{n_1, n_2, \ldots, n_k} p_1^{n_1} p_2^{n_2} \cdots p_k^{n_k} \tag{5.68}$$

for every choice of nonnegative integers $n_1, n_2, \ldots, n_k$ with $n_1 + n_2 + \cdots + n_k = n$ and $f(n_1, n_2, \ldots, n_k) = 0$ otherwise. ♦

Next, we consider the joint distributions of continuous random variables.

Definition 5.4.3. *Joint Density Function.*

Let X and Y be two continuous random variables on the same probability space. If there exists an integrable nonnegative function $f(x, y)$ on $\mathbb{R}^2$ such that

$$\mathrm{P}(a < X < b, c < Y < d) = \int_c^d \int_a^b f(x, y) dx dy \qquad (5.69)$$

for all real numbers a, b, c, d, then f is called a joint or bivariate probability density function[12] of X and Y or of the pair (X, Y), and X and Y are said to be jointly continuous.

Similarly, for a set of n continuous random variables on the same probability space, with n a positive integer greater than 2, if there exists an integrable nonnegative function $f(x_1, x_2, \ldots, x_n)$ on $\mathbb{R}^n$ such that, for any coordinate rectangle[13] R of $\mathbb{R}^n$,

$$P((X_1, X_2, \ldots, X_n) \in R) = \int \cdots \int_R f(x_1, x_2, \ldots, x_n) dx_1 \ldots dx_n, \quad (5.70)$$

then f is called a joint or multivariate probability density function of $X_1, X_2, \ldots, X_n$ or of the point or vector $(X_1, X_2, \ldots, X_n)$, and $X_1, X_2, \ldots, X_n$ are said to be jointly continuous.

Similarly as for discrete variables, in the continuous bivariate case, $\int_{-\infty}^{\infty} f(x, y) dx = f_Y(y)$ is the (marginal) density of Y, and $\int_{-\infty}^{\infty} f(x, y) dy = f_X(x)$ is the (marginal) density of X. In the multivariate case, integrating the joint density over any k of its arguments from $-\infty$ to ∞, we get the (marginal) joint density of the remaining $n - k$ random variables.

The relationship between the p.d.f. and the d.f. is analogous to the one for a single random variable: For a continuous bivariate distribution,

$$F(x, y) = \mathrm{P}(X \leq x, Y \leq y) = \int_{-\infty}^{y} \int_{-\infty}^{x} f(s, t) ds dt, \qquad (5.71)$$

and

$$f(x, y) = \frac{\partial^2 F(x, y)}{\partial x \partial y}, \qquad (5.72)$$

wherever the derivative on the right exists and is continuous. Similar relations exist for multivariate distributions.

An important class of joint distributions is obtained by generalizing the notion of a uniform distribution on an interval to higher dimensions:

---

[12] The same ambiguities arise as in the one-dimensional case. (See footnote [4] on page 115.)

[13] That is, a Cartesian product of n intervals, one from each coordinate axis.

Definition 5.4.4. ***Uniform Distribution on Various Regions.*** *Let D be a region of* $\mathbb{R}^n$, *with n-dimensional volume V. Then the point* $(X_1, X_2, \ldots, X_n)$ *is said to be chosen at random or uniformly distributed on D, if its distribution is given by the density function*[14]

$$f(x_1, x_2, \ldots, x_n) = \begin{cases} \frac{1}{V} & \text{if } (x_1, x_2, \ldots, x_n) \in D \\ 0 & \text{otherwise} \end{cases}. \tag{5.73}$$

Example 5.4.4. ***Uniform Distribution on the Unit Square.***

Let D be the closed unit square of $\mathbb{R}^2$, that is, $D = \{(x, y) : 0 \leq x \leq 1, 0 \leq y \leq 1\}$. Then the random point (X, Y) is uniformly distributed on D, if its distribution is given by the density function

$$f(x, y) = \begin{cases} 1 & \text{if } (x, y) \in D \\ 0 & \text{otherwise} \end{cases}. \tag{5.74}$$

Clearly, the marginal densities are the uniform densities on the $[0, 1]$ intervals of the x and y axes, respectively.   ♦

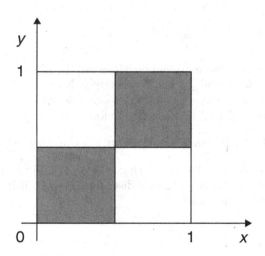

Fig. 5.12. D is the shaded area

Example 5.4.5. ***Uniform Distribution on Part of the Unit Square.***

Let D be the union of the lower left quarter and of the upper right quarter of the unit square of $\mathbb{R}^2$, that is, $D = \{(x, y) : 0 \leq x \leq 1/2, 0 \leq y \leq 1/2\} \cup \{(x, y) : 1/2 \leq x \leq 1, 1/2 \leq y \leq 1\}$ as shown in Figure 5.12.

---

[14] Note that it makes no difference for this assignment of probabilities whether we consider the region D open or closed or, more generally, whether we include or omit any set of points of dimension less than n.

Then, clearly, the area of D is $1/2$, and so the density function of a random point (X, Y), uniformly distributed on D, is given by

$$f(x, y) = \begin{cases} 2 \text{ if } (x, y) \in D \\ 0 \text{ otherwise} \end{cases}. \tag{5.75}$$

The surprising thing about this distribution is that the marginal densities are again the uniform densities on the $[0, 1]$ intervals of the x and y axes, just as in the previous example, although the joint density is very different and not even continuous on the unit square.   ♦

Example 5.4.6. Uniform Distribution on a Diagonal of the Unit Square.

Let D be again the unit square of $\mathbb{R}^2$, that is, $D = \{(x, y) : 0 \leq x \leq 1, 0 \leq y \leq 1\}$, and let the random point (X, Y) be uniformly distributed on the diagonal $y = x$ between the vertices $(0, 0)$ and $(1, 1)$, that is, on the line segment $L = \{(x, y) : y = x, 0 \leq x \leq 1\}$. In other words, assign probabilities to regions A in the plane by

$$P((X, Y) \in A) = \frac{\text{length}(A \cap L)}{\sqrt{2}}. \tag{5.76}$$

Clearly, here again, the marginal densities are the uniform densities on the $[0, 1]$ intervals of the x and y axes, respectively. Note, however, that X and Y are not jointly continuous (nor discrete) and do not have a joint density function, in spite of X and Y being continuous separately.   ♦

Example 5.4.7. Uniform Distribution on the Unit Disk.

Let D be the open unit disk of $\mathbb{R}^2$, that is, $D = \{(x, y) : x^2 + y^2 < 1\}$. Then the random point (X, Y) is uniformly distributed on D, if its distribution is given by the density function

$$f(x, y) = \begin{cases} 1/\pi \text{ if } (x, y) \in D \\ 0 \quad \text{otherwise} \end{cases}. \tag{5.77}$$

The marginal density of X is obtained from its definition $f_X(x) = \int_{-\infty}^{\infty} f(x, y) dy$. Now, for any fixed $x \in (-1, 1)$, $f(x, y) \neq 0$ if and only if $-\sqrt{1 - x^2} < y < \sqrt{1 - x^2}$, and so for such x

$$\int_{-\infty}^{\infty} f(x, y) dy = \int_{-\sqrt{1-x^2}}^{\sqrt{1-x^2}} \frac{1}{\pi} dy = \frac{2}{\pi}\sqrt{1 - x^2}. \tag{5.78}$$

Thus,

$$f_X(x) = \begin{cases} \frac{2}{\pi}\sqrt{1-x^2} & \text{if } x \in (-1,1) \\ 0 & \text{otherwise} \end{cases} . \tag{5.79}$$

By symmetry, the marginal density of Y is the same, just with x replaced by y :

$$f_Y(y) = \begin{cases} \frac{2}{\pi}\sqrt{1-y^2} & \text{if } y \in (-1,1) \\ 0 & \text{otherwise} \end{cases} . \tag{5.80}$$

$\blacklozenge$

Frequently, as for single random variables, we know the general form of a joint distribution except for an unknown coefficient, which we determine from the requirement that the total probability must be 1.

Example 5.4.8. A Distribution on a Triangle.

Let D be the triangle in $\mathbb{R}^2$ given by $D = \{(x,y) : 0 < x, 0 < y, x+y < 1\}$, and let (X,Y) have the density function

$$f(x,y) = \begin{cases} Cxy^2 & \text{if } (x,y) \in D \\ 0 & \text{otherwise} \end{cases} . \tag{5.81}$$

Find the value of C and compute the probability $P(X < Y)$.

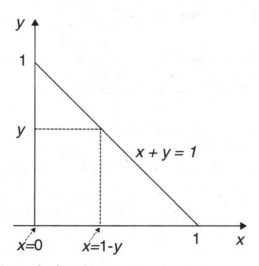

Fig. 5.13. The range of x for a given y

Then, by Figure 5.13,

$$1 = \iint_{\mathbb{R}^2} f(x,y)dxdy = \iint_D Cxy^2 dxdy = \int_0^1 \int_0^{1-y} Cxy^2 dxdy$$

$$= C\int_0^1 \frac{1}{2}(1-y)^2 y^2 dy = C\int_0^1 \frac{1}{2}(y^2 - 2y^3 + y^4)dy$$

$$= C\frac{1}{2}\left(\frac{1}{3} - \frac{1}{2} + \frac{1}{5}\right) = \frac{C}{60}. \tag{5.82}$$

Thus $C = 60$.

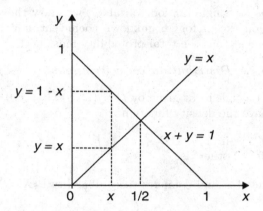

Fig. 5.14. The integration limits for $P(X < Y)$

To compute the probability $P(X < Y)$, we have to integrate f over those values (x, y) of (X, Y) for which $x < y$ holds, that is, for the half of the triangle D above the $y = x$ line. (See Figure 5.14.) Thus

$$P(X < Y) = 60\int_0^{1/2}\int_x^{1-x} xy^2 dydx = 60\int_0^{1/2} x\left[\frac{y^3}{3}\right]_x^{1-x} dx$$

$$= 20\int_0^{1/2} x\left[(1-x)^3 - x^3\right]dx = 20\int_0^{1/2}(x - 3x^2 + 3x^3 - 2x^4)\,dx$$

$$= 20\left[\frac{1}{2}\left(\frac{1}{2}\right)^2 - \left(\frac{1}{2}\right)^3 + \frac{3}{4}\left(\frac{1}{2}\right)^4 - \frac{2}{5}\left(\frac{1}{2}\right)^5\right] = \frac{11}{16}. \tag{5.83}$$

♦

The second part of the above example is an instance of the following general principle: If (X, Y) is continuous with joint p.d.f. f and A is any set[15] in $\mathbb{R}^2$, then

$$P((X, Y) \in A) = \iint_A f(x, y)dxdy. \tag{5.84}$$

In particular, if the set A is defined by a function g so that $A = \{(x, y) : g(x, y) \leq a\}$, for some constant a, then

$$P(g(X, Y) \leq a) = \iint_{\{g(x,y) \leq a\}} f(x, y)dxdy. \tag{5.85}$$

Relations similar to Equations 5.84 and 5.85 hold for discrete random variables as well; we just have to replace the integrals by sums.

Equation 5.85 shows how to obtain the d.f. of a new random variable $Z = g(X, Y)$. This is illustrated in the following example.

Example 5.4.9. Distribution of the Sum of the Coordinates of a Point.

Let the random point (X, Y) be uniformly distributed on the unit square $D = \{(x, y) : 0 \leq x \leq 1, 0 \leq y \leq 1\}$, as in Example 5.4.4. Find the d.f. of $Z = X + Y$.

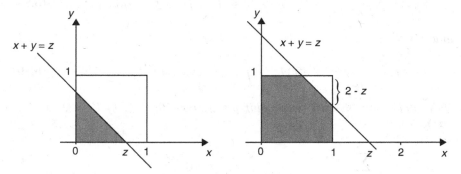

Fig. 5.15. The region $\{x + y \leq z\} \cap D$, depending on the value of z

---

[15] More precisely, A is any set in $\mathbb{R}^2$ such that $\{s : (X(s), Y(s)) \in A\}$ is an event.

By Equation 5.85 (see Figure 5.15),

$$F_Z(z) = P(X + Y \le z) = \iint_{\{x+y \le z\}} f(x,y)dxdy = \iint_{\{x+y \le z\} \cap D} dxdy$$

$$= \text{Area of } D \text{ under the line } x + y = z$$

$$= \begin{cases} 0 & \text{if } z < 0 \\ \frac{z^2}{2} & \text{if } 0 \le z < 1 \\ 1 - \frac{(2-z)^2}{2} & \text{if } 1 \le z < 2 \\ 1 & \text{if } 2 \le z \end{cases} \tag{5.86}$$

and so the p.d.f. of Z is

$$f_Z(z) = F_Z'(z) = \begin{cases} 0 & \text{if } z < 0 \\ z & \text{if } 0 \le z < 1 \\ 2 - z & \text{if } 1 \le z < 2 \\ 0 & \text{if } 2 \le z. \end{cases} \tag{5.87}$$

◆

The method of the foregoing example can be generalized as follows:

Theorem 5.4.2. *Distribution of the Sum of Two Random Variables.*
If X and Y are continuous with joint density f, then the d.f. and the density of $Z = X + Y$ are given by

$$F_Z(z) = \int_{-\infty}^{\infty} \int_{-\infty}^{z-x} f(x,y)dydx = \int_{-\infty}^{\infty} \int_{-\infty}^{z-y} f(x,y)dxdy \tag{5.88}$$

and

$$f_Z(z) = \int_{-\infty}^{\infty} f(x, z-x)dx = \int_{-\infty}^{\infty} f(z-y, y)dy. \tag{5.89}$$

If X and Y are discrete with joint p.f. f, then the p.f. of $Z = X + Y$ is given by

$$f_Z(z) = \sum_{x=-\infty}^{\infty} f(x, z-x) = \sum_{y=-\infty}^{\infty} f(z-y, y). \tag{5.90}$$

Proof. In the continuous case,

$$F_Z(z) = P(X + Y \le Z) = \iint_{x+y \le z} f(x,y)dxdy = \int_{-\infty}^{\infty} \int_{-\infty}^{z-x} f(x,y)dydx \tag{5.91}$$

and, by the fundamental theorem of calculus,

$$f_Z(z) = F'_Z(z) = \int_{-\infty}^{\infty} \left(\frac{\partial}{\partial z} \int_{-\infty}^{z-x} f(x,y)dy \right) dx = \int_{-\infty}^{\infty} f(x, z-x)dx.$$
(5.92)

In the discrete case,

$$f_Z(z) = \sum_{x+y=z} f(x,y) = \sum_{x=-\infty}^{\infty} f(x, z-x)dx.$$
(5.93)

In each formula, the second form can be obtained by interchanging the roles of x and y. ∎

Exercises

Exercise 5.4.1.

Roll two dice as in Example 5.4.1. Find the joint probability function of $U = X + Y$ and $V = X - Y$.

Exercise 5.4.2.

Roll two dice as in Example 5.4.1. Find the joint probability function of $U = \max(X, Y)$ and $V = \min(X, Y)$.

Exercise 5.4.3.

Roll six dice. Find the probabilities of obtaining:

1. Each of the six possible numbers once,
2. One 1, two 2's, and three 3's.

Exercise 5.4.4.

Let the random point (X, Y) be uniformly distributed on the triangle $D = \{(x,y) : 0 \leq x \leq y \leq 1\}$. Find the marginal densities of X and Y and plot their graphs.

Exercise 5.4.5.

Let the random point (X, Y) be uniformly distributed on the unit disk $D = \{(x,y) : x^2 + y^2 < 1\}$. Find the d.f. and the p.d.f. of the point's distance $Z = \sqrt{X^2 + Y^2}$ from the origin.

Exercise 5.4.6.

Let (X, Y) be continuous with density $f(x, y) = Ce^{-x-2y}$ for $x \geq 0, y \geq 0$ and 0 otherwise. Find:

1. The value of the constant C,
2. The marginal densities of X and Y,
3. The joint d.f. $F(x, y)$,
4. $\mathrm{P}(X < Y)$.

Exercise 5.4.7.

Let (X, Y) be continuous with density $f(x, y) = Cxy^2$ on the triangle $D = \{(x, y) : 0 \leq x \leq y \leq 1\}$ and 0 otherwise. Find:

1. The value of the constant C,
2. The marginal densities of X and Y,
3. The joint d.f. $F(x, y)$,
4. $\mathrm{P}(X > Y^2)$.

Exercise 5.4.8.

Let the random point (X, Y) be uniformly distributed on the square $D = \{(x, y) : -1 \leq x \leq 1, -1 \leq y \leq 1\}$. Find the d.f. and the p.d.f. of $Z = X + Y$.

Exercise 5.4.9.

Show that, for any random variables X and Y and any real numbers $x_1 < x_2$ and $y_1 < y_2$,

$$\mathrm{P}(x_1 < X \leq x_2, y_1 < Y \leq y_2) = F(x_2, y_2) - F(x_1, y_2) + F(x_1, y_1) - F(x_2, y_1).$$

Exercise 5.4.10.

Let the random point (X, Y) be uniformly distributed on the unit square $D = \{(x, y) : 0 \leq x \leq 1, 0 \leq y \leq 1\}$. Find the d.f. and the density of $Z = X - Y$.

Exercise 5.4.11.

Find formulas analogous to those in Theorem 5.4.2 for:

1. $Z = X - Y$,
2. $Z = 2X - Y$,
3. $Z = XY$.

5.5 Independence of Random Variables

The notion of independence of events can easily be extended to random variables, by applying the product rule to their joint distributions.

Definition 5.5.1. *Independence of Two Random Variables.*

Two random variables X and Y are said to be independent of each other if, for all intervals A and B,

$$P(X \in A, Y \in B) = P(X \in A)P(Y \in B). \tag{5.94}$$

Equivalently, we can reformulate the defining condition in terms of F or f:

Theorem 5.5.1. *Alternative Conditions for Independence of Two Random Variables.* *Two random variables X and Y are independent of each other if and only if their joint d.f. is the product of their marginal d.f.'s:*

$$F(x,y) = F_X(x)F_Y(y) \text{ for all } x, y. \tag{5.95}$$

Two discrete or absolutely continuous random variables X and Y are independent of each other if and only if their joint p.f. or p.d.f. is the product of their marginal p.f.'s or p.d.f.'s[16]*:*

$$f(x,y) = f_X(x)f_Y(y) \text{ for all } x, y. \tag{5.96}$$

Proof. If in Definition 5.5.1 we choose $A = (-\infty, x]$ and $B = (-\infty, y]$, then we get Equation 5.95. Conversely, if Equation 5.95 holds, then Equation 5.94 follows for any intervals from Theorem 5.1.2.

For discrete variables, Equation 5.96 follows from Definition 5.5.1 by substituting the one-point intervals $A = [x, x]$ and $B = [y, y]$, and for continuous variables by differentiating Equation 5.95. Conversely, we can obtain Equation 5.95 from Equation 5.96 by summation or integration. ∎

Example 5.5.1. Two Discrete Examples.

In Example 5.4.2 we obtained the following table (Table 5.5) for the joint p.f. f and the marginals of two discrete random variables X and Y:

$y \backslash x$	3	4	Any x
1	1/4	1/2	3/4
2	0	1/4	1/4
Any y	1/4	3/4	1

Table 5.5. The joint p.f. and marginals of two discrete dependent random variables

---

[16] More precisely, two absolutely continuous r.v.'s are independent if and only if there exist versions of the densities for which Equation 5.96 holds. (See footnote [4] on page 115.)

These variables are *not independent*, because $f(x,y) \neq f_X(x)f_Y(y)$ for all x, y. For instance, $f(3,1) = \frac{1}{4}$ but $f_X(3)f_Y(1) = \frac{1}{4} \cdot \frac{3}{4} = \frac{3}{16}$. (Note that we need to establish only one instance of $f(x,y) \neq f_X(x)f_Y(y)$ to *disprove* independence, but to *prove* independence we need to show $f(x,y) = f_X(x)f_Y(y)$ for all x, y.)

We can easily construct a table for an f with the same x, y values and the same marginals that represents the distribution of *independent* X and Y. All we have to do is to make each entry $f(x,y)$ equal to the product of the corresponding numbers on the margins (Table 5.6):

$y \backslash x$	3	4	Any x
1	3/16	9/16	3/4
2	1/16	3/16	1/4
Any y	1/4	3/4	1

Table 5.6. The joint p.f. and marginals of two discrete independent random variables

These examples show that there are usually many possible joint distributions for given marginals, but only one of those represents independent random variables.   ♦

Example 5.5.2. Independent Uniform Random Variables.

Let the random point (X, Y) be uniformly distributed on the rectangle $D = \{(x,y) : a \leq x \leq b, c \leq y \leq d\}$. Then

$$f(x,y) = \begin{cases} \frac{1}{(b-a)(d-c)} & \text{if } (x,y) \in D \\ 0 & \text{otherwise} \end{cases} \tag{5.97}$$

and the marginal densities are obtained by integration as

$$f_X(x) = \int_{-\infty}^{\infty} f(x,y)dy = \begin{cases} \int_c^d \frac{dy}{(b-a)(d-c)} = \frac{1}{(b-a)} & \text{if } a \leq x \leq b \\ 0 & \text{otherwise} \end{cases} \tag{5.98}$$

and

$$f_Y(y) = \int_{-\infty}^{\infty} f(x,y)dx = \begin{cases} \int_a^b \frac{dx}{(b-a)(d-c)} = \frac{1}{(d-c)} & \text{if } c \leq y \leq d \\ 0 & \text{otherwise} \end{cases}. \tag{5.99}$$

Hence X and Y are uniformly distributed on their respective intervals and are independent, because $f(x,y) = f_X(x)f_Y(y)$ for all x, y, as the preceding formulas show.

Clearly, the converse of our result is also true: If X and Y are uniformly distributed on their respective intervals and are independent, then $f_X(x)f_Y(y)$ yields the p.d.f. 5.97 of a point (X,Y) uniformly distributed on the corresponding rectangle.   ♦

Example 5.5.3. Uniform (X,Y) on the Unit Disk.

Let the random point (X,Y) be uniformly distributed on the unit disk $D = \{(x,y) : x^2 + y^2 < 1\}$. In Example 5.4 we obtained

$$f(x,y) = \begin{cases} 1/\pi & \text{if } (x,y) \in D \\ 0 & \text{otherwise} \end{cases}, \tag{5.100}$$

$$f_X(x) = \begin{cases} \frac{2}{\pi}\sqrt{1-x^2} & \text{if } x \in (-1,1) \\ 0 & \text{otherwise} \end{cases} \tag{5.101}$$

and

$$f_Y(y) = \begin{cases} \frac{2}{\pi}\sqrt{1-y^2} & \text{if } y \in (-1,1) \\ 0 & \text{otherwise} \end{cases}. \tag{5.102}$$

Now, clearly, $f(x,y) \neq f_X(x)f_Y(y)$ for all x,y, and so X and Y are not independent.

Note that this result is in agreement with the nontechnical meaning of dependence: From the shape of the disk, it follows that some values of X more or less determine the corresponding values of Y (and vice versa). For instance, if X is close to ± 1, then Y must be close to 0, and so X and Y are not expected to be independent of each other.   ♦

Example 5.5.4. Constructing a Triangle.

Suppose we pick two random points X and Y independently and uniformly on the interval $[0,1]$. What is the probability that we can construct a triangle from the resulting three segments as its sides?

A triangle can be constructed if and only if the sum of any two sides is longer than the third side. In our case, this condition means that each side must be shorter than $\frac{1}{2}$. (Prove this!) Thus X and Y must satisfy either

$$0 < X < \frac{1}{2}, \quad 0 < Y - X < \frac{1}{2}, \quad \frac{1}{2} < Y < 1, \tag{5.103}$$

or

$$0 < Y < \frac{1}{2}, \quad 0 < X - Y < \frac{1}{2}, \quad \frac{1}{2} < X < 1. \tag{5.104}$$

By Example 5.5.2 the given selection of the two points X and Y on a line is equivalent to the selection of the single point (X,Y) with a uniform distribution on the unit square of the plane. Now, the two sets of inequalities describe the two triangles at the center, shown in Figure 5.16, and the required probability is their combined area: $\frac{1}{4}$.

♦

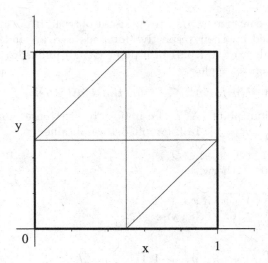

Fig. 5.16.

Example 5.5.5. Buffon's Needle Problem.

In 1777 a French scientist Comte de Buffon published the following problem: Suppose a needle is thrown at random on a floor marked with equidistant parallel lines. What is the probability that the needle will cross one of the lines?

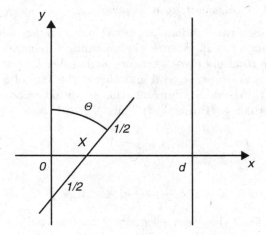

Fig. 5.17.

Let the distance between the lines be d and the length of the needle l. (See Fig. 5.17.) Choose a coordinate system in which the lines are vertical and one of the lines is the y-axis. Let the center of the needle have the random coordinates (X, Y). Clearly, the Y-coordinate is irrelevant to the problem, and we may assume that the center lies on the x-axis. Because of the periodicity, we may also assume that the center falls in the first strip, that is, that $0 \leq X \leq d$. Now, if the needle makes a random angle Θ with the y-axis, then it will cross the y-axis if and only if $0 \leq X \leq (l/2) \sin \Theta$, and it will cross the $y = d$ line if and only if $d - (l/2) \sin \Theta \leq X \leq d$. The random throw of the needle implies that X and Θ are uniform r.v.'s on the $[0, d]$ and the $[0, \pi]$ intervals, respectively, which is, by Example 5.5.2, equivalent to the random point (Θ, X) being uniform on the $[0, \pi] \times [0, d]$ rectangle in the (θ, x) plane. The needle will cross one of the lines if and only if the random point (Θ, X) falls into either one of the D-shaped regions in Figure 5.18. Since the area of the rectangle is πd and the area of each D-shaped region is $\int_0^\pi (l/2) \sin \theta d\theta = l$, the required probability is

$$P(\text{the needle will cross a line}) = \frac{2l}{\pi d}. \tag{5.105}$$

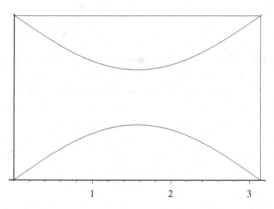

Fig. 5.18. The (θ, x) plane, with the $[0, \pi] \times [0, d]$ rectangle and the $y = (l/2) \sin \theta$ and $y = d - (l/2) \sin \theta$ curves

In principle, this experiment can be used for the determination of π. However, it is difficult to arrange completely random throws of a needle, and it would be a very inefficient way to obtain π; we have much better methods. On the other hand, computer simulations of needle throws have been performed to obtain approximations to π, just to illustrate the result.♦

Example 5.5.6. Bertrand's Paradox.

In 1889, another Frenchman, Joseph Bertrand, considered the following problem:

Consider a circle of radius r and select one of its chords at random. What is the probability that the selected chord will be longer than the side of an equilateral triangle inscribed into the given circle.

Bertrand presented three solutions with "paradoxically" different results:

1. Since the length of a chord is uniquely determined by the position of its center, we just choose a point for it at random, that is, with a uniform distribution inside the circle. We can see from Figure 5.19 that the chord will be longer than the side of an inscribed equilateral triangle if and only if its center falls inside the circle of radius $r/2$ concentric with the given circle. Thus

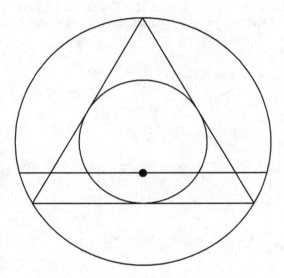

Fig. 5.19.

P(the chord is longer than the side of the triangle) $= \dfrac{\pi\left(r/2\right)^2}{\pi r^2} = \dfrac{1}{4}.$

(5.106)

2. By symmetry, we may consider only horizontal chords, and then we may assume that their center is uniformly distributed on the vertical diameter of the given circle. The chord will be longer than the side of the triangle if and only if its center falls on the thick vertical segment in Figure 5.20. Thus

P(the chord is longer than the side of the triangle) $= \dfrac{1}{2}.$ (5.107)

3. We may also choose a random chord by fixing one of its endpoints and choosing the other one at random on the circle, that is, uniformly distributed on the perimeter. Let the fixed point be on top, as shown in Figure 5.21. Clearly, the chord will be longer than the side of the triangle

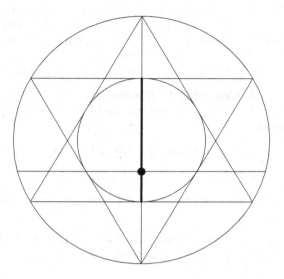

Fig. 5.20.

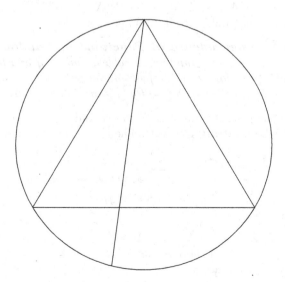

Fig. 5.21.

if and only if its random end falls on the bottom one third of the circle. Thus

$$P(\text{the chord is longer than the side of the triangle}) = \frac{1}{3}. \quad (5.108)$$

The resolution of the paradox lies in realizing that the statement of the problem is ambiguous. Choosing a chord at random is not well defined; each of the three choices presented above is a reasonable but different way of doing so.   ◆

Next, we present some theorems about independence of random variables.

Theorem 5.5.2. *A Constant Is Independent of Any Random Variable. Let $X = c$, where c is any constant, and let Y be any r.v. Then X and Y are independent.*

Proof. Let $X = c$, and let Y be any r.v. Then Equation 5.95 becomes

$$P(c \leq x, Y \leq y) = P(c \leq x)P(Y \leq y), \quad (5.109)$$

and this equation is true because for $x \geq c$ and any y, it reduces to $P(Y \leq y) = P(Y \leq y)$, and for $x < c$ it reduces to $0 = 0$.   ∎

Theorem 5.5.3. *No Nonconstant Random Variable Is Independent of Itself. Let X be any nonconstant random variable and let $Y = X$. Then X and Y are dependent.*

Proof. Let A and B be two disjoint intervals for which $P(X \in A) > 0$ and $P(X \in B) > 0$ hold. Since X is not constant, such intervals clearly exist. If $Y = X$, then $P(X \in A, Y \in B) = 0$, but $P(X \in A)P(Y \in B) > 0$, and so Equation 5.94 does not hold for all intervals A and B.   ∎

Theorem 5.5.4. *Independence of Functions of Random Variables. Let X and Y be independent random variables, and let g and h be any real-valued measurable functions (see the footnote on page 105) on range(X) and range(Y), respectively. Then $g(X)$ and $h(Y)$ are independent.*

Proof. We give the proof for discrete X and Y only.

Let A and B be arbitrary intervals. Then

$$P(g(X) \in A, h(Y) \in B) = \sum_{\{x:g(x)\in A\}} \sum_{\{y:h(y)\in B\}} P(X = x, Y = y)$$

$$= \sum_{\{x:g(x)\in A\}} \sum_{\{y:h(y)\in B\}} P(X = x)P(Y = y)$$

$$= \sum_{\{x:g(x)\in A\}} P(X = x) \sum_{\{y:h(y)\in B\}} P(Y = y)$$

$$= P(g(X) \in A)P(h(Y) \in B). \quad (5.110)$$

∎

We can extend the definition of independence to several random variables as well, but we need to distinguish different types of independence, depending on the number of variables involved:

Definition 5.5.2. *Independence of Several Random Variables.*

Let $X_1, X_2, \ldots, X_n$, for $n = 2, 3, \ldots$, be arbitrary random variables. They are (totally) independent, if

$$P(X_1 \in A_1, X_2 \in A_2, \ldots, X_n \in A_n) = P(X_1 \in A_1)P(X_2 \in A_2) \cdots P(X_n \in A_n)$$
(5.111)

for all intervals $A_1, A_2, \ldots, A_n$.
They are pairwise independent if

$$P(X_i \in A_i, X_j \in A_j) = P(X_i \in A_i)P(X_j \in A_j)$$
(5.112)

for all $i \neq j$ and all intervals A_i, A_j.

Note that in the case of total independence, it is not necessary to require the product rule for all subsets of the n random variables (as we had to for general events), because the product rule for any number less than n follows from Equation 5.111 by setting $A_i = \mathbb{R}$ for all values of i that we want to omit. On the other hand, pairwise independence is a weaker requirement than total independence: Equation 5.112 does not imply Equation 5.111. Also, we could have defined various types of independence between total and pairwise, but such types generally do not occur in practice.

We have the following theorems for several random variables, analogous to Theorem 5.5.1 and Theorem 5.5.4, which we state without proof.

Theorem 5.5.5. *Alternative Conditions for Independence of Several Random Variables.* *Any random variables $X_1, X_2, \ldots, X_n$, for $n = 2, 3, \ldots$, are independent of each other if and only if their joint d.f. is the product of their marginal d.f.'s:*

$$F(x_1, x_2, \ldots, x_n) = F_1(x_1)F_2(x_2) \cdots F_n(x_n) \text{ for all } x_1, x_2, \ldots, x_n. \quad (5.113)$$

Also, any discrete or absolutely continuous random variables $X_1, X_2, \ldots, X_n$, for $n = 2, 3, \ldots$, are independent of each other if and only if their joint p.f. or p.d.f. is the product of their marginal p.f.'s or (appropriate versions of their) p.d.f.'s:

$$f(x_1, x_2, \ldots, x_n) = f_1(x_1)f_2(x_2) \cdots f_n(x_n) \text{ for all } x_1, x_2, \ldots, x_n. \quad (5.114)$$

Theorem 5.5.6. *Independence of Functions of Random Variables.* *Let $X_1, X_2, \ldots, X_n$, for $n = 2, 3, \ldots$, be independent random variables, and let the g_i be real-valued measurable functions on range(X_i) for $i = 1, 2, \ldots, n$. Then $g_1(X_1), g_2(X_2), \ldots, g_n(X_n)$ are independent.*

Theorem 5.5.6 could be further generalized in an obvious way by taking the g_i to be functions of several, non-overlapping variables. For example, in the case of three random variables, we have the following theorem:

Theorem 5.5.7. *Independence of* $g(X, Y)$ *and* Z. *If* Z *is independent of* (X, Y), *then* Z *is independent of* $g(X, Y)$, *too, for any measurable function* g.

Proof. We give the proof for jointly continuous X, Y, and Z only.
    For arbitrary t and z,

$$
\begin{aligned}
\mathrm{P}\left(g\left(X,Y\right) \leq t, Z \leq z\right) &= \int_{-\infty}^{z} \iint_{g(x,y)\leq t} f\left(x,y,\varsigma\right) dx dy d\varsigma \\
&= \int_{-\infty}^{z} \iint_{g(x,y)\leq t} f_{X,Y}\left(x,y\right) f_{Z}\left(\varsigma\right) dx dy d\varsigma \\
&= \iint_{g(x,y)\leq t} f_{X,Y}\left(x,y\right) dx dy \int_{-\infty}^{z} f_{Z}\left(\varsigma\right) d\varsigma \\
&= \mathrm{P}\left(g\left(X,Y\right) \leq t\right) \mathrm{P}\left(Z \leq z\right). \qquad (5.115)
\end{aligned}
$$

By Theorem 5.5.1, Equation 5.115 proves the independence of $g(X, Y)$ and Z.
∎

In some applications, we need to find the distribution of the maximum or of the minimum of several independent random variables. This can be done as follows:

Theorem 5.5.8. *Distribution of Maximum and Minimum of Several Random Variables.* *Let* $X_1, X_2, \ldots, X_n$, *for* $n = 2, 3, \ldots$, *be independent, identically distributed (abbreviated i.i.d.) random variables with common d.f.* F_X, *and let* $Y = \max\{X_1, X_2, \ldots, X_n\}$ *and* $Z = \min\{X_1, X_2, \ldots, X_n\}$.[17] *Then the distribution functions of* Y *and* Z *are given by*

$$F_Y(y) = [F_X(y)]^n \text{ for all } y \in \mathbb{R} \qquad (5.116)$$

and

$$F_Z(z) = 1 - [1 - F_X(z)]^n \text{ for all } z \in \mathbb{R}. \qquad (5.117)$$

Proof. For any $y \in \mathbb{R}$, $Y = \max\{X_1, X_2, \ldots, X_n\} \leq y$ holds if and only if, for every i, $X_i \leq y$. Thus, we have

$$
\begin{aligned}
F_Y(y) &= \mathrm{P}\left(X_1 \leq y, X_2 \leq y, \ldots, X_n \leq y\right) \\
&= \mathrm{P}\left(X_1 \leq y\right)\mathrm{P}\left(X_2 \leq y\right) \cdots \mathrm{P}\left(X_n \leq y\right) = [F_X(y)]^n. \qquad (5.118)
\end{aligned}
$$

---

[17] Note that the max and the min have to be taken pointwise, that is, for each sample point s, we have to consider the max and the min of $\{X_1(s), X_2(s), \ldots, X_n(s)\}$, and so Y and Z will in general be different from each of the X_i.

Similarly,

$$
\begin{aligned}
F_Z(z) &= \mathrm{P}\left(Z \leq z\right) = 1 - \mathrm{P}\left(Z > z\right)\\
&= 1 - \mathrm{P}\left(X_1 > z, X_2 > z, \ldots, X_n > z\right)\\
&= 1 - \mathrm{P}\left(X_1 > z\right)\mathrm{P}\left(X_2 > z\right)\cdots\mathrm{P}\left(X_n > z\right)\\
&= 1 - \left[1 - F_X(z)\right]^n.
\end{aligned}
\tag{5.119}
$$

∎

Example 5.5.7. Maximum of Two Independent Uniformly Distributed Points.

Let X_1 and X_2 be independent, uniform random variables on the interval $[0, 1]$. Find the d.f. and the p.d.f. of $Y = \max\{X_1, X_2\}$.

By Equation 5.19,

$$
F_X\left(x\right) = \begin{cases} 0 \text{ if } & x < 0\\ x \text{ if } & 0 \leq x < 1\\ 1 \text{ if } & x \geq 1 \end{cases},
\tag{5.120}
$$

and so, by Theorem 5.5.8,

$$
F_Y\left(y\right) = \begin{cases} 0 \text{ if } & y < 0\\ y^2 \text{ if } & 0 \leq y < 1\\ 1 \text{ if } & y \geq 1 \end{cases}.
\tag{5.121}
$$

Hence the p.d.f. of Y is given by

$$
f_Y\left(y\right) = \begin{cases} 2y \text{ if } & 0 \leq y < 1\\ 0 \text{ if } y < 0 \text{ or } y \geq 1 \end{cases},
\tag{5.122}
$$

which shows that the probability of $Y = \max\{X_1, X_2\}$ falling in a subinterval of length dy is no longer constant over $[0, 1]$, as for X_1 or X_2, but increases linearly.

The two functions above can also be seen in Figure 5.22. The sample space is the set of points $s = (x_1, x_2)$ of the unit square, and, for any sample point s, $X_1\left(s\right) = x_1$ and $X_2\left(s\right) = x_2$. The sample points are uniformly distributed on the unit square, and so the areas of subsets give the corresponding probabilities. Since for any sample point s above the diagonal $x_1 < x_2$ holds, $Y\left(s\right) = x_2$ there and, similarly, below the diagonal $Y\left(s\right) = x_1$. Thus, the set $\{s : Y\left(s\right) \leq y\}$ is the shaded square of area y^2, and the thin strip of width dy, to the right and above the square, has an area $\approx 2y\,dy$.   ♦

Another, very important function of two independent random variables is their sum. We have the following theorem for its distribution:

Theorem 5.5.9. Sum of Two Independent Random Variables. *Let X and Y be independent random variables and $Z = X + Y$. If X and Y are discrete, then the p.f. of Z is given by*

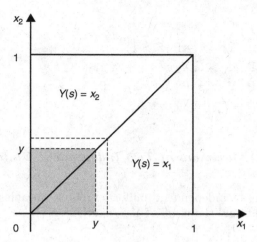

Fig. 5.22. The d.f and the p.d.f. of $Y = \max\{X_1, X_2\}$ for two i.i.d. uniform r.v.'s on $[0,1]$

$$f_Z(z) = \sum_{x+y=z} f_X(x)f_Y(y) = \sum_x f_X(x)f_Y(z-x), \qquad (5.123)$$

where, for a given z, the summation is extended over all possible values of X and Y for which $x + y = z$, if such values exist. Otherwise $f_Z(z)$ is taken to be 0. The expression on the right is called the convolution of f_X and f_Y.

If X and Y are continuous with densities f_X and f_Y, then the density of $Z = X + Y$ is given by

$$f_Z(z) = \int_{-\infty}^{\infty} f_X(x)f_Y(z-x)dx, \qquad (5.124)$$

where the integral is again called the convolution of f_X and f_Y.

Proof. These results follow from Theorem 5.4.2 by substituting $f(x,y) = f_X(x)f_Y(y)$. However, in the continuous case, we also give another, more direct and visual proof: If X and Y are independent and continuous with densities f_X and f_Y, then Z falls between z and $z+dz$ if and only if the point (X,Y) falls in the oblique strip between the lines $x+y = z$ and $x+y = z+dz$, as shown in Figure 5.23. The area of the shaded parallelogram is $dxdz$, and the probability of (X,Y) falling into it is[18]

$$P(x \leq X < x + dx, z \leq Z < z + dz) \sim f(x,y)dxdz = f_X(x)f_Y(z-x)dxdz. \qquad (5.125)$$

---

[18] Recall that the symbol $\sim$ means that the ratio of the expressions on either side of it tends to 1 as dx and dz tend to 0.

Hence the probability of the strip is obtained by integrating this expression over all x as

$$P(z \leq Z < z + dz) \sim \left[\int_{-\infty}^{\infty} f_X(x) f_Y(z - x) dx \right] dz, \qquad (5.126)$$

and, since $P(z \leq Z < z + dz) \sim f_Z(z) dz$, Equation 5.126 implies Equation 5.124. ∎

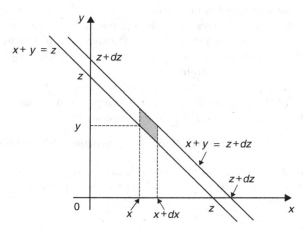

Fig. 5.23. The probability of (X, Y) falling in the oblique strip is dz times the convolution

The convolution formulas for two special classes of random variables are worth mentioning separately:

Corollary 5.5.1. *If the possible values of X and Y are the natural numbers $i, j = 0, 1, 2, \ldots$, then the p.f. of $Z = X + Y$ is given by*

$$f_Z(k) = \sum_{i=0}^{k} f_X(i) f_Y(k - i) \text{ for } k = 0, 1, 2, \ldots, \qquad (5.127)$$

and if X and Y are continuous nonnegative random variables, then the p.d.f. of $Z = X + Y$ is given by

$$f_Z(z) = \int_{0}^{z} f_X(x) f_Y(z - x) dx. \qquad (5.128)$$

Example 5.5.8. Sum of Two Binomial Random Variables.

Let X and Y be independent, binomial r.v.'s with parameters n_1, p and n_2, p, respectively. Then $Z = X + Y$ is binomial with parameters $n_1 + n_2, p$ because, by Equation 5.127 and Equation 3.37,

$$f_Z(k) = \sum_{i=0}^{k} \binom{n_1}{i} p^i q^{n_1-i} \binom{n_2}{k-i} p^{k-i} q^{n_2-k+i}$$

$$= \sum_{i=0}^{k} \binom{n_1}{i} \binom{n_2}{k-i} p^k q^{n_1+n_2-k}$$

$$= \binom{n_1+n_2}{k} p^k q^{n_1+n_2-k} \text{ for } k = 0, 1, 2, \ldots, n_1 + n_2. \qquad (5.129)$$

This result should be obvious without any computation as well, since X counts the number of successes in n_1 independent trials and Y the number of successes in n_2 trials, independent of each other and of the first n_1 trials, and so $Z = X + Y$ counts the number of successes in $n_1 + n_2$ independent trials, all with the same probability p.

On the other hand, for sampling without replacement, the trials are not independent, and the analogous sum of two independent hypergeometric random variables does not turn out to be hypergeometric.   ♦

Example 5.5.9. *Sum of Two Uniform Random Variables.*

Let X and Y be i.i.d. random variables, uniform on $[0,1]$ as in Example 5.4.9. This time, however, we are going to use Equation 5.128 to obtain the density of $Z = X + Y$.

The common density of X and Y is

$$f(x) = \begin{cases} 1 \text{ if } 0 \le x \le 1 \\ 0 \text{ otherwise} \end{cases} \qquad (5.130)$$

Hence $f_Y(z-x) = 1$ if $0 \le z - x \le 1$ or, equivalently, if $z - 1 \le x \le z$ and is 0 otherwise. Thus the density of Z is the convolution from Equation 5.128:

$$f_Z(z) = \int_0^z f_X(x) f_Y(z-x) dx = \int_{[0,1] \cap [z-1,z]} 1 dx = \begin{cases} 0 & \text{if } z < 0 \\ z & \text{if } 0 \le z < 1 \\ 2 - z & \text{if } 1 \le z < 2 \\ 0 & \text{if } 2 \le z. \end{cases}$$

$$(5.131)$$

This result is the same, of course, as the corresponding one in Example 5.4.9. ♦

Theorem 5.5.10. *Product and Ratio of Two Independent Random Variables.* *Let X and Y be independent, continuous, positive random variables with given densities f_X and f_Y, with $f_X(x) = 0$ for $x < 0$ and $f_Y(y) = 0$ for $y < 0$:*

1. *The density function of $Z = XY$ is given by*

$$f_Z\left(z\right) = \begin{cases} \int_0^\infty f_X\left(\frac{z}{y}\right)f_Y\left(y\right)\frac{1}{y}dy & \text{if } z > 0 \\ 0 & \text{if } z \leq 0 \end{cases} \qquad (5.132)$$

and alternatively by

$$f_Z\left(z\right) = \begin{cases} \int_0^\infty f_X\left(x\right)f_Y\left(\frac{z}{x}\right)\frac{1}{x}dx & \text{if } z > 0 \\ 0 & \text{if } z \leq 0. \end{cases} \qquad (5.133)$$

2. *The density function of $Z = X/Y$ is given by*

$$f_Z\left(z\right) = \begin{cases} \int_0^\infty f_X\left(zy\right)f_Y\left(y\right)ydy & \text{if } z > 0 \\ 0 & \text{if } z \leq 0 \end{cases} \qquad (5.134)$$

and alternatively by

$$f_Z\left(z\right) = \begin{cases} \int_0^\infty f_X\left(x\right)f_Y\left(\frac{x}{z}\right)\frac{x}{z^2}dx & \text{if } z > 0 \\ 0 & \text{if } z \leq 0. \end{cases} \qquad (5.135)$$

Proof.

1. For $z > 0$

$$F_Z\left(z\right) = \mathrm{P}\left(XY \leq z\right) = \iint_{xy \leq z} f_X\left(x\right)f_Y\left(y\right)dxdy$$

$$= \int_0^\infty \left[\int_0^{z/y} f_X\left(x\right)dx\right]f_Y\left(y\right)dy = \int_0^\infty F_X\left(\frac{z}{y}\right)f_Y\left(y\right)dy, \qquad (5.136)$$

and so, by the chain rule,

$$f_Z\left(z\right) = F_Z'\left(z\right) = \int_0^\infty f_X\left(\frac{z}{y}\right)f_Y\left(y\right)\frac{1}{y}dy. \qquad (5.137)$$

If $z \leq 0$, then, by the assumed positivity of X and Y, $\mathrm{P}\left(XY \leq z\right) = 0$ and $f_Z\left(z\right) = 0$. The alternative formula can be obtained by interchanging x and y.

2. For $z > 0$

$$F_Z\left(z\right) = \mathrm{P}\left(\frac{X}{Y} \leq z\right) = \iint_{x/y \leq z} f_X\left(x\right)f_Y\left(y\right)dxdy$$

$$= \int_0^\infty \left[\int_0^{zy} f_X\left(x\right)dx\right]f_Y\left(y\right)dy = \int_0^\infty F_X\left(zy\right)f_Y\left(y\right)dy, \qquad (5.138)$$

and so

$$f_Z\left(z\right) = F_Z'\left(z\right) = \int_0^\infty f_X(zy)f_Y(y)y\,dy.$$ (5.139)

Alternatively,

$$F_Z\left(z\right) = \mathrm{P}\left(\frac{X}{Y} \le z\right) = \iint_{x/y \le z} f_X(x)f_Y(y)dxdy$$

$$= \int_0^\infty f_X(x)\left[\int_{x/z}^\infty f_Y(y)dy\right]dx = \int_0^\infty f_X(x)\left[1 - F_Y\left(\frac{x}{z}\right)\right]dx,$$ (5.140)

and so

$$f_Z\left(z\right) = F_Z'\left(z\right) = \int_0^\infty f_X(x)f_Y\left(\frac{x}{z}\right)\frac{x}{z^2}dx.$$ (5.141)

If $z \le 0$, then clearly $\mathrm{P}\left(X/Y \le z\right) = 0$ and $f_Z\left(z\right) = 0$. ∎

Example 5.5.10. Ratio of Two Exponential Random Variables.

Let X and Y be two exponential random variables with unequal parameters λ_1 and λ_2, respectively. Find the density of their ratio $Z = X/Y$.

Now, $f_X(x) = \lambda_1 e^{-\lambda_1 x}$ and $f_Y(y) = \lambda_2 e^{-\lambda_2 y}$ for $x, y > 0$. Thus, from Equation 5.134 and using integration by parts, we obtain the density of $Z = X/Y$, for $z > 0$, as

$$f_Z\left(z\right) = \int_0^\infty \lambda_2 e^{-\lambda_2 y}\lambda_1 e^{-\lambda_1 z y}y\,dy = \lambda_1\lambda_2\int_0^\infty e^{-(\lambda_2 + \lambda_1 z)y}y\,dy$$

$$= \lambda_1\lambda_2\left[\frac{e^{-(\lambda_2 + \lambda_1 z)y}}{-(\lambda_2 + \lambda_1 z)}y\bigg|_0^\infty - \int_0^\infty \frac{e^{-(\lambda_2 + \lambda_1 z)y}}{-(\lambda_2 + \lambda_1 z)}dy\right]$$

$$= \lambda_1\lambda_2\left[0 - \frac{e^{-(\lambda_2 + \lambda_1 z)y}}{(\lambda_2 + \lambda_1 z)^2}\bigg|_0^\infty\right] = \frac{\lambda_1\lambda_2}{(\lambda_2 + \lambda_1 z)^2}.$$ (5.142)

Exercises

Exercise 5.5.1.

Two cards are dealt from a regular deck of 52 cards without replacement. Let X denote the number of spades and Y the number of hearts obtained. Are X and Y independent?

Exercise 5.5.2.

We roll two dice once. Let X denote the number of 1's and Y the number of 6's obtained. Are X and Y independent?

Exercise 5.5.3.

Let the random point (X, Y) be uniformly distributed on $D = \{(x, y) : 0 \leq x \leq 1/2, 0 \leq y \leq 1/2\} \cup \{(x, y) : 1/2 \leq x \leq 1, 1/2 \leq y \leq 1\}$ as in 5.4.5. Are X and Y independent?

Exercise 5.5.4.

Let X and Y be continuous random variables with density

$$f(x, y) = \begin{cases} xe^{-x(y+1)} & \text{if } x > 0, y > 0 \\ 0 & \text{otherwise.} \end{cases} \tag{5.143}$$

Are X and Y independent?

Exercise 5.5.5.

Recall that the *indicator function* or *indicator random variable* I_A of an event A in any sample space S is defined by

$$I_A(s) = \begin{cases} 1 \text{ if } s \in A \\ 0 \text{ if } s \in \overline{A}. \end{cases} \tag{5.144}$$

1. Prove that $I_{A \cup B} = I_A + I_B - I_{AB}$.
2. Prove that A and B are independent events if and only if I_A and I_B are independent random variables.

Exercise 5.5.6.

Let the random point (X, Y) be uniformly distributed on the unit disk as in 5.4. Show that the polar coordinates $R \in [0, 1]$ and $\Theta \in [0, 2\pi]$ of the point are independent. (*Hint*: Determine the joint d.f. $F_{R,\Theta}(r, \theta)$ and the marginals $F_R(r) = F_{R,\Theta}(r, 2\pi)$ and $F_\Theta(\theta) = F_{R,\Theta}(1, \theta)$ from a picture, and use Equation 5.95.)

Exercise 5.5.7.

Alice and Bob visit the school library, each at a random time uniformly distributed between 2PM and 6PM, independently of each other, and stay there for an hour. What is the probability that they meet?

Exercise 5.5.8.

A point X is chosen at random on the interval $[0,1]$ and independently another point Y on the interval $[1,2]$. What is the probability that we can construct a triangle from the resulting three segments $[0,X], [X,Y], [Y,2]$ as sides?

Exercise 5.5.9.

We choose a point at random on the perimeter of a circle and then independently another point at random in the interior of the circle. What is the probability that the two points will be nearer to each other than the radius of the circle?

Exercise 5.5.10.

Let X be a discrete uniform r.v. on the set $\{000, 011, 101, 110\}$ of four binary integers, and let X_i denote the ith digit of X, for $i = 1, 2, 3$. Show that X_1, X_2, X_3 are independent pairwise but not totally independent.

Can you generalize this example to more than three random variables?

Exercise 5.5.11.

Let X and Y be i.i.d. uniform on $(0,1)$:

1. Find the joint density of $Z = XY$.
2. Find the joint density of $Z = X/Y$.

Exercise 5.5.12.

What is the probability that in ten independent tosses of a fair coin, we get two heads in the first four tosses and five heads altogether?

Exercise 5.5.13.

Consider light bulbs with independent, exponentially distributed lifetimes with parameter $\lambda = \frac{1}{100 \text{ days}}$:

1. Find the probability that such a bulb survives to 200 days.
2. Find the probability that such a bulb dies before 40 days.
3. Find the probability that the bulb with the longest lifetime in a batch of 10 survives to 200 days.
4. Find the probability that the bulb with the shortest lifetime in a batch of 10 dies before 40 days.

Exercise 5.5.14.

Let $X_1, X_2, \ldots, X_n$, for $n = 2, 3, \ldots$, be i.i.d. random variables with common d.f. F_X. Find a formula for the joint d.f. $F_{Y,Z}$ of $Y = \max\{X_1, X_2, \ldots, X_n\}$ and $Z = \min\{X_1, X_2, \ldots, X_n\}$ in terms of F_X.

Exercise 5.5.15.

Show that the p.d.f. of the sum $S = T_1 + T_2$ of two i.i.d exponential r.v.'s with parameter λ is given by

$$f_S(s) = \begin{cases} 0 & \text{if } s < 0 \\ \lambda^2 s e^{-\lambda s} & \text{if } s \geq 0 \end{cases} \tag{5.145}$$

Exercise 5.5.16.

Find the p.d.f. of the sum $S = T_1 + T_2$ of two independent exponential r.v.'s with parameters λ and $\mu \neq \lambda$, respectively.

Exercise 5.5.17.

Show that the p.d.f. of the difference $Z = T_1 - T_2$ of two i.i.d exponential r.v.'s with parameter λ is $f_Z(z) = \frac{\lambda}{2} e^{-\lambda |z|}$.

Exercise 5.5.18.

Let X_i for $i = 1, 2, \ldots$ be i.i.d. random variables, uniform on $[0, 1]$, and let f_n denote the p.d.f. of $S_n = \sum_{i=1}^{n} X_i$ for $n \geq 1$:

1. Show that $f_{n+1}(z) = \int_{z-1}^{z} f_n(x) \, dx$.
2. Evaluate $f_3(z)$ and sketch its graph.

Exercise 5.5.19.

Let X and Y be i.i.d. random variables, uniform on $[0, 1]$. Find the density of $Z = X - Y$.

5.6 Conditional Distributions

In many applications, we need to consider the distribution of a random variable under certain conditions. For conditions with nonzero probabilities, we can just apply the definition of conditional probabilities to events associated with random variables. Thus, we make the following definition:

Definition 5.6.1. *Conditional Distributions for Conditions with Nonzero Probabilities.*

Let A be any event with $\mathrm{P}(A) \neq 0$ and X any random variable. Then we define the conditional distribution function of X under the condition A by

$$F_{X|A}(x) = \mathrm{P}(X \leq x | A) \text{ for all } x \in \mathbb{R}. \tag{5.146}$$

If X is a discrete random variable, then we define the conditional probability function of X under the condition A by

$$f_{X|A}(x) = P(X = x|A) \text{ for all } x \in \mathbb{R}. \tag{5.147}$$

If X is a continuous random variable and there exists a nonnegative function $f_{X|A}$ that is integrable over R and for which

$$\int_{-\infty}^{x} f_{X|A}(t)dt = F_{X|A}(x), \text{ for all } x, \tag{5.148}$$

then $f_{X|A}$ is called the conditional density function of X under the condition A.

If Y is a discrete random variable and $A = \{Y = y\}$, then we write

$$F_{X|Y}(x,y) = P(X \leq x|Y = y) \text{ for all } x \in \mathbb{R} \text{ and all possible values } y \text{ of } Y \tag{5.149}$$

and call $F_{X|Y}$ the conditional distribution function of X given Y.

If both X and Y are discrete, then the conditional probability function of X given Y is defined by

$$f_{X|Y}(x,y) = P(X = x|Y = y) \text{ for all possible values } x \text{ and } y \text{ of } X \text{ and } Y. \tag{5.150}$$

If X is continuous, Y is discrete, $A = \{Y = y\}$, and $f_{X|A}$ in Equation 5.148 exists, then $f_{X|A}$ is called the conditional density function of X given $Y = y$ and is denoted by $f_{X|Y}(x,y)$ for all $x \in \mathbb{R}$ and all possible values y of Y.

If X is a continuous random variable with a conditional density function $f_{X|A}$, then, by the fundamental theorem of calculus, Equation 5.148 gives that

$$f_{X|A}(x) = F'_{X|A}(x), \tag{5.151}$$

wherever $F'_{X|A}$ is continuous. At such points, we also have

$$f_{X|A}(x)dx \sim P(x \leq X < x + dx \mid A) = \frac{P(\{x \leq X < x + dx\} \cap A)}{P(A)}. \tag{5.152}$$

By the definitions of conditional probabilities and joint distributions, Equation 5.150 for discrete X and Y can also be written as

$$f_{X|Y}(x,y) = \frac{f(x,y)}{f_Y(y)} \text{ for all possible values } x \text{ and } y \text{ of } X \text{ and } Y, \tag{5.153}$$

where $f(x,y)$ is the joint p.f. of X and Y and $f_Y(y)$ the marginal p.f. of Y.

Example 5.6.1. Sum and Absolute Difference of Two Dice.

In Example 5.4.1 we considered the random variables $U = X + Y$ and $V = |X - Y|$, where X and Y were the numbers obtained with rolling two dice. Now, we want to find the values of the conditional probability functions $f_{U|V}$ and $f_{V|U}$. For easier reference, we first reproduce the table of the joint probability function $f(u, v)$ and the marginals here (Table 5.7):

$u \backslash v$	0	1	2	3	4	5	$f_U(u)$
2	1/36	0	0	0	0	0	1/36
3	0	2/36	0	0	0	0	2/36
4	1/36	0	2/36	0	0	0	3/36
5	0	2/36	0	2/36	0	0	4/36
6	1/36	0	2/36	0	2/36	0	5/36
7	0	2/36	0	2/36	0	2/36	6/36
8	1/36	0	2/36	0	2/36	0	5/36
9	0	2/36	0	2/36	0	0	4/36
10	1/36	0	2/36	0	0	0	3/36
11	0	2/36	0	0	0	0	2/36
12	1/36	0	0	0	0	0	1/36
$f_V(v)$	6/36	10/36	8/36	6/36	4/36	2/36	1

Table 5.7. The joint and marginal probability functions of $U = X + Y$ and $V = |X - Y|$, for the numbers X and Y showing on two dice

According to Equation 5.153, the table of the conditional probability function $f_{U|V}(u, v)$ can be obtained from the table above by dividing each $f(u, v)$ value by the marginal probability below it, and similarly, the table of the conditional probability function $f_{V|U}(u, v)$ can be obtained by dividing each $f(u, v)$ value by the marginal probability to the right of it:

$u \backslash v$	0	1	2	3	4	5
2	1/6	0	0	0	0	0
3	0	1/5	0	0	0	0
4	1/6	0	1/4	0	0	0
5	0	1/5	0	1/3	0	0
6	1/6	0	1/4	0	1/2	0
7	0	1/5	0	1/3	0	1
8	1/6	0	1/4	0	1/2	0
9	0	1/5	0	1/3	0	0
10	1/6	0	1/4	0	0	0
11	0	1/5	0	0	0	0
12	1/6	0	0	0	0	0

Table 5.8. The conditional probability function of $U = X + Y$ given $V = |X - Y|$, for the numbers X and Y showing on two dice

$u\backslash v$	0	1	2	3	4	5
2	1	0	0	0	0	0
3	0	1	0	0	0	0
4	1/3	0	2/3	0	0	0
5	0	1/2	0	1/2	0	0
6	1/5	0	2/5	0	2/5	0
7	0	1/3	0	1/3	0	1/3
8	1/5	0	2/5	0	2/5	0
9	0	1/2	0	1/2	0	0
10	1/3	0	2/3	0	0	0
11	0	1	0	0	0	0
12	1	0	0	0	0	0

Table 5.9. The conditional probability function of $V = |X - Y|$ given $U = X + Y$, for the numbers X and Y showing on two dice

The conditional probabilities in these tables make good sense. For instance, if $V = |X - Y| = 1$, then $U = X + Y$ can be only $3 = 1 + 2 = 2 + 1, 5 = 2 + 3 = 3 + 2, 7 = 3 + 4 = 4 + 3, 9 = 4 + 5 = 5 + 4$, or $11 = 5 + 6 = 6 + 5$. Since each of these five possible U values can occur under the condition $V = 1$ in exactly two ways, their conditional probabilities must be $1/5$ each, as shown in the second column of Table 5.8.

Similarly, if $U = X + Y = 3$, then we must have $(X, Y) = (1, 2)$ or $(X, Y) = (2, 1)$, and in either case $V = |X - Y| = 1$. Thus, $f_{V|U}(3, 1) = 1$ as shown for $(u, v) = (3, 1)$ in Table 5.9.   ♦

For a continuous random variable Y, $P(A|Y = y)$ and the conditional density $f_{X|Y}(x, y)$ are undefined because $P(Y = y) = 0$. Nevertheless we can define $P(A|Y = y)$ as a limit with Y falling in an infinitesimal interval at y, rather than being equal to y. For $f_{X|Y}(x, y)$ we can use Equation 5.153 as a model, with f and f_Y reinterpreted as densities.

Definition 5.6.2. *Conditional Probabilities and Densities for Given Values of a Continuous Random Variable.*

For a continuous random variable Y and any event A, we define

$$P(A|Y = y) = \lim_{h \to 0^+} P(A|y \leq Y < y + h), \tag{5.154}$$

if the limit exists. In particular, if $A = \{X \leq x\}$, for any random variable X and any real x, then the conditional p.d.f. of X, given $Y = y$, is defined as

$$F_{X|Y}(x, y) = \lim_{h \to 0^+} P(X \leq x|y \leq Y < y + h), \tag{5.155}$$

if the limit exists, and, if X is discrete, then the conditional p.f. of X, given $Y = y$, is defined as

$$f_{X|Y}(x, y) = \lim_{h \to 0^+} P(X = x|y \leq Y < y + h), \tag{5.156}$$

if the limit exists.

Furthermore, for continuous random variables X and Y with joint density $f(x,y)$ and Y having marginal density $f_Y(y)$, we define the conditional density $f_{X|Y}$ by

$$f_{X|Y}(x,y) = \begin{cases} \frac{f(x,y)}{f_Y(y)} & \text{if } f_Y(y) \neq 0 \\ 0 & \text{otherwise} \end{cases} \tag{5.157}$$

for all real x and y.

Example 5.6.2. Conditional Density for (X,Y) Uniform on Unit Disk.

Let (X,Y) be uniform on the unit disk $D = \{(x,y) : x^2 + y^2 < 1\}$ as in Example 5.4.7. Hence

$$f_{X|Y}(x,y) = \begin{cases} \frac{f(x,y)}{f_Y(y)} = \frac{1}{2\sqrt{1-y^2}} & \text{if } (x,y) \in D \\ 0 & \text{otherwise} \end{cases}. \tag{5.158}$$

For a fixed $y \in (-1,1)$, this expression is constant over the x-interval $\left(-\sqrt{1-y^2}, \sqrt{1-y^2}\right)$, and therefore, not unexpectedly, it is the density of the uniform distribution over that interval.   ♦

Note that $f_{X|Y}$ can also be interpreted as a limit. Indeed,

$$\lim_{h \to 0^+} P\left(x \leq X < x + dx \,|\, y \leq Y < y + h\right)$$

$$= \lim_{h \to 0^+} \frac{P\left(x \leq X < x + dx, \, y \leq Y < y + h\right)}{P\left(y \leq Y < y + h\right)}$$

$$\sim \lim_{h \to 0^+} \frac{f(x,y)\,h\,dx}{f_Y(y)\,h} = \frac{f(x,y)\,dx}{f_Y(y)} = f_{X|Y}(x,y)\,dx, \tag{5.159}$$

wherever $f(x,y)$ and $f_Y(y)$ exist and are continuous and $f_Y(y) \neq 0$. On the other hand, $P(A|Y = y)$ can be interpreted also *without* a limit as

$$P(A|Y = y) = \frac{P(A)\,f_{Y|A}(y)}{f_Y(y)}, \tag{5.160}$$

wherever $f_{Y|A}(y)$ and $f_Y(y)$ exist and are continuous and $f_Y(y) \neq 0$, because then

$$\lim_{h \to 0^+} P\left(A|y \leq Y < y + h\right)$$

$$= \lim_{h \to 0^+} \frac{P\left(A \cap \{y \leq Y < y + h\}\right)}{P\left(y \leq Y < y + h\right)} = \lim_{h \to 0^+} \frac{P(A)\,P\left(y \leq Y < y + h \,|\, A\right)}{P\left(y \leq Y < y + h\right)}$$

$$= \lim_{h \to 0^+} \frac{P(A)\,f_{Y|A}(y)\,h}{f_Y(y)\,h} = \frac{P(A)\,f_{Y|A}(y)}{f_Y(y)}. \tag{5.161}$$

Equation 5.160 is valid also when $f_{Y|A}(y)$ and $f_Y(y)$ exist and are continuous and $f_Y(y) \neq 0$, but $P(A) = 0$. For, in this case, $A \cap \{y \leq Y < y + h\} \subset A$, and so $P(A \cap \{y \leq Y < y + h\}) = 0$, which implies $P(A|y \leq Y < y + h) = 0$ and $P(A|Y = y) = 0$ as well. Thus, Equation 5.160 reduces to $0 = 0$.

Equation 5.160 can be written in multiplicative form as

$$P(A|Y = y) f_Y(y) = P(A) f_{Y|A}(y). \tag{5.162}$$

This equation is valid also when $f_Y(y) = 0$, as well, because in that case $f_{Y|A}(y) = 0$ as well. This fact follows from Equation 5.152 with Y in place of X:

$$\begin{aligned} f_{Y|A}(y)dy &\sim \frac{P(\{y \leq Y < y + dy\} \cap A)}{P(A)} \\ &\leq \frac{P(y \leq Y < y + dy)}{P(A)} \sim \frac{f_Y(y)\,dy}{P(A)} = 0. \end{aligned} \tag{5.163}$$

Similarly, Equation 5.157 too can be written in multiplicative form as

$$f_{X|Y}(x, y) f_Y(y) = f(x, y). \tag{5.164}$$

This equation is valid when $f_Y(y) = 0$, as well, because $f_Y(y) = 0$ implies $f(x, y) = 0$. Interchanging x and y, we also have

$$f_{Y|X}(x, y) f_X(x) = f(x, y). \tag{5.165}$$

Returning to $f_{X|Y}$, we can see that, for any fixed y such that $f_Y(y) \neq 0$, it is a density as a function of x. Consequently, it can be used to define conditional probabilities for X, given $Y = y$, as

$$P(a < X < b|Y = y) = \int_a^b f_{X|Y}(x, y)\,dx = \frac{1}{f_Y(y)} \int_a^b f(x, y)\,dx \tag{5.166}$$

and, in particular, the *conditional distribution function of X, given $Y = y$,* as

$$F_{X|Y}(x, y) = \int_{-\infty}^x f_{X|Y}(t, y)\,dt = \frac{1}{f_Y(y)} \int_{-\infty}^x f(t, y)\,dt. \tag{5.167}$$

Using Definition 5.6.2, we can generalize the Theorem of Total Probability (Theorem 4.5.2) as follows:

Theorem 5.6.1. *Theorem of Total Probability, Continuous Versions.* *For a continuous random variable Y and any event A, if $f_{Y|A}$ and f_Y exist for all y, then*

$$P(A) = \int_{-\infty}^\infty P(A|Y = y) f_Y(y)\,dy \tag{5.168}$$

and if X and Y are both continuous and $f_{X|Y}$ and f_Y exist for all x, y, then

$$f_X(x) = \int_{-\infty}^\infty f_{X|Y}(x, y) f_Y(y)\,dy. \tag{5.169}$$

Proof. Integrating both sides of Equation 5.162 from $-\infty$ to ∞, we obtain Equation 5.168, because $\int_{-\infty}^{\infty} f_{Y|A}(y)dy = 1$ from Equation 5.148.

Similarly, integrating both sides of Equation 5.164 with respect to y from $-\infty$ to ∞, we obtain Equation 5.169.   ∎

We have new versions of Bayes' theorem as well:

Theorem 5.6.2. Bayes' Theorem, Continuous Versions. *For a continuous random variable Y and any event A with nonzero probability, if $P(A|Y = y)$ and f_Y exist for all y, then*

$$f_{Y|A}(y) = \frac{P(A|Y = y)\, f_Y(y)}{\int_{-\infty}^{\infty} P(A|Y = y)\, f_Y(y)\, dy}. \tag{5.170}$$

Here f_Y is called the prior density of Y, and $f_{Y|A}$ its posterior density, referring to the fact that these are the densities of Y before and after the observation of A.

Furthermore, if X and Y are both continuous, $f_{X|Y}$ and f_Y exist for all x, y, and $f_X(x) \neq 0$, then

$$f_{Y|X}(y, x) = \frac{f_{X|Y}(x, y)\, f_Y(y)}{\int_{-\infty}^{\infty} f_{X|Y}(x, y)\, f_Y(y)\, dy}. \tag{5.171}$$

Again, f_Y is called the prior density of Y, and $f_{Y|X}$ its posterior density.

Proof. From Equation 5.162 we get, when P$(A) \neq 0$,

$$f_{Y|A}(y) = \frac{\mathrm{P}(A|Y = y)\, f_Y(y)}{\mathrm{P}(A)}. \tag{5.172}$$

Substituting the expression for P(A) here from Equation 5.168, we obtain Equation 5.170.

Similarly, from Equations 5.164 and 5.165, we obtain, when $f_X(x) \neq 0$,

$$f_{Y|X}(y, x) = \frac{f_{X|Y}(x, y)\, f_Y(y)}{f_X(x)}, \tag{5.173}$$

and substituting the expression for $f_X(x)$ here from Equation 5.169, we obtain Equation 5.171.   ∎

Example 5.6.3. Bayes Estimate of a Bernoulli Parameter.

Suppose that X is a Bernoulli random variable with an unknown parameter P, which is uniformly distributed on the interval $[0, 1]$. In other words,[19]

$$f_{X|P}(x, p) = p^x (1 - p)^{1-x} \text{ for } x = 0, 1 \tag{5.174}$$

---

[19] We assume $0^0 = 1$ where necessary.

and

$$f_P(p) = \begin{cases} 1 \text{ for } p \in [0,1] \\ 0 \text{ otherwise} \end{cases}. \tag{5.175}$$

We make an observation of X and want to find the posterior density $f_{P|X}(p,x)$ of P. (This problem is a very simple example of the so-called Bayesian method of statistical estimation. It will be generalized to several observations instead of just one in Example 7.4.4.)

By Equation 5.171,

$$f_{P|X}(p,x) = \begin{cases} \frac{p^x(1-p)^{1-x}}{\int_0^1 p^x(1-p)^{1-x}dp} & \text{for } p \in [0,1] \text{ and } x = 0,1 \\ 0 & \text{otherwise} \end{cases}. \tag{5.176}$$

For $x = 1$ we have $\int_0^1 p^x (1-p)^{1-x} dp = \int_0^1 p\, dp = \frac{1}{2}$, and for $x = 0$, similarly, $\int_0^1 p^x (1-p)^{1-x} dp = \int_0^1 (1-p)\, dp = \frac{1}{2}$. Hence,

$$f_{P|X}(p,x) = \begin{cases} 2p & \text{for } p \in [0,1] \text{ and } x = 1 \\ 2(1-p) & \text{for } p \in [0,1] \text{ and } x = 0 \\ 0 & \text{otherwise} \end{cases}. \tag{5.177}$$

Thus, the observation changes the uniform prior density into a triangular posterior density that gives more weight to p-values near the observed value of X.   ♦

Before closing this section, we want to present one more theorem, which follows from the definitions at once:

Theorem 5.6.3. *Conditions for Independence of Random Variables.*

If A is any event with $P(A) \neq 0$ and X any random variable, then A and X are independent of each other if and only if

$$F_{X|A}(x) = F_X(x) \text{ for all } x \in \mathbb{R}. \tag{5.178}$$

If X and Y are any random variables, then they are independent of each other if and only if

$$F_{X|Y}(x,y) = F_X(x) \tag{5.179}$$

for all $x \in \mathbb{R}$ and, for discrete Y, at all possible values y of Y and, for continuous Y, at all y values where $f_{X|Y}(x,y)$ exists.

If A is any event with $P(A) \neq 0$ and X any discrete random variable, then A and X are independent of each other if and only if

$$f_{X|A}(x) = f_X(x) \text{ for all } x \in \mathbb{R}. \tag{5.180}$$

If X and Y are any random variables, both discrete or both absolutely continuous, then they are independent of each other if and only if

$$f_{X|Y}(x, y) = f_X(x) \tag{5.181}$$

for all $x \in \mathbb{R}$ and all y values where $f_Y(y) \neq 0$.

In closing this section, let us mention that all the conditional functions considered above can easily be generalized to more than two random variables, as will be seen in some exercises and later chapters.

Exercises

Exercise 5.6.1.

Roll four dice. Let X denote the number of 1's and Y the number of 6's obtained. Find the values of the p.f. $f_{X|Y}(x, y)$ and display them in a 5×5 table.

Exercise 5.6.2.

Roll two dice. Let X and Y denote the numbers obtained and let $Z = X + Y$:

1. Find the values of the p.f. $f_{X|Z}(x, z)$ and display them in a 6×11 table.
2. Find the values of the conditional joint p.f. $f_{(X,Y)|Z}(x, y, z)$ for $z = 2$, and show that X and Y are independent under this condition.
3. Find the values of the conditional joint p.f. $f_{(X,Y)|Z}(x, y, z)$ for $z = 3$, and show that X and Y are not independent under this condition.

Exercise 5.6.3.

As in Example 5.5.4, pick two random points X and Y independently and uniformly on the interval $[0, 1]$, and let A denote the event that we can construct a triangle from the resulting three segments as its sides. Find the probability $P(A|X = x)$ as a function of x and the conditional density function $f_{X|A}(x)$.

Exercise 5.6.4.

As in Example 5.6.3, let X be a Bernoulli random variable with an unknown parameter P, which is uniformly distributed on the interval $(0, 1)$. Suppose we make two independent observations X_1 and X_2 of X, so that

$$f_{(X_1, X_2)|P}(x_1, x_2, p) = p^{x_1 + x_2}(1 - p)^{2 - x_1 - x_2} \text{ for } x_1, x_2 = 0, 1. \tag{5.182}$$

Find and graph $f_{P|(X_1, X_2)}(p, x_1, x_2)$ for all four possible values of (x_1, x_2).

Exercise 5.6.5.

Let (X, Y) be uniform on the triangle $D = \{(x, y) : 0 < x, 0 < y, x + y < 1\}$. Find the conditional densities $f_{X|Y}(x, y)$ and $f_{Y|X}(x, y)$.

Exercise 5.6.6.

Let $D = \{(x, y) : 0 < x, 0 < y, x + y < 1\}$ and (X, Y) have density

$$f(x, y) = \begin{cases} 60xy^2 & \text{if } (x, y) \in D \\ 0 & \text{otherwise} \end{cases} . \tag{5.183}$$

(See Example 5.4.8.) Find the conditional densities $f_{X|Y}(x, y)$ and $f_{Y|X}(x, y)$.

Exercise 5.6.7.

Let (X, Y) be uniform on the open unit square $D = \{(x, y) : 0 < x < 1, 0 < y < 1\}$ and $Z = X + Y$ (see Example 5.4.9):

1. Find the conditional distribution functions $F_{X|Z}(x, z)$ and $F_{Y|Z}(y, z)$ and the conditional densities $f_{X|Z}(x, z)$ and $f_{Y|Z}(y, z)$.
2. Let A be the event $\{Z < 1\}$. Find $F_{X|A}(x)$ and $f_{X|A}(x)$.

6. Expectation, Variance, and Moments

6.1 Expected Value

Just as probabilities are idealized relative frequencies, so are expected values analogous idealizations of averages of random variables. Before presenting the formal definition, let us consider an example.

Example 6.1.1. Average of Dice Rolls.

Suppose we roll a die $n = 18$ times, and observe the following outcomes: $2, 4, 2, 1, 5, 5, 4, 3, 4, 2, 6, 6, 3, 4, 1, 2, 5, 6$. The average of these numbers can be computed as

$$\text{Average} = \frac{2 \cdot 1 + 4 \cdot 2 + 2 \cdot 3 + 4 \cdot 4 + 3 \cdot 5 + 3 \cdot 6}{18} =$$

$$= \frac{2}{18} \cdot 1 + \frac{4}{18} \cdot 2 + \frac{2}{18} \cdot 3 + \frac{4}{18} \cdot 4 + \frac{3}{18} \cdot 5 + \frac{3}{18} \cdot 6$$

$$= \sum_{i=1}^{6} f_i \cdot i = \frac{65}{18} = 3.611\ldots, \tag{6.1}$$

where f_i stands for the relative frequency of the outcome i.

Now ideally, since for a fair die the six outcomes are equally likely, we should have obtained each number three times, but that is not what usually happens. For large n, however, the relative frequencies are approximately equal to the corresponding probabilities $p_i = 1/6$, and the average becomes close to

$$\sum_{i=1}^{6} p_i \cdot i = \sum_{i=1}^{6} \frac{1}{6} \cdot i = \frac{1}{6} \sum_{i=1}^{6} i = \frac{21}{6} = 3.5. \tag{6.2}$$

♦

© Springer International Publishing Switzerland 2016
G. Schay, *Introduction to Probability with Statistical Applications*,
DOI 10.1007/978-3-319-30620-9_6

It is the first sum in Equation 6.2 that we use as the paradigm for our general idealized average:

Definition 6.1.1. *Expected Value.* *For any discrete random variable X, writing $p_i =$P$(X = x_i)$, we define the expected value, mean, or expectation of X as*

$$E(X) = \sum p_i x_i, \tag{6.3}$$

provided, in the case of an infinite sum, that the sum is absolutely convergent[1]. *The summation runs over all i for which x_i is a possible value of X.*

For any continuous random variable X with density $f(x)$, we define the expected value, mean, or expectation of X as

$$E(X) = \int_{-\infty}^{\infty} x f(x) dx, \tag{6.4}$$

provided that the improper integral is absolutely convergent.

Remarks:

1. We did not give the definition for general random variables; that is a topic taken up in graduate courses. We shall assume, without further mention, that the random variables we discuss are either discrete or absolutely continuous.
2. Because of the occurrence of infinite sums and integrals, $E(X)$ does not exist for some random variables, as will be illustrated shortly. These cases are rare, however, in real-life applications.
3. The expected value of a random variable X is not necessarily a possible value of X, despite its name (see, for instance, Example 6.1.1), but in many cases it can be used to predict, before the experiment is performed, that a value of X close to $E(X)$ can be expected.
4. The expected value of a random variable X depends only on the distribution of X and not on any other properties of X. Thus, if two different random variables have the same distribution, then they have the same expectation as well. For instance, if X is the number of H's in n tosses of a fair coin and Y is the number of T's, then $E(X) = E(Y)$.
5. In the discrete case, $E(X)$ can also be written as

$$E(X) = \sum_{x:f(x)>0} x f(x), \tag{6.5}$$

where f is the p.f. of X.
6. $E(X)$ is often abbreviated as μ or μ_X.

---

[1] Requiring absolute convergence is necessary, because if the sum were merely conditionally convergent, then the value of $E(X)$ would depend on the order of the terms. Similarly, in the continuous case, if the integral were merely conditionally convergent, then $E(X)$ would depend on the manner in which the limits of the integral tend to $\pm\infty$.

Example 6.1.2. Bernoulli Random Variable.

Recall that X is a Bernoulli random variable with parameter p (see Definition 5.1.4), if it has two possible values: 1 and 0 and $P(X = 1) = p$ and $P(X = 0) = q = 1 - p$.

Hence, $E(X) = 1p + 0q = p$.   ♦

Example 6.1.3. Expected Value of Uniform Random Variables.

Let X be uniform over an interval $[a, b]$, that is, have p.d.f.

$$f(x) = \begin{cases} \dfrac{1}{b-a} & \text{if} \quad a < x < b \\ 0 & \text{if } x \le a \text{ or } x \ge b \end{cases}.$$ (6.6)

Then its expected value is given by

$$E(X) = \int_{-\infty}^{\infty} xf(x)dx = \int_{a}^{b} \frac{x}{b-a}dx = \frac{1}{b-a}\frac{x^2}{2}\Big|_{a}^{b} = \frac{a+b}{2}.$$ (6.7)

♦

Example 6.1.4. Expected Value of Discrete Uniform Random Variables.

Let X be uniform over the set $\{x_1, x_2, \ldots, x_n\}$. Then its expected value is given by

$$E(X) = \frac{1}{n}\sum_{x=1}^{n} x_i.$$ (6.8)

♦

Example 6.1.5. Expected Value of Exponential Random Variables.

Let T be an exponential r.v. with parameter λ. (See Definition 5.2.3.) Then its p.d.f. is $f(t) = \lambda e^{-\lambda t}$ for $t \ge 0$, and so

$$E(T) = \int_{-\infty}^{\infty} tf(t)dt = \int_{0}^{\infty} t\lambda e^{-\lambda t}dt.$$ (6.9)

Integrating by parts with $u = t$ and $dv = \lambda e^{-\lambda t}dt$, we get

$$E(T) = -te^{-\lambda t}\Big|_{0}^{\infty} + \int_{0}^{\infty} e^{-\lambda t}dt = 0 - \frac{e^{-\lambda t}}{\lambda}\Big|_{0}^{\infty} = \frac{1}{\lambda}.$$ (6.10)

♦

In Examples 6.1.3 and 6.1.1, $E(X)$ was at the center of the distribution. This property of $E(X)$ is true in general, as explained in the following observation and in the subsequent theorem.

The expected value is a measure of the center of a probability distribution, because the defining formulas are exactly the same as the corresponding ones for the center of mass in mechanics for masses on the x-axis (or, more generally, for the x-coordinates of masses in space), with p_i for the mass of a point at x_i and $f(x)$ for the mass density for a smeared-out mass distribution. Thus, if we were to cut out the graph of the p.f. or p.d.f. of a r.v. X from cardboard, then it would be balanced if supported under the point $x = E(X)$. In a similar vein, the following theorem confirms that $E(X)$ yields the obvious center for a symmetric distribution.

Theorem 6.1.1. *The Center of Symmetry Equals* $E(X)$. *If the distribution of a random variable is symmetric about a point α, that is, the p.f. or the p.d.f. satisfies $f(\alpha - x) = f(\alpha + x)$ for all x, and $E(X)$ exists, then $E(X) = \alpha$.*

Proof. We give the proof for continuous X only; for discrete X the proof is similar and is left as an exercise.

We can write

$$E(X) = \int_{-\infty}^{\infty} x f(x) dx = \int_{-\infty}^{\infty} (x - \alpha + \alpha) f(x) dx$$
$$= \int_{-\infty}^{\infty} (x - \alpha) f(x) dx + \int_{-\infty}^{\infty} \alpha f(x) dx, \tag{6.11}$$

where the first integral on the right will be shown to be 0, and so

$$E(X) = \int_{-\infty}^{\infty} \alpha f(x) dx = \alpha \int_{-\infty}^{\infty} f(x) dx = \alpha. \tag{6.12}$$

The integral of $(x - \alpha) f(x)$ may be evaluated as follows:

$$\int_{-\infty}^{\infty} (x - \alpha) f(x) dx = \int_{-\infty}^{\alpha} (x - \alpha) f(x) dx + \int_{\alpha}^{\infty} (x - \alpha) f(x) dx, \tag{6.13}$$

where in the first integral on the right, we substitute $u = \alpha - x$ and in the second integral $u = x - \alpha$. Hence

$$\int_{-\infty}^{\infty} (x - \alpha) f(x) dx = \int_{\infty}^{0} u f(\alpha - u) du + \int_{0}^{\infty} u f(\alpha + u) du$$
$$= \int_{0}^{\infty} u \left[f(\alpha + u) - f(\alpha - u) \right] du = 0, \tag{6.14}$$

where the last step follows from the symmetry assumption. ∎

If a random variable is bounded from below, say by 0, and we know its expected value, then only a small fraction of its values can fall far out on the right, that is, the expected value yields a bound for the right tail of the distribution:

Theorem 6.1.2. Markov's Inequality. *If X is a nonnegative random variable with expected value μ and a is any positive number, then*

$$P(X \geq a) \leq \frac{\mu}{a}. \tag{6.15}$$

Proof. We prove the statement only for continuous X with density f. Then

$$\mu = \int_0^\infty x f(x) dx = \int_0^a x f(x) dx + \int_a^\infty x f(x) dx$$

$$\geq \int_a^\infty x f(x) dx \geq a \int_a^\infty f(x) dx = a P(X \geq a), \tag{6.16}$$

from which Equation 6.15 follows at once.   ∎

The main use of Theorem 6.1.2 is in proving another inequality in the next section, for not necessarily positive random variables, which, in turn, will be used for a proof of the so-called law of large numbers.

In addition to providing a measure of the center of a probability distribution, the expected value has many other uses, as will be discussed later. For now, we just describe its occurrence in gambling games.

Example 6.1.6. Total Gain in Dice Rolls.

Consider the same game as in Example 6.1.1 with the same outcomes, and assume that whenever the die shows the number i, we win i dollars. In that case our total gain will be \$65, which can be written as $18 \times average$. Similarly, in the ideal situation, the total gain would be $18 \times 3.5 = \$63$.   ♦

Thus, in general games, our ideal gain is $nE(X)$, where n is the number of times we play. (Mathematically, this result follows from Theorem 6.1.5 below.) Hence, $E(X)$ is a measure of the fairness of a game, and a game is called *fair* if $E(X) = 0$.

The dice game described above is very unfair, and we may ask the question how much should we be required to bet each time to make the game fair. Clearly, the answer is \$3.50, that is, if we lose this bet each time and win i dollars with probability $\frac{1}{6}$ for $i = 1, 2, \ldots, 6$, then

$$E(X - 3.50) = \sum_{i=1}^{6} \frac{1}{6}(i - 3.50) = 0, \tag{6.17}$$

and the game is turned into a fair one.

In general, if we have an unfair game with $E(X) > 0$, then paying an entrance fee of $E(X)$ dollars each time will turn the game into a fair one. (This follows from Corollary 6.1.1 below.)

Example 6.1.7. Roulette.

In Nevada roulette, a wheel with 38 numbered pockets is spun around, and a ball is rolled around its rim in the opposite direction until it falls at random into one of the pockets. On a table the numbers of the pockets are laid out, and the players can bet on various combinations to come up, with predetermined payouts. Eighteen of the numbers are black and 18 are red, while two are green. One of the possible betting combinations is that of betting on red with a $1 payout for every $1 bet (i.e., if red comes up, you keep your bet and get another dollar, and if black or green comes up, you lose your bet). Let us compute the expected gain from such a bet.

If we denote the amount won or lost in a single play of $1 by X, then $P(X = 1) = \frac{18}{38}$ and $P(X = -1) = \frac{20}{38}$. Thus,

$$E(X) = \frac{18}{38} \cdot 1 + \frac{20}{38} \cdot (-1) \approx -.0526 = -5.26 \text{ cents.} \qquad (6.18)$$

This result means that in the long run, the players will lose about 5.26 cents on every dollar bet.

The house advantage is set up to be about 5% for the other possible betting combinations as well.   ◆

Example 6.1.8. The Saint Petersburg Paradox.

Some gamblers in Saint Petersburg, Russia, in the eighteenth century devised a betting scheme for even money bets, as for betting on red in roulette. First you bet 1 unit, and if you win, you quit. If you lose, you bet 2 units on the next game. If you win this game, then you are ahead by 1 unit, because you have lost 1 and won 2, and you quit. If you lose again, then you bet 4 on the third game. If you win this time, then you are again ahead by 1, since you have lost 1+2, but have won 4. If you lose, you bet 8, and so on. Thus, the claim was that, following this scheme, you are assured of winning 1 unit.

The expected gain is also 1 unit.

If X denotes the net gain and n the number of plays till you win and stop, then, according to the above discussion, $X = 1$ for any n. If $p < 1$ denotes the probability of winning in any trial and $q = 1 - p$ is the probability of losing, then $P(\text{first win occurs on the } n\text{th play}) = q^{n-1}p$. Hence, by the sum formula for a geometric series,

$$E(X) = \sum_{n=1}^{\infty} q^{n-1}p \cdot 1 = \frac{p}{1-q} = 1. \qquad (6.19)$$

On the other hand, roulette is an unfavorable game, so how can it be possible to beat it? The answer is simple: it cannot be beaten. In this game, you need an infinite amount of money to be assured of winning, since it is quite possible that you may need to bet 2^n units, with n arbitrarily large.

If the bet size is capped, however, either by the house or by the player's capital, then the scheme has no advantage over any other scheme. Indeed, if the maximum bet size is 2^N, then

$$E(X) = \sum_{n=1}^{N} q^{n-1}p \cdot 1 - \sum_{n=N+1}^{\infty} q^{n-1}p \cdot (2^N - 1)$$

$$= \frac{p\left(1 - q^N\right)}{1 - q} - \frac{pq^N(2^N - 1)}{1 - q} = 1 - 2^N q^N. \tag{6.20}$$

This result is exactly what we would expect, since 1 is the expected value for an overall win and 2^N is the last bet in the case of a string of losses, which has probability q^N, and so $2^N q^N$ is the expected loss. Notice that in the case of a fair game, $q = \frac{1}{2}$ and $E(X) = 1 - 2^N \left(\frac{1}{2}\right)^N = 0$, that is, a fair game remains fair under this doubling scheme as well.

Another variant of the Saint Petersburg scheme provides an example of a random variable with infinite expectation. For the sake of simplicity, we assume that we are betting on H in independent tosses of a fair coin. Again, we play until the first H comes up, but this time we bet even more: $(n+1)2^{n-1}$ units on the nth toss if the first $n-1$ tosses resulted in T, for $n = 1, 2, \ldots$. If the first H occurs on the nth toss, which has probability $\frac{1}{2^n}$, then the gain is (see Exercise 6.1.5)

$$(n+1)2^{n-1} - \sum_{i=1}^{n-1} (i+1)2^{i-1} = 2^n \tag{6.21}$$

and so

$$E(X) = \sum_{n=1}^{\infty} \frac{1}{2^n} \cdot 2^n = \sum_{n=1}^{\infty} 1 = \infty. \tag{6.22}$$

♦

Next, we present a surprising, but very useful, theorem, which enables us to compute the expectation of a function $Y = g(X)$ of a r.v. X without going through the laborious process of first finding the distribution of Y.

Theorem 6.1.3. *Expectation of a Function of a Random Variable.*
Let X be any random variable, and define a new random variable as $Y = g(X)$, where g is any[2] function on the range of X.
If X is discrete then, writing $p_i = \mathrm{P}\,(X = x_i)$, we have

$$E(Y) = \sum p_i g\,(x_i), \tag{6.23}$$

provided the sum is absolutely convergent. (The summation runs over all i for which x_i is a possible value of X.)

---

[2] Actually, g must be a so-called measurable function. This restriction is discussed in more advanced texts; all functions encountered in elementary calculus courses are of this type.

If X is continuous with p.d.f. f_X, then

$$E(Y) = \int_{-\infty}^{\infty} g(x) f_X(x) dx, \tag{6.24}$$

provided the integral is absolutely convergent.

Before giving the proof, let us compare the evaluation of $E(Y)$ by the theorem with its evaluation from the definition, on a simple example.

Example 6.1.9. Expectation of $g(X) = |X|$ for a Discrete X.

Let the p.f. of X be given by

$$f_X(x) = \begin{cases} 1/8 \text{ if } & x = -1 \\ 3/8 \text{ if } & x = 0 \\ 3/8 \text{ if } & x = 1 \\ 1/8 \text{ if } & x = 2 \end{cases} \tag{6.25}$$

and let $Y = |X|$. Then

$$f_Y(y) = \begin{cases} 3/8 \text{ if } & y = 0 \\ 1/2 \text{ if } & y = 1 \\ 1/8 \text{ if } & y = 2 \end{cases}. \tag{6.26}$$

Hence, by Definition 6.1.1,

$$E(Y) = \frac{3}{8} \cdot 0 + \frac{1}{2} \cdot 1 + \frac{1}{8} \cdot 2 = \frac{3}{4}. \tag{6.27}$$

On the other hand, by Theorem 6.1.3,

$$E(Y) = \frac{1}{8} \cdot |-1| + \frac{3}{8} \cdot |0| + \frac{3}{8} \cdot |1| + \frac{1}{8} \cdot |2| = \frac{3}{4}. \tag{6.28}$$

Thus, we see that the difference between the two evaluations is that the two terms in Equation 6.28 that contain $|-1|$ and $|1|$ are combined into a single term in Equation 6.27. This is the sort of thing that happens in the general discrete case as well. In the evaluation of f_Y, we combine the probabilities of various x-values (see the proof below), which we can treat separately when using the theorem. In more complicated cases, it can be difficult to find the x-values that need to be combined, but treating them separately is very straightforward.   ◆

Proof (of Theorem 6.1.3). In the discrete case, we evaluate $\sum p_i g(x_i)$ in two stages: first, we sum over all x_i for which $g(x_i)$ is a fixed value y_k of Y, and then sum over all k for which y_k is a possible value of Y. Thus, assuming absolute convergence,

$$\sum p_i g(x_i) = \sum_k \sum_{i:g(x_i)=y_k} p_i g(x_i) = \sum_k \left(\sum_{i:g(x_i)=y_k} p_i \right) y_k$$

$$= \sum_k P(Y = y_k) y_k = \sum_k f_Y(y_k) y_k = E(Y). \tag{6.29}$$

For continuous X, the general proof is beyond the scope of this book and is therefore omitted[3]. However, if g is one-to-one and differentiable, then the proof is easy, and goes like this:

By Definition 6.1.1,

$$E(Y) = \int_{-\infty}^{\infty} y f_Y(y) dy, \tag{6.30}$$

where $f_Y(y)$ is given by the methods of Section 5.3 as

$$f_Y(y) = \begin{cases} f_X\left(g^{-1}(y)\right) \left|\frac{d}{dy}g^{-1}(y)\right| = \frac{f_X(x)}{|g'(x)|} & \text{if } y = g(x) \\ & \text{for some } x \in \text{range}(X) \\ 0 & \text{otherwise.} \end{cases} \tag{6.31}$$

Thus, changing variables in Equation 6.30 from $y = g(x)$ to $x = g^{-1}(y)$, we get

$$E(Y) = \int_{-\infty}^{\infty} g(x) \frac{f_X(x)}{|g'(x)|} \left|\frac{dy}{dx}\right| dx = \int_{-\infty}^{\infty} g(x) f_X(x) dx, \tag{6.32}$$

as stated in Equation 6.24.   ∎

Example 6.1.10. Average Area of Circles.

Assume that we draw a circle with a random radius R, uniformly distributed between 0 and some constant a. What is the expected value of the area $Y = \pi R^2$ of such a circle?

Now,

$$f(r) = \begin{cases} 1/a & \text{if } 0 < r < a \\ 0 & \text{otherwise} \end{cases} \tag{6.33}$$

and $g(r) = \pi r^2$. Thus, substituting into Equation 6.24 gives

$$E(Y) = \int_0^a \pi r^2 \frac{1}{a} dr = \left.\frac{\pi r^3}{3a}\right|_0^a = \frac{\pi a^2}{3}. \tag{6.34}$$

It is quite surprising that, though the mean radius is half of the maximal radius, the mean area is one third of the maximal area.   ◆

Theorem 6.1.3 has a frequently used application to linear functions:

Corollary 6.1.1. Expectation of a Linear Function. *For any random variable X such that $E(X)$ exists and for any constants a and b,*

$$E(aX + b) = aE(X) + b. \tag{6.35}$$

---

[3] The proof would require taking limits of approximations of the given continuous r.v. by discrete r.v.'s with a finite number of values.

Proof. We give the proof for continuous X only; for discrete X the proof is similar and is left as an exercise.

In Equation 6.24 let $g(x) = ax + b$. Then

$$E(aX+b) = \int_{-\infty}^{\infty} (ax + b) f(x)dx = a \int_{-\infty}^{\infty} xf(x)dx + b \int_{-\infty}^{\infty} f(x)dx = aE(X)+b. \quad (6.36)$$

∎

Example 6.1.11. Average Temperature.

Assume that at noon on April 15 at a certain place, the temperature C is a random variable (i.e., it varies randomly from year to year) with an unknown distribution but with known mean $E(C) = 15°$ Celsius. If $F = 1.8C + 32$ is the corresponding mean temperature in Fahrenheit degrees, then, by Corollary 6.1.1,

$$E(F) = 1.8E(C) + 32 = 59° \text{ Fahrenheit.} \quad (6.37)$$

Thus, the expected temperature transforms in the same way as the individual values do.   ♦

Example 6.1.12. Expected Value of a Geometric Random Variable.

Let X be geometric with parameter p. (See Definition 5.1.7.) We can obtain $E(X)$ by computing $E(X - 1)$ in two ways:

By Theorem 6.1.3,

$$E(X - 1) = \sum_{k=2}^{\infty}(k - 1)pq^{k-1} = \sum_{j=1}^{\infty} jpq^j = q \sum_{j=1}^{\infty} jpq^{j-1} = qE(X) \quad (6.38)$$

and, by Corollary 6.1.1,

$$E(X - 1) = E(X) - 1. \quad (6.39)$$

Thus,

$$qE(X) = E(X) - 1, \quad (6.40)$$

$$(1 - q)E(X) = 1, \quad (6.41)$$

and so

$$E(X) = \frac{1}{p}. \quad (6.42)$$

♦

A theorem analogous to Theorem 6.1.3 holds also for functions of several variables:

Theorem 6.1.4. *Expectation of a Function of Several Random Variables.* *Let $X_1, X_2, \ldots, X_n$ be any random variables, and define a new random variable as $Y = g(X_1, X_2, \ldots, X_n)$, where g is any function on $\mathbb{R}^n$. If f denotes the joint p.f. or p.d.f. of $X_1, X_2, \ldots, X_n$, then in the discrete case,*

$$E(Y) = \sum \cdots \sum g(x_1, x_2, \ldots, x_n) f(x_1, x_2, \ldots, x_n), \tag{6.43}$$

where the summations run over all possible values $x_1, x_2, \ldots, x_n$ of $X_1, X_2, \ldots X_n$, and in the continuous case,

$$E(Y) = \int \cdots \int_{\mathbb{R}^n} g(x_1, x_2, \ldots, x_n) f(x_1, x_2, \ldots, x_n) dx_1 dx_2 \cdots dx_n, \tag{6.44}$$

provided the sum and the integral are absolutely convergent.

We omit the proof. (In the discrete case, it would be similar to the proof of Theorem 6.1.3, and in the continuous case, it would present the same difficulties.)

Example 6.1.13. *Expectation of the Distance of a Random Point from the Center of a Circle.*

Let the random point (X, Y) be uniformly distributed on $D = \{(x, y) : x^2 + y^2 < 1\}$. (See Example 5.4.7.) Let $R = \sqrt{X^2 + Y^2}$ and find $E(R)$.
Then

$$E(R) = \frac{1}{\pi} \iint_D \sqrt{x^2 + y^2} \, dx \, dy. \tag{6.45}$$

Changing over to polar coordinates, we get

$$E(R) = \frac{1}{\pi} \int_0^{2\pi} \int_0^1 r^2 \, dr \, d\theta = \frac{1}{\pi} 2\pi \left. \frac{r^3}{3} \right|_0^1 = \frac{2}{3}. \tag{6.46}$$

$\blacklozenge$

Theorem 6.1.4 has the following very important consequence:

Theorem 6.1.5. *Expectation of a Sum of Two Random Variables.* *For any two random variables X and Y whose expectations exist,*

$$E(X + Y) = E(X) + E(Y). \tag{6.47}$$

Proof. We give the proof for continuous (X, Y) only; for discrete (X, Y) the proof is similar and is left as an exercise.
By Theorem 6.1.4, with $X_1 = X$, $X_2 = Y$, and $g(X, Y) = X + Y$, we have

$$\begin{aligned}
E(X + Y) &= \int_{-\infty}^{\infty} \int_{-\infty}^{\infty} (x + y) \, f(x, y) \, dx \, dy \\
&= \int_{-\infty}^{\infty} x \left(\int_{-\infty}^{\infty} f(x, y) dy \right) dx + \int_{-\infty}^{\infty} y \left(\int_{-\infty}^{\infty} f(x, y) dx \right) dy \\
&= \int_{-\infty}^{\infty} x f_X(x) \, dx + \int_{-\infty}^{\infty} y f_Y(y) \, dy = E(X) + E(Y). \quad (6.48)
\end{aligned}$$

$\blacksquare$

Repeated application of Theorem 6.1.5 and Equation 6.35 leads to:

Corollary 6.1.2. *Expectation of a Linear Function of Several Random Variables.* *For any positive integer n and any random variables $X_1, X_2, \ldots, X_n$ with finite expectations and constants $a_1, a_2, \ldots, a_n$,*

$$E\left(\sum_{i=1}^{n} a_i X_i\right) = \sum_{i=1}^{n} a_i E(X_i). \tag{6.49}$$

Example 6.1.14. *Expectation of Binomial Random Variables.*

Recall that a random variable X is called binomial with parameters n and p, (see Definition 5.1.5) if it has p.f.

$$f(x; n, p) = \binom{n}{x} p^x q^{n-x} \text{ for } x = 0, 1, \ldots, n. \tag{6.50}$$

Now, X counts the number of successes in n trials (or the number of good items selected in sampling *with* replacement). It can be written as a sum of n identical (and independent, but that is irrelevant here) Bernoulli random variables X_i with parameter p. Indeed, let $X_i = 1$ if the ith trial results in success and 0 otherwise. Then $X = \sum_{i=1}^{n} X_i$, because the number of 1's in the sum is exactly the number of successes, and the rest of the terms equal 0. Hence

$$E(X) = E\left(\sum_{i=1}^{n} X_i\right) = \sum_{i=1}^{n} E(X_i) = \sum_{i=1}^{n} p = np. \tag{6.51}$$

This result can, of course, be obtained directly from the definitions as well (see Exercise 6.1.15), but the present method is much simpler and explains the reason behind the formula. ◆

Example 6.1.15. *Hypergeometric Random Variable.*

A hypergeometric random variable X counts the number of successes, that is, the number of good items picked, if we select a sample of size n *without* replacement from a mixture of N good and bad items. (See Example 4.2.4.) If p stands for the fraction of good items in the lot and $q = 1 - p$ the fraction of bad items, then the p.f. of X is

$$f(x; n, N, p) = \frac{\binom{Np}{x}\binom{Nq}{n-x}}{\binom{N}{n}} \text{ for } \max(0, n - Nq) \leq x \leq \min(n, Np). \tag{6.52}$$

A direct evaluation of $E(X)$ would be quite difficult from here, but we can do the same thing that we did in the binomial case. Again, if X_i is a Bernoulli random variable for each i, such that $X_i = 1$ if the ith trial (i.e., the ith choice) results in success and $X_i = 0$ otherwise, then $X = \sum_{i=1}^{n} X_i$. Now $P(X_i = 1) = p$ for every i, because if we do not know the outcomes of the previous choices, then the probability of success on the ith trial is the same as for the first trial. Thus Equation 6.51 also applies now and gives the same result $E(X) = np$. ◆

Theorem 6.1.6. *Expectation of the Product of Two Independent Random Variables.* For any two independent random variables X and Y whose expectations exist,

$$E(XY) = E(X)E(Y). \tag{6.53}$$

Proof. We give the proof for continuous (X, Y) only; for discrete (X, Y) we would just have to replace the integrals by sums.

By the assumed independence, $f(x, y) = f_X(x)f_Y(y)$. By Theorem 6.1.4, with $X_1 = X, X_2 = Y$, and $g(X, Y) = XY$, we have

$$E(XY) = \int_{-\infty}^{\infty} \int_{-\infty}^{\infty} xyf(x,y)dxdy = \int_{-\infty}^{\infty} \int_{-\infty}^{\infty} xyf_X(x)f_Y(y)dxdy$$

$$= \int_{-\infty}^{\infty} xf_X(x)dx \int_{-\infty}^{\infty} yf_Y(y)dy = E(X)E(Y). \tag{6.54}$$

∎

Note that in the preceding proof, the assumption of independence was crucial. For dependent random variables, Equation 6.53 usually does not hold.

A similar proof leads to the analogous theorem for more than two random variables:

Theorem 6.1.7. *Expectation of the Product of Several Independent Random Variables.* For any positive integer n and any independent random variables $X_1, X_2, \ldots, X_n$ whose expectations exist,

$$E\left(\prod_{i=1}^{n} X_i\right) = \prod_{i=1}^{n} E(X_i). \tag{6.55}$$

Exercises

Exercise 6.1.1.

From a regular deck of 52 playing cards, we pick one at random. Let the r.v. X equal the number on the card if it is a numbered one (Ace counts as 1) and 10 if it is a face card. Find $E(X)$.

Exercise 6.1.2.

Four indistinguishable balls are distributed randomly into three distinguishable boxes. (See Example 3.5.3.) Let X denote the number of balls that end up in the first box. Find $E(X)$.

Exercise 6.1.3.

Find $E(T)$ for a r.v. T with density

$$f(t) = \begin{cases} 0 & \text{if } t < 0 \\ \lambda^2 t e^{-\lambda t} & \text{if } t \geq 0 \end{cases}. \tag{6.56}$$

(This is the density of the sum of two independent exponential r.v.'s with parameter λ.)

Exercise 6.1.4.

In the game of roulette (Example 6.1.7), a winning bet on any single number pays 35:1. Find $E(X)$, where X denotes the gain from a bet of \$1 on a single number.

Exercise 6.1.5.

Prove Equation 6.21. Hint: Let $g(x) = \sum_{i=1}^{n-1} x^i = \frac{x^n - x}{x-1}$. First, compute $g'(x)$ from both expressions for $g(x)$ and set $x = 2$.

Exercise 6.1.6.

A random variable X with p.d.f. $f(x) = \frac{1}{\pi} \frac{1}{1+x^2}$ for any real x is called a *Cauchy r.v.* Show that:

1. This f is indeed a p.d.f.,
2. $E(X)$ does not exist, because the integral of $xf(x)$ is not absolutely convergent.

Exercise 6.1.7.

Prove Theorem 6.1.1 for discrete X.

Exercise 6.1.8.

Toss a fair coin repeatedly, until HH or TT comes up. Let X be the number of tosses required. Find $E(X)$. (See Exercise 5.1.7 and Example 6.1.12.)

Exercise 6.1.9.

Let X be an exponential r.v. with parameter λ. (See Definition 5.2.3.) Find $E(X^2)$.

Exercise 6.1.10.

Let X be uniform over the interval $(0, 1)$. Find $E(|X - \frac{1}{2}|)$.

Exercise 6.1.11.

Let X be uniform over the interval $(0, 1)$. Show that $E(1/X)$ does not exist.

Exercise 6.1.12.

Prove Equation 6.35 for discrete X.

Exercise 6.1.13.

Prove Equation 6.35, for continuous X and $a \neq 0$, directly from Example 5.3.1 without using Theorem 6.1.3.

Exercise 6.1.14.

Prove Theorem 6.1.5 for discrete X.

Exercise 6.1.15.

Prove $E(X) = np$ for a binomial r.v. directly from Equation 6.50 and Definition 6.1.1.

Exercise 6.1.16.

Let the random point (X, Y) be uniformly distributed on $D = \{(x, y) : x^2 + y^2 < 1\}$, and let $Z = X^2 + Y^2$. Find $E(Z)$.

Exercise 6.1.17.

Let the random point (X, Y) be uniformly distributed on $D = \{(x, y) : x^2 + y^2 < 1\}$. Does Equation 6.53 hold in this case?

Exercise 6.1.18.

Let X be a discrete uniform r.v. on the set $\{-1, 0, 1\}$, and let $Y = X^2$. Show that X and Y are not independent but $E(XY) = E(X)E(Y)$ nevertheless.

Exercise 6.1.19.

Let the random point (X, Y) be uniformly distributed on the unit square $D = \{(x, y) : 0 \le x \le 1, 0 \le y \le 1\}$, as in Example 5.4.4, and let $Z = X^2 + Y^2$. Find $E(Z)$.

Exercise 6.1.20.

Let the random point (X, Y) be uniformly distributed on the unit square $D = \{(x, y) : 0 \le x \le 1, 0 \le y \le 1\}$, as in Example 5.4.4, and let $Z = X + Y$. Find $E(Z)$.

Exercise 6.1.21.

Give an alternative proof for the expectation of a geometric r.v. X (Example 6.1.12), based on the observation that $E(X) = \sum_{k=0}^{\infty} kpq^{k-1} = p\frac{d}{dq}\sum_{k=0}^{\infty} q^k$ for $0 < q < 1$.

Exercise 6.1.22.

Let X be a hypergeometric random variable (the number of good items in a sample, see Example 6.1.15), and let $Y = n - X$ be the number of bad items in the same sample. Find $E(X - Y)$.

6.2 Variance and Standard Deviation

As we have seen, the expected value gives some information about a distribution by providing a measure of its center. Another characteristic of a distribution is the standard deviation, which gives a measure of its average width.

The first idea most people have for an average width of the distribution of a random variable X is the mean of the deviations $X - \mu$ from the mean $\mu = E(X)$, that is, the quantity $E(X - \mu)$. Unfortunately, however, $E(X - \mu) = E(X) - \mu = 0$ for every r.v. that has an expectation, and so this would be a useless definition. We must do something to avoid the cancelations of the positive and negative deviations.

So next, one would try $E(|X - \mu|)$. Though this definition does provide a good measure of the average width, it is generally difficult to compute and does not have the extremely useful properties and the amazingly fruitful applications that our next definition has.

Definition 6.2.1. *Variance and Standard Deviation. Let X be any random variable with mean $\mu = E(X)$. We define its variance and standard deviation as*

$$Var(X) = E\left((X - \mu)^2\right) \tag{6.57}$$

and

$$SD(X) = \sqrt{Var(X)}, \tag{6.58}$$

provided $E\left((X - \mu)^2\right)$ exists as a finite quantity.

Note that $(X - \mu)^2 \geq 0$, and so here the cancelations implicit in $E(X - \mu)$ are avoided. Moreover, the squaring of $X - \mu$ introduces a change of units, and the square root in $SD(X)$ undoes this. For instance, if X is a length, then $Var(X)$ is an area, but $SD(X)$ is a length again.

$SD(X)$ is often abbreviated as σ or σ_X.

Example 6.2.1. Roll of a Die.

Let X denote the number obtained in a roll of a die, that is, $P(X = i) = \frac{1}{6}$ for $i = 1, 2, \ldots 6$. Then $\mu = 3.5$, and

$$Var(X) = \sum_{i=1}^{6} \frac{1}{6} \cdot (i - 3.5)^2 \tag{6.59}$$

$$= \frac{1}{6} \left[(-2.5)^2 + (-1.5)^2 + (-0.5)^2 + (0.5)^2 + (1.5)^2 + (2.5)^2 \right]$$

$$\approx 2.9167$$

and

$$SD(X) \approx 1.7078. \tag{6.60}$$

In Figure 6.1 we show the graph of the p.f. with μ and $\mu \pm \sigma$ marked on the x-axis.

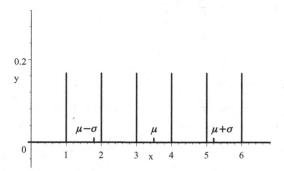

Fig. 6.1. Graph of the probability function of a discrete uniform random variable over $\{1, 2, \ldots 6\}$ with μ and $\mu \pm \sigma$ indicated

As can be seen, the distance between $\mu - \sigma$ and $\mu + \sigma$ is indeed a reasonable measure of the average width of the graph. ◆

Example 6.2.2. Variance and Standard Deviation of a Discrete Uniform Random Variable.

Let X be discrete uniform over the set $\{x_1, x_2, \ldots, x_n\}$. Then, writing $\bar{x}$ for its expected value, we can obtain its variance as

$$Var(X) = \frac{1}{n} \sum_{x=1}^{n} (x_i - \bar{x})^2, \tag{6.61}$$

and its standard deviation as

$$SD(X) = \left[\frac{1}{n} \sum_{x=1}^{n} (x_i - \overline{x})^2 \right]^{1/2}.$$ (6.62)

◆

Example 6.2.3. Variance and Standard Deviation of a Uniform Random Variable.

Let X be uniform over the interval $[a, b]$, that is, have p.d.f.

$$f(x) = \begin{cases} \dfrac{1}{b-a} & \text{if} \quad a < x < b \\ 0 & \text{if } x \le a \text{ or } x \ge b \end{cases}.$$ (6.63)

Then $\mu = \frac{a+b}{2}$ and

$$Var(X) = \frac{1}{b-a} \int_a^b (x - \mu)^2 \, dx = \frac{(b-a)^2}{12}$$ (6.64)

and

$$SD(X) = \frac{b-a}{2\sqrt{3}}.$$ (6.65)

In Figure 6.2 we show the graph of the uniform p.d.f. over the $[0, 1]$ interval, with μ and $\mu \pm \sigma$ marked on the x-axis.   ◆

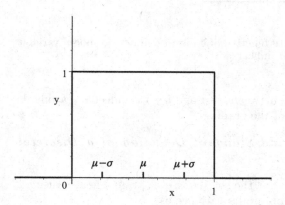

Fig. 6.2. Graph of the p.d.f. of a uniform random variable over $[0, 1]$ with μ and $\mu \pm \sigma$ indicated

Next, we present several useful theorems.

Theorem 6.2.1. *Zero Variance.* *For any random variable X such that $Var(X)$ exists, $Var(X) = 0$ if and only if $P(X = c) = 1$ for some constant c.*

Proof. We give the proof for discrete X only.

If $P(X = c) = 1$ for some constant c, then $f(x) = 0$ for all $x \neq c$, and so

$$\mu = E(X) = \sum_{x:f(x)>0} xf(x) = c \cdot 1 = c. \tag{6.66}$$

Similarly,

$$Var(X) = E\left((X - \mu)^2\right) = \sum_{x:f(x)>0} (x - c)^2 f(x) = 0. \tag{6.67}$$

Conversely, assume that $Var(X) = 0$. Then every term on the left of

$$\sum_{x:f(x)>0} (x - \mu)^2 f(x) = 0 \tag{6.68}$$

is nonnegative and must therefore be 0. So if $f(x) > 0$, then we must have $x - \mu = 0$, that is, $x = \mu$. For $x \neq \mu, (x - \mu)^2 \neq 0$, and so $f(x) = 0$ must hold. Since $\sum_{x:f(x)>0} f(x) = 1$ and the only possible nonzero $f(x)$ is $f(\mu)$, we get $f(\mu) = 1$ or, in other words, $P(X = \mu) = 1$.   ∎

Theorem 6.2.2. *Variance and Standard Deviation of a Linear Function of a Random Variable.* *If X is a random variable such that $Var(X)$ exists, then, for any constants a and b,*

$$Var(aX + b) = a^2 Var(X) \tag{6.69}$$

and

$$SD(aX + b) = |a| \, SD(X). \tag{6.70}$$

Proof. By Equation 6.35, $E(aX + b) = a\mu + b$, and so

$$Var(aX + b) = E\left[(aX + b - a\mu - b)^2\right] = E\left[a^2(X - \mu)^2\right]$$
$$= a^2 E\left[(X - \mu)^2\right] = a^2 Var(X). \tag{6.71}$$

Equation 6.70 follows from here by taking square roots.   ∎

Example 6.2.4. Standardization.

In some applications we transform random variables to a standard scale in which all random variables are centered at 0 and have standard deviations equal to 1. For any given r.v. X, for which μ and σ exist, we define its standardization as the new r.v.

$$Z = \frac{X - \mu}{\sigma}. \tag{6.72}$$

Then indeed, by Equation 6.35,

$$E(Z) = E\left(\frac{X}{\sigma} - \frac{\mu}{\sigma}\right) = \frac{1}{\sigma}E(X) - \frac{\mu}{\sigma} = 0 \tag{6.73}$$

and, by Equation 6.70,

$$SD(Z) = \left|\frac{1}{\sigma}\right| SD(X) = 1. \tag{6.74}$$

$\blacklozenge$

Theorem 6.2.3. *An Alternative Formula for Computing the Variance.* *If X is a random variable such that $Var(X)$ exists, then*

$$Var(X) = E(X^2) - \mu^2. \tag{6.75}$$

Proof.

$$Var(X) = E\left((X - \mu)^2\right) = E\left(X^2 - 2\mu X + \mu^2\right)$$
$$= E(X^2) - 2\mu E(X) + \mu^2 = E(X^2) - \mu^2. \tag{6.76}$$

$\blacksquare$

Example 6.2.5. *Variance and Standard Deviation of an Exponential Random Variable.*

Let T be an exponential r.v. with parameter λ. We use Equation 6.75 to compute the variance. Then

$$E(T^2) = \int_{-\infty}^{\infty} t^2 f(t) dt = \int_{0}^{\infty} t^2 \lambda e^{-\lambda t} dt. \tag{6.77}$$

Integrating by parts twice as in Example 6.1.5, we obtain

$$E(T^2) = \frac{2}{\lambda^2}. \tag{6.78}$$

Hence,

$$Var(T) = \frac{2}{\lambda^2} - \frac{1}{\lambda^2} = \frac{1}{\lambda^2} \tag{6.79}$$

and so

$$SD(T) = \frac{1}{\lambda}. \tag{6.80}$$

$\blacklozenge$

Theorem 6.2.4. *Variance of the Sum of Two Independent Random Variables. For any two independent random variables X and Y whose variances exist,*

$$Var(X + Y) = Var(X) + Var(Y). \tag{6.81}$$

Proof. Writing $E(X) = \mu_X$ and $E(Y) = \mu_Y$, we have $E(X + Y) = \mu_X + \mu_Y$, and so

$$
\begin{aligned}
Var(X + Y) &= E\left((X + Y - (\mu_X + \mu_Y))^2\right) \\
&= E\left(((X - \mu_X) + (Y - \mu_Y))^2\right) \\
&= E\left((X - \mu_X)^2 + 2(X - \mu_X)(Y - \mu_Y) + (Y - \mu_Y)^2\right) \\
&= E\left((X - \mu_X)^2\right) + 2E((X - \mu_X)(Y - \mu_Y)) + E\left((Y - \mu_Y)^2\right) \\
&= Var(X) + 0 + Var(Y) = Var(X) + Var(Y). \tag{6.82}
\end{aligned}
$$

The reason that the middle term is 0 is the independence of X and Y, which implies

$$
\begin{aligned}
E((X - \mu_X)(Y - \mu_Y)) &= E(XY - \mu_X Y - \mu_Y X + \mu_X \mu_Y) \\
&= E(X)E(Y) - \mu_X E(Y) - \mu_Y E(X) + \mu_X \mu_Y = 0. \tag{6.83}
\end{aligned}
$$

∎

Theorem 6.2.4 can easily be generalized to more than two random variables:

Theorem 6.2.5. *Variance of Sums of Pairwise Independent Random Variables. For any positive integer n and any pairwise independent random variables $X_1, X_2, \ldots, X_n$ whose variances exist,*

$$Var\left(\sum_{i=1}^{n} X_i\right) = \sum_{i=1}^{n} Var(X_i). \tag{6.84}$$

We omit the proof. It would be similar to that of Theorem 6.2.4, and because each mixed term involves the product of only two factors, we do not need to assume total independence; pairwise independence is enough.

It is this additivity of the variance that makes it, together with the SD, such a useful quantity, a property that other measures of the spread of a distribution, like $E(|X - \mu|)$, lack.

The preceding results have a corollary that is very important in statistical sampling:

Corollary 6.2.1. Square-Root Law. *For any positive integer n, consider n pairwise independent, identically distributed random variables $X_1, X_2, \ldots, X_n$ with mean μ and standard deviation σ. Let S_n denote their sum and $\overline{X}_n$ their average, that is, let*

$$S_n = \sum_{i=1}^{n} X_i \tag{6.85}$$

and

$$\overline{X}_n = \frac{1}{n} \sum_{i=1}^{n} X_i. \tag{6.86}$$

Then

$$E(S_n) = n\mu \quad and \quad SD(S_n) = \sqrt{n}\sigma, \tag{6.87}$$

and

$$E(\overline{X}_n) = \mu \quad and \quad SD(\overline{X}_n) = \frac{\sigma}{\sqrt{n}}. \tag{6.88}$$

Example 6.2.6. Variance and SD of a Bernoulli Random Variable.

If X is a Bernoulli random variable with parameter p, then $E(X) = p$, and

$$Var(X) = E\left((X-p)^2\right) = p(1-p)^2 + (1-p)(0-p)^2 = p - p^2 = pq, \tag{6.89}$$

and

$$SD(X) = \sqrt{pq}. \tag{6.90}$$

◆

Example 6.2.7. Variance and SD of a Binomial Random Variable.

Again, as in Example 6.1.14, we write the binomial r.v. X with parameters n and p as a sum of n identical and pairwise independent (this time, the independence is crucial) Bernoulli random variables X_i with parameter p. Then $X = S_n = \sum_{i=1}^{n} X_i$, and so, by the square-root law,

$$Var(X) = nVar(X_i) = npq, \tag{6.91}$$

$$SD(X) = \sqrt{npq}. \tag{6.92}$$

and

$$SD(\overline{X}_n) = \sqrt{\frac{pq}{n}}. \tag{6.93}$$

◆

There is another important general relation that we should mention here. It gives bounds for the probability of the tails of a distribution expressed in terms of multiples of the standard deviation. That such a relation exists should not be surprising, because both quantities—standard deviation and tail probability—are measures of the width of a distribution.

Theorem 6.2.6. *Chebyshev's Inequality.* *For any random variable X with mean μ and variance σ^2 and for any positive number k,*

$$P(|X - \mu| \geq k\sigma) \leq \frac{1}{k^2}. \tag{6.94}$$

Proof. Clearly,

$$P(|X - \mu| \geq k\sigma) = P((X - \mu)^2 \geq k^2\sigma^2) \tag{6.95}$$

and, applying Markov's inequality (Theorem 6.1.2) to the nonnegative random variable $(X - \mu)^2$ with $a = k^2\sigma^2$, we get

$$P(|X - \mu| \geq k\sigma) \leq \frac{E((X - \mu)^2)}{a} = \frac{\sigma^2}{k^2\sigma^2} = \frac{1}{k^2}. \tag{6.96}$$

$\blacksquare$

Theorem 6.2.6 should be used to estimate tail probabilities only if we do not know anything about a distribution. If we know the d.f., then we should use that to get a precise value for $P(|X - \mu| \geq k\sigma)$, which is usually much smaller than $\frac{1}{k^2}$. (See the example below.)

Example 6.2.8. Tail Probabilities of an Exponential Random Variable.

Let T be an exponential r.v. with parameter $\lambda = 1$. Then $\mu = \sigma = 1$ and $F(t) = 1 - e^{-t}$ for $t \geq 0$. Also,

$$P(|T - 1| \geq k) = P(T - 1 \geq k) = 1 - F(1 + k) = e^{-1-k} \text{ for } k \geq 1. \tag{6.97}$$

Thus

$$P(|T - 1| \geq k) \approx \begin{cases} 0.14 \text{ if } k = 1 \\ 0.05 \text{ if } k = 2 \\ 0.02 \text{ if } k = 3 \end{cases}, \tag{6.98}$$

while Chebyshev's inequality gives

$$P(|T - 1| \geq k) \leq \begin{cases} 1 \quad \text{ if } k = 1 \\ 0.25 \text{ if } k = 2 \\ 0.11 \text{ if } k = 3 \end{cases}. \tag{6.99}$$

$\blacklozenge$

The most important use of Chebyshev's inequality is in the proof of a limit theorem, known as the law of large numbers[4]:

Theorem 6.2.7. *Law of Large Numbers.* *For any positive integer n, let $X_1, X_2, \ldots, X_n$ be i.i.d. random variables with mean μ and standard deviation σ. Then, for any $\varepsilon > 0$, their mean $\overline{X}_n$ satisfies the relation*

$$\lim_{n \to \infty} P\left(\left|\overline{X}_n - \mu\right| < \varepsilon\right) = 1. \tag{6.100}$$

Proof. By Corollary 6.2.1, for any i.i.d. $X_1, X_2, \ldots, X_n$ with mean μ and standard deviation σ, their average $\overline{X}_n$ has $E\left(\overline{X}_n\right) = \mu$ and $SD(\overline{X}_n) = \frac{\sigma}{\sqrt{n}}$. Thus, applying Chebyshev's inequality to $\overline{X}_n$ with $\varepsilon = k\left(\sigma/\sqrt{n}\right)$, we obtain $P\left(\left|\overline{X}_n - \mu\right| \geq \varepsilon\right) = P\left(\left|\overline{X}_n - \mu\right| \geq k\left(\sigma/\sqrt{n}\right)\right) \leq 1/k^2 = \sigma^2/\left(n\varepsilon^2\right)$. Since $\sigma^2/\left(n\varepsilon^2\right) \to 0$ as $n \to \infty$, the left side is squeezed to 0 as $n \to \infty$. ∎

Remarks.

1. The Relation 6.100 is true even if σ does not exist for the X_i.
2. In the special case of the X_i being Bernoulli random variables with parameter p, the mean $\overline{X}_n$ is the relative frequency of successes in n trials and $\mu = p$. So, in that case, the law of large numbers says that, as $n \to \infty$, the relative frequency of successes will be arbitrarily close to the probability of success with probability one. (Note that this is only a probability statement about p. We cannot use this theorem as a definition of probability, that is, we cannot say that the relative frequency becomes the probability p; we can make statements only about the *probability* of this event, even in the stronger version in the footnote.)
3. The SDs of S_n and $\overline{X}_n$ are sometimes called their *standard errors* (SEs).

Exercises

Exercise 6.2.1.

Find two random variables X and Y whose variances do not exist but the variance of their sum does.

Exercise 6.2.2.

1. Let X and Y be two independent random variables whose variances exist. Show that $Var(X - Y) = Var(X + Y)$ in this case.
2. Is the above relation necessarily true if X and Y are not independent?

---

[4] In fact, there is a stronger version of this law, $P\left(\lim_{n \to \infty} \overline{X}_n = \mu\right) = 1$, under appropriate conditions, but we do not prove this strong law of large numbers here. Actually, the precise name of Theorem 6.2.7 is the weak law of large numbers.

Exercise 6.2.3.

Let X and Y be two independent random variables whose variances exist. For any constants a, b, c, express $Var(aX + bY + c)$ in terms of $Var(X)$ and $Var(Y)$.

Exercise 6.2.4.

Show that the converse of Theorem 6.2.4 is false: For X and Y as in Exercise 6.1.18, the relation $Var(X+Y) = Var(X)+Var(Y)$ holds, although X and Y are not independent.

Exercise 6.2.5.

Prove that if for a r.v. X, both $E(X) = \mu$ and $SD(X) = \sigma$ exist and c is any constant, then:

1. $E\left((X - c)^2\right) = \sigma^2 + (\mu - c)^2$

2. $\min_c E\left((X - c)^2\right) = Var(X)$, that is, the mean of squared deviations is minimum if the deviations are taken from the mean.

Exercise 6.2.6.

Let X and Y be two independent random variables, both with density $f(x) = 3x^2$ for $x \in [0,1]$ and 0 otherwise. Find the expected value and the variance of

1. X,
2. $X - Y$,
3. XY,
4. X^2,
5. $(X + Y)^2$.

Exercise 6.2.7.

Let X and Y be two independent exponential random variables, both with parameter λ. Find the expected value and the variance of

1. $X + 2Y$,
2. $X - 2Y$,
3. XY,
4. X^2,
5. $(X + Y)^2$.

Exercise 6.2.8.

Let X be a binomial random variable with $E(X) = 5$. Find the least upper bound of $SD(X)$ as a function of n.

Exercise 6.2.9.

Toss a fair coin n times, and let X denote the number of H's and Y the number of T's obtained. Does $E(XY) = E(X)E(Y)$ hold in this case?

6.3 Moments and Generating Functions

The notions of expected value and variance of a r.v. X can be generalized to higher powers of X:

Definition 6.3.1. _Moments._ _For any positive integer k, we call $E\left(X^k\right)$ the kth moment of X and $E\left((X - \mu)^k\right)$ the kth central moment of X, if they exist. (The name "moment" is borrowed from physics.)_

Thus, $E(X)$ is the first moment and $Var(X)$ is the second central moment of X. Other than these two, only the third and fourth central moments have some probabilistic significance: they can be used to measure the skewness and the flatness of a distribution.

The use of moments is analogous to the use of higher derivatives in calculus. There, the higher derivatives have no independent geometrical meaning, but are needed in Taylor expansions. Similarly, the higher moments are significant only in the Taylor expansions of certain functions obtained from probability distributions: the moment generating function, the probability generating function, and the characteristic function.

The moment generating function is closely related to the Laplace transform, which may be familiar from differential equation courses, and has similar properties. Its main use is the simplification it brings to finding the distributions of sums of i.i.d. random variables, which would, in most cases, be hopeless with the convolution formula when the number of terms gets large.

Definition 6.3.2. _Moment Generating Function._ _The moment generating function (m.g.f.) ψ or ψ_X of any random variable X is defined by_

$$\psi(t) = E\left(e^{tX}\right). \tag{6.101}$$

Clearly, the m.g.f. may not exist for certain random variables or for certain values of t. For most distributions that we are interested in, $\psi(t)$ will exist for all real t or on some interval.

Also, note that the m.g.f., being an expectation, depends only on the distribution of X and not on any other property of X. That is, if two r.v.'s have the same distribution, then they have the same m.g.f. as well. For this reason, it is correct to speak of the _m.g.f. of a distribution_ rather than that of the corresponding r.v.

Example 6.3.1. Binomial Distribution.

If X is binomial with parameters n and p, then

$$\psi(t) = E\left(e^{tX}\right) = \sum_{x=0}^{n} \binom{n}{x} p^x q^{n-x} e^{tx}$$

$$= \sum_{x=0}^{n} \binom{n}{x} \left(pe^t\right)^x q^{n-x} = \left(pe^t + q\right)^n. \tag{6.102}$$

In particular, if $n = 1$, then X is Bernoulli and its m.g.f. is

$$\psi(t) = pe^t + q. \tag{6.103}$$

◆

Example 6.3.2. Geometric Distribution.

If X is geometric with parameter p, then

$$\psi(t) = E\left(e^{tX}\right) = \sum_{x=1}^{\infty} pq^{x-1}e^{tx}$$

$$= \sum_{x=1}^{\infty} p\left(qe^t\right)^x q^{-1} = \frac{pe^t}{1 - qe^t}. \tag{6.104}$$

◆

Example 6.3.3. Uniform Distribution.

If X is uniform on $[a, b]$, then

$$\psi(t) = \int_a^b \frac{e^{tx}}{b - a} dx = \left. \frac{e^{tx}}{(b - a)\,t} \right|_a^b$$

$$= \frac{e^{bt} - e^{at}}{(b - a)\,t}. \tag{6.105}$$

◆

Example 6.3.4. Exponential Distribution.

If X is exponential with parameter $\lambda > 0$, then

$$\psi(t) = \int_0^\infty e^{tx} \lambda e^{-\lambda x} dx = \lambda \int_0^\infty e^{(t-\lambda)x} dx$$

$$= \left. \frac{\lambda}{t - \lambda} e^{(t-\lambda)x} \right|_0^\infty = \frac{\lambda}{\lambda - t} \text{ if } t < \lambda. \tag{6.106}$$

Clearly, $\psi(t)$ does not exist for $t \geq \lambda$.   ◆

Let us see now how the m.g.f. and the moments are connected.

Theorem 6.3.1. ψ **Generates Moments.** *If the m.g.f.* ψ *of a random variable X exists for all t in a neighborhood of 0, then all the moments of X exist, and*

$$\psi(t) = \sum_{k=0}^{\infty} E\left(X^k\right) \frac{t^k}{k!}, \tag{6.107}$$

that is, the moments are the coefficients of the Maclaurin series of ψ. Also, the function ψ is then infinitely differentiable at 0, and

$$\psi^{(k)}(0) = E\left(X^k\right) \text{ for } k = 0, 1, 2, \ldots. \tag{6.108}$$

Proof. We omit the technical details and just outline the proof. Since the Maclaurin series of e^{tX} is

$$e^{tX} = \sum_{k=0}^{\infty} \frac{(tX)^k}{k!}, \tag{6.109}$$

which is convergent for all real t, we have

$$\psi(t) = E\left(e^{tX}\right) = E\left(\sum_{k=0}^{\infty} \frac{(tX)^k}{k!}\right)$$
$$= \sum_{k=0}^{\infty} E\left(\frac{(tX)^k}{k!}\right) = \sum_{k=0}^{\infty} E\left(X^k\right) \frac{t^k}{k!}. \tag{6.110}$$

Equation 6.108 follows from Equation 6.107 by differentiating both sides k times and setting $t = 0$. ∎

Example 6.3.5. Mean and Variance of Exponential X.

If X is exponential with parameter $\lambda > 0$, then expanding the m.g.f. from Example 6.3.4 into a geometric series, we obtain

$$\psi(t) = \frac{\lambda}{\lambda - t} = \frac{1}{1 - t/\lambda} = \sum_{k=0}^{\infty} \frac{t^k}{\lambda^k} \text{ if } t < \lambda. \tag{6.111}$$

Comparing the coefficients of t and t^2 in the sum here with those of Equation 6.107 results in

$$E(X) = \frac{1}{\lambda} \tag{6.112}$$

and

$$\frac{E\left(X^2\right)}{2} = \frac{1}{\lambda^2}. \tag{6.113}$$

Hence

$$Var(X) = E\left(X^2\right) - [E(X)]^2 = \frac{2}{\lambda^2} - \frac{1}{\lambda^2} = \frac{1}{\lambda^2}, \tag{6.114}$$

just as in Chapter 5.

We could, of course, also have obtained these results by using Equation 6.108 rather than Equation 6.107. ◆

The most important properties of the m.g.f. are stated in the next three theorems.

Theorem 6.3.2. *The Multiplicative Property of Moment Generating Functions.* *For any positive integer n, let $X_1, X_2, \ldots, X_n$ be independent random variables with m.g.f. $\psi_1, \psi_2, \ldots, \psi_n$, respectively, and let $Y = \sum_{i=1}^{n} X_i$. Then $\psi_Y(t)$ exists for all t for which each $\psi_i(t)$ exists, and*

$$\psi_Y(t) = \prod_{i=1}^{n} \psi_i(t). \tag{6.115}$$

Proof.

$$\psi_Y(t) = E\left(e^{tY}\right) = E\left(e^{t\sum X_i}\right) = E\left(\prod_{i=1}^{n} e^{tX_i}\right)$$

$$= \prod_{i=1}^{n} E\left(e^{tX_i}\right) = \prod_{i=1}^{n} \psi_i(t). \tag{6.116}$$

■

The next two theorems will be stated without proof. Their proofs can be found in more advanced texts.

Theorem 6.3.3. *Uniqueness of the Moment Generating Function.* *If the moment generating functions of two random variables are equal on a neighborhood of 0, then their distributions are also equal.*

Theorem 6.3.4. *Limits of Sequences of Moment Generating Functions.* *Let $X_1, X_2, \ldots$ be a sequence of random variables with m.g.f.'s $\psi_1, \psi_2, \ldots$ and d.f.'s $F_1, F_2, \ldots$. If $\lim_{i \to \infty} \psi_i(t) = \psi(t)$ for all t in a neighborhood of 0, then $\lim_{i \to \infty} F_i(x) = F(x)$ exists for all x, and $\psi(t)$ is the m.g.f. of a r.v. whose d.f. is F.*

Example 6.3.6. *Sum of Binomial Random Variables.*

We rederive the result of Example 5.5.8, using m.g.f.'s.

Let X and Y be independent, binomial r.v.'s with parameters n_1, p and n_2, p, respectively. Then $Z = X + Y$ is binomial with parameters $n_1 + n_2, p$. By Example 6.3.1

$$\psi_X(t) = \left(pe^t + q\right)^{n_1} \text{ and } \psi_Y(t) = \left(pe^t + q\right)^{n_2}. \tag{6.117}$$

Hence, by Theorem 6.3.2, the m.g.f. of $Z = X + Y$ is given by

$$\psi_Z(t) = \left(pe^t + q\right)^{n_1 + n_2}. \tag{6.118}$$

This function is the m.g.f. of a binomial r.v. with parameters $n_1 + n_2, p$, and so, by the uniqueness theorem, $Z = X + Y$ is binomial with parameters $n_1 + n_2, p$.   ◆

Equation 6.107 is a particular case of the *generating function of a sequence*. In general, for any sequence $a_0, a_1, \ldots$, we call $G(s) = \sum_{k=0}^{\infty} a_k s^k$ its generating function. (As known from calculus, the infinite sum here is convergent on a finite or infinite interval centered at 0 or just at the point 0 itself.) Thus the m.g.f. is the generating function of the sequence $\langle E(X^k)/k! \rangle$ and not of the sequence of moments, despite its name.

For a discrete random variable, the probability function provides another sequence, in addition to the moments. The corresponding generating function for nonnegative, integer-valued random variables plays an important role in many applications.

Definition 6.3.3. *Probability Generating Function. The probability generating function* (p.g.f.) G *or* G_X *of any nonnegative integer-valued random variable X is defined by*

$$G(s) = E(s^X) = \sum_{x=0}^{\infty} f(x) s^x, \tag{6.119}$$

where f is the p.f. of X.

If we put $s = 1$ in Equation 6.119, then the sum on the right becomes the sum of the probabilities, and so we obtain

$$G(1) = 1. \tag{6.120}$$

Hence, the power series in Equation 6.119 is convergent for all $|s| \leq 1$.

If we know the generating function G, then we can obtain the probability function f from Equation 6.119, either by expanding $G(s)$ into a power series and extracting the coefficients or by using the formula

$$f(k) = \frac{G^{(k)}(0)}{k!} \text{ for } k = 0, 1, \ldots . \tag{6.121}$$

The p.g.f. is closely related to the m.g.f. If we put $s = e^t$ in Equation 6.119, then we obtain

$$\psi(t) = G(e^t). \tag{6.122}$$

Thus, the p.g.f. has similar properties to those of the m.g.f., and, especially, *it has the corresponding multiplicative and uniqueness properties*. The p.g.f. is, however, defined only for nonnegative integer-valued random variables, whereas the m.g.f. exists for all random variables whose moments exist. The p.g.f. is used to derive certain specific distributions, mainly in problems involving difference equations, like the gambler's ruin (Example 4.5.5), which we shall revisit below, and the m.g.f. is used to derive general theorems like the CLT in Section 7.3.

Example 6.3.7. The Gambler's Ruin.

In Example 4.5.5 we asserted that the difference equation

$$P(A_m) = P(A_{m+1}) \cdot \frac{1}{2} + P(A_{m-1}) \cdot \frac{1}{2} \tag{6.123}$$

is known to have the general solution $P(A_m) = a + bm$, where a and b are arbitrary constants. While it is easy to see by direct substitution that $P(A_m) = a + bm$ is a solution, it is not obvious that there are no other solutions. We now use the p.g.f. to prove this fact.

Multiplying both sides of Equation 6.123 by s^m and summing over m from 1 to ∞, we get

$$\sum_{m=1}^{\infty} P(A_m)s^m = \frac{1}{2s} \sum_{m=1}^{\infty} P(A_{m+1})s^{m+1} + \frac{s}{2} \sum_{m=1}^{\infty} P(A_{m-1})s^{m-1}. \tag{6.124}$$

With the notations $p_m = P(A_m)$ and $G(s) = \sum_{m=0}^{\infty} P(A_m)s^m$, the above equation can be written as

$$G(s) - p_0 = \frac{1}{2s} [G(s) - p_1 s - p_0] + \frac{s}{2} G(s), \tag{6.125}$$

and, solving for $G(s)$, we obtain

$$G(s) = \frac{(p_1 - 2p_0) s + p_0}{(1-s)^2}. \tag{6.126}$$

As known from calculus, the expression on the right can be decomposed into partial fractions as

$$G(s) = \frac{a}{1-s} + \frac{bs}{(1-s)^2}, \tag{6.127}$$

with appropriate constants a and b. These partial fractions are well-known sums of a geometric series and one derived from a geometric series[5], and so

$$G(s) = \sum_{m=0}^{\infty} as^m + \sum_{m=0}^{\infty} bms^m = \sum_{m=0}^{\infty} (a + bm) s^m. \tag{6.128}$$

Comparing this result with the definition of $G(s)$, we can see that $p_m = a + bm$ must hold for all m. In particular, $p_0 = a$ and $p_1 = a + b$, and so $a = p_0$ and $b = p_1 - p_0$.   ◆

The moments can also be obtained directly from the p.g.f.. For instance,

$$G'(s) = \sum_{x=0}^{\infty} f(x) x s^{x-1}, \tag{6.129}$$

---

[5] The second sum can be derived by differentiation from the geometric sum: $\sum_{m=0}^{\infty} ms^m = s \sum_{m=0}^{\infty} ms^{m-1} = s \frac{d}{ds} \sum_{m=0}^{\infty} s^m = s \frac{d}{ds} \frac{1}{1-s} = \frac{s}{(1-s)^2}$.

and so[6]

$$G'(1) = \sum_{x=0}^{\infty} f(x)\, x = E(X).$$ (6.130)

Similarly,

$$G''(s) = \sum_{x=0}^{\infty} f(x)\, x\,(x-1)\, s^{x-2},$$ (6.131)

and

$$G''(1) = \sum_{x=0}^{\infty} f(x)\, x\,(x-1) = E\left(X^2\right) - E(X).$$ (6.132)

Hence

$$E\left(X^2\right) = G''(1) + G'(1),$$ (6.133)

and

$$Var(X) = G''(1) + G'(1) - G'(1)^2.$$ (6.134)

As mentioned at the beginning of this section, there is yet another widely used function related to the generating functions described above:

Definition 6.3.4. *Characteristic Function.* *The characteristic function ϕ or ϕ_X of any random variable X is defined by*

$$\phi(t) = E\left(e^{itX}\right).$$ (6.135)

This function has properties similar to those of the m.g.f. and has the advantage that, unlike the m.g.f., it exists for every random variable X, since e^{itX} is a bounded function. On the other hand, its use requires complex analysis, and therefore we shall not discuss it further.

Exercises

Exercise 6.3.1.

Show that, for independent random variables, the third central moments are additive. That is, writing $m_3(X) = E\left((X - \mu_X)^3\right)$, we have, for independent X and Y, $m_3(X + Y) = m_3(X) + m_3(Y)$.

---

[6] Since the power series of $G(s)$ may not be convergent for $s > 1$, we consider $G'(1)$ and $G''(1)$ to be left derivatives.

Exercise 6.3.2.

Show that, for independent random variables, the fourth central moments are not additive. That is, writing $m_4(X) = E\left((X - \mu_X)^4\right)$, for independent X and Y, $m_4(X + Y) \neq m_4(X) + m_4(Y)$ in general.

Exercise 6.3.3.

Express the m.g.f. ψ_Y of $Y = aX + b$ in terms of ψ_X.

Exercise 6.3.4.

Use the m.g.f. from Example 6.3.2 to show that for a geometric r.v., $Var(X) = \frac{q}{p^2}$.

Exercise 6.3.5.

For any random variable X, the function $\psi_{X-\mu}$ is called the *central moment generating function* of X. Find $\psi_{X-\mu}$ for an X having the binomial n, p distribution, and use $\psi_{X-\mu}$ to find $Var(X)$.

Exercise 6.3.6.

Find the m.g.f. and the p.g.f. of a discrete uniform r.v. with possible values $1, 2, \ldots, n$. Simplify your answers.

Exercise 6.3.7.

Let X and Y be i.i.d. random variables with m.g.f. ψ. Express the m.g.f. ψ_Z of $Z = Y - X$ in terms of ψ.

Exercise 6.3.8.

Let X be a continuous r.v. with density $f(x) = \frac{1}{2}e^{-|x|}$ for $-\infty < x < \infty$:

1. Show that $\psi(t) = \frac{1}{1-t^2}$.
2. Use this ψ to find a formula for the moments of X.

Exercise 6.3.9.

Find the p.g.f. of a binomial n, p random variable.

Exercise 6.3.10.

Find the p.g.f. of a geometric random variable with parameter p.

Exercise 6.3.11.

We roll three dice. Use the p.g.f. to find the probability p_k that the sum of the points showing is k for $k = 3, 4$, and 5. (Hint: cf. Exercise 6.3.6.)

6.4 Covariance and Correlation

The expected value and the variance provided useful summary information about single random variables. The new notions of covariance and correlation, to be introduced in this section, provide information about the relationship between two or more random variables.

Definition 6.4.1. *Covariance. Given random variables X and Y with expected values μ_X and μ_Y, their covariance is defined as*

$$Cov\left(X,Y\right) = E\left(\left(X - \mu_X\right)\left(Y - \mu_Y\right)\right), \tag{6.136}$$

whenever the expected value on the right exists.

Example 6.4.1. *Covariance of (X,Y) Uniform on a Triangle.*

Let (X,Y) be uniform on the triangle $D = \{(x,y) : 0 \le x \le y \le 1\}$. Then $f(x,y) = 2$ on D, and

$$\mu_X = \int_0^1 \int_0^y 2x\,dx\,dy = \int_0^1 y^2\,dy = \frac{1}{3}, \tag{6.137}$$

$$\mu_Y = \int_0^1 \int_0^y 2y\,dx\,dy = \int_0^1 2y^2\,dy = \frac{2}{3}, \tag{6.138}$$

and

$$
\begin{aligned}
Cov\left(X,Y\right) &= \int_0^1 \int_0^y 2\left(x - \frac{1}{3}\right)\left(y - \frac{2}{3}\right)dx\,dy \\
&= \int_0^1 \left(y^2 - \frac{2}{3}y\right)\left(y - \frac{2}{3}\right)dy = \frac{1}{36}. \tag{6.139}
\end{aligned}
$$

♦

We can see from the definition that the covariance is positive if $(X - \mu_X)$ and $(Y - \mu_Y)$ tend to have the same sign, as in the example above, and it is negative if they tend to have opposite signs. If the sign combinations are equally balanced, then $Cov\left(X,Y\right) = 0$. The latter happens, in particular, whenever X and Y are independent, but it can happen in other cases, too.

Theorem 6.4.1. *An Alternative Formula for the Covariance. If X and Y are random variables such that $E\left(X\right), E\left(Y\right)$, and $E\left(XY\right)$ exist, then*

$$Cov\left(X,Y\right) = E\left(XY\right) - E\left(X\right)E\left(Y\right). \tag{6.140}$$

Proof. From Definition 6.4.1,

$$
\begin{aligned}
Cov\left(X,Y\right) &= E\left(XY - \mu_X Y - \mu_Y X + \mu_X \mu_Y\right) \\
&= E\left(XY\right) - \mu_X E\left(Y\right) - \mu_Y E\left(X\right) + \mu_X \mu_Y \\
&= E\left(XY\right) - E\left(X\right)E\left(Y\right). \tag{6.141}
\end{aligned}
$$

∎

Theorem 6.4.2. Independence Implies Zero Covariance. *For independent random variables X and Y whose expectations exist, $Cov(X,Y) = 0$.*

Proof. By Theorem 6.1.6 the two terms on the right of Equation 6.140 are equal in this case. ∎

As mentioned above, the converse is not true; the covariance may be zero for dependent random variables as well, as shown by the next example.

Example 6.4.2. Covariance of (X,Y) Uniform on a Disk.

Let (X,Y) be uniform on the unit disk $D = \{(x,y) : x^2 + y^2 < 1\}$. Then, clearly, $\mu_X = \mu_Y = 0$ and

$$
\begin{aligned}
Cov(X,Y) &= \int_{-1}^{1} \int_{-\sqrt{1-x^2}}^{\sqrt{1-x^2}} \frac{1}{\pi} xy\, dy\, dx \\
&= \int_{-1}^{1} \frac{2x}{\pi} \sqrt{1-x^2}\, dx = 0.
\end{aligned}
\tag{6.142}
$$

♦

In order to shed more light on what the covariance measures, it is useful to standardize the variables, so that their magnitude should not influence the value obtained. Thus we make a new definition:

Definition 6.4.2. Correlation Coefficient. *We define the correlation coefficient of any random variables X and Y with nonzero variances and existing covariance as*

$$
\rho(X,Y) = E\left(\frac{X - \mu_X}{\sigma_X} \cdot \frac{Y - \mu_Y}{\sigma_Y}\right).
\tag{6.143}
$$

We have the following obvious theorem:

Theorem 6.4.3. Alternative Formulas for the Correlation Coefficient. *If $\rho(X,Y)$ exists, then*

$$
\begin{aligned}
\rho(X,Y) &= Cov\left(\frac{X - \mu_X}{\sigma_X}, \frac{Y - \mu_Y}{\sigma_Y}\right) = \frac{Cov(X,Y)}{\sigma_X \sigma_Y} \\
&= \frac{E(XY) - E(X)E(Y)}{\sigma_X \sigma_Y}.
\end{aligned}
\tag{6.144}
$$

Example 6.4.3. Correlation Coefficient of Discrete Uniform (X,Y).

Consider the distribution that assigns probability $1/n$ to each of n data points (x_i, y_i). Then

$$
\rho(X,Y) = \frac{E(XY) - E(X)E(Y)}{\sigma_X \sigma_Y} = \frac{\frac{1}{n}\sum x_i y_i - \overline{x}\,\overline{y}}{\sigma_X \sigma_Y},
\tag{6.145}
$$

where

$$\overline{x} = \frac{1}{n} \sum x_i, \ \overline{y} = \frac{1}{n} \sum y_i, \tag{6.146}$$

and

$$\sigma_X = \left[\frac{1}{n} \sum (x_i - \overline{x})^2 \right]^{1/2}, \ \sigma_Y = \left[\frac{1}{n} \sum (y_i - \overline{y})^2 \right]^{1/2}. \tag{6.147}$$

◆

We are going to show that $\rho(X, Y)$ falls between -1 and $+1$, with ρ taking on the values ± 1, if and only if there is a linear relation $Y = aX + b$ between X and Y with probability 1. (ρ is $+1$ if a is positive and -1 if a is negative.) Thus, $|\rho|$ measures how close the points (X, Y) fall to a straight line in the plane. If ρ is 0, then X and Y are said to be *uncorrelated*, which means that there is no bunching of the points around a line. We say that $\rho(X, Y)$ *measures the strength of the linear association between X and Y.*

To prove the foregoing statements, we first present a general theorem about expectations.

Theorem 6.4.4. *Schwarz Inequality. For any random variables X and Y such that the expectations below exist,*

$$[E(XY)]^2 \le E(X^2) E(Y^2). \tag{6.148}$$

Furthermore, the two sides are equal if and only if $P(aX + bY = 0) = 1$ for some constants a and b, not both 0.

Proof. First assume that $E(Y^2) > 0$. Then, for any real number λ,

$$0 \le E\left((X - \lambda Y)^2\right) = \lambda^2 E(Y^2) - 2\lambda E(XY) + E(X^2), \tag{6.149}$$

and the right hand side is a quadratic function of λ whose graph is a parabola facing upward. The minimum occurs at

$$\lambda = \frac{E(XY)}{E(Y^2)} \tag{6.150}$$

and at that point the Inequality 6.149 becomes

$$0 \le \left[\frac{E(XY)}{E(Y^2)} \right]^2 E(Y^2) - 2 \frac{E(XY)}{E(Y^2)} E(XY) + E(X^2)$$

$$= E(X^2) - \frac{[E(XY)]^2}{E(Y^2)}, \tag{6.151}$$

which is equivalent to the Inequality 6.148.

In case $E\left(Y^2\right) = 0$, by Theorem 6.2.1, $P(Y = 0) = 1$, and then we also have $P(XY = 0) = 1$ and $E\left(XY\right) = 0$. Thus Inequality 6.148 is valid with both sides equal to 0.

To prove the second statement of the theorem, first assume that $P(aX + bY = 0) = 1$ for some constants a and b, not both 0. If $a = 0$, then this condition reduces to $P(Y = 0) = 1$, which we have just discussed. If $a \neq 0$, then, by Theorem 6.2.1, $E\left((aX + bY)^2\right) = 0$, and so

$$a^2 E\left(X^2\right) + 2ab E\left(XY\right) + b^2 E\left(Y^2\right) = 0, \tag{6.152}$$

or, equivalently,

$$\left(\frac{b}{a}\right)^2 E\left(Y^2\right) + 2\frac{b}{a} E\left(XY\right) + E\left(X^2\right) = 0. \tag{6.153}$$

Now, this is a quadratic equation for $\frac{b}{a}$ and we know that it has a single solution. (If it had two solutions, then both X and Y would have to be 0 with probability 1: a trivial case.) Thus its discriminant must be zero, that is, we must have

$$(2E\left(XY\right))^2 - 4E\left(X^2\right) E\left(Y^2\right) = 0, \tag{6.154}$$

which reduces to

$$[E\left(XY\right)]^2 = E\left(X^2\right) E\left(Y^2\right). \tag{6.155}$$

If we assume Equation 6.155, then the last argument can be traced backwards, and we can conclude that $P(aX + bY = 0) = 1$ must hold for some constants a and b, not both 0.   ∎

If we apply Theorem 6.4.4 to $\frac{X-\mu_X}{\sigma_X}$ and $\frac{Y-\mu_Y}{\sigma_Y}$ in place of X and Y, we obtain the following relation for the correlation coefficient:

Corollary 6.4.1. *For any random variables X and Y such that $\rho\left(X,Y\right)$ exists,*

$$-1 \leq \rho\left(X,Y\right) \leq 1. \tag{6.156}$$

Furthermore, $\rho\left(X,Y\right) = \pm 1$ if and only if $P(Y = aX + b) = 1$ for some constants $a \neq 0$ and b, with $\text{sign}(\rho\left(X,Y\right)) = \text{sign}(a)$.

Thus, the correlation coefficient puts a numerical value on the strength of the *linear association* between X and Y, that is, the closer ρ is to ± 1, the closer the random points (X,Y) bunch around a straight line and vice versa. Note that the line cannot be vertical or horizontal, because then ρ would not exist. The correlation coefficient conveys no useful information if the points bunch around any curve other than a straight line. For example, if (X,Y) is uniform on a circle, then ρ is zero, even though the points are on a curve.

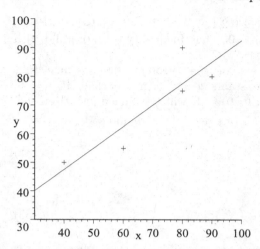

Fig. 6.3. Scatter plot for the exam scores with the least squares line superimposed

Example 6.4.4. Correlation Between Two Exams.

Suppose five students take two exams. Let X and Y denote the grades of a randomly selected student, as given in the X and Y columns of the following table. The rest of the table is included for the computation of ρ.

Student	X	Y	X^2	Y^2	XY
A	40	50	1600	2500	2000
B	60	55	3600	3025	3300
C	80	75	6400	5625	6000
D	90	80	8100	6400	7200
E	80	90	6400	8100	7200
Ave.	70	70	5220	5130	5140

Hence $\mu_X = \mu_Y = 70, \sigma_X = \sqrt{5220 - 70^2} \approx 17.889, \sigma_Y = \sqrt{5130 - 70^2} \approx$ 15.166, and $\rho \approx \frac{5140 - 70^2}{17.889 \cdot 15.166} \approx 0.88$.

The grades of each student are shown above as points in a so-called *scatter plot* (see Figure 6.3), together with the line of the best fit in the least squares sense or, briefly, *the least squares line or regression line of Y on X*, given by $y = 70 + \frac{3}{4}(x - 70)$. (The general formula will be given below in Theorem 6.4.5, and regression will be discussed in Section 8.8.) It should not be surprising that the points bunch around a straight line, because we would expect good students to do well on both exams, bad students to do poorly, and mediocre students to be in the middle, both times. On the other hand, the points need not fall exactly on a line, since there is usually some randomness in the scores; people do not always perform at the same level. Furthermore, the slope of the line does not have to be 1, because the two exams may differ in difficulty.

The value 0.88 for ρ shows that the points are fairly close to a line. If ρ were 1, then they would all fall on a line, and if ρ were 0, then the points would seem to bunch the same way around any line through their center of gravity, with no preferred direction.

♦

Example 6.4.5. Correlation of (X, Y) Uniform on a Triangle.

Let (X, Y) be uniform on the triangle $D = \{(x, y) : 0 \leq x \leq y \leq 1\}$ as in Example 6.4.1. Then

$$E\left(X^2\right) = \int_0^1 \int_0^y 2x^2 dx dy = \int_0^1 \frac{2}{3}y^3 dy = \frac{1}{6}, \tag{6.157}$$

$$E\left(Y^2\right) = \int_0^1 \int_0^y 2y^2 dx dy = \int_0^1 2y^3 dy = \frac{1}{2}, \tag{6.158}$$

and

$$\sigma_X^2 = \frac{1}{6} - \left(\frac{1}{3}\right)^2 = \frac{1}{18}, \text{ and } \sigma_Y^2 = \frac{1}{2} - \left(\frac{2}{3}\right)^2 = \frac{1}{18}. \tag{6.159}$$

Thus,

$$\rho(X, Y) = \frac{Cov(X, Y)}{\sigma_X \sigma_Y} = \frac{1/36}{1/18} = \frac{1}{2} \tag{6.160}$$

This result shows that the points of the triangle D are rather loosely grouped around a line, as can also be seen in Figure 6.4. However, this line is not unique: the line joining the origin and the centroid would do just as well as the one shown.

♦

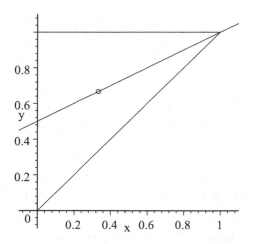

Fig. 6.4. The triangle D with the least squares line and the point of averages drawn in

In addition to being a measure of the linear association between two random variables, the correlation coefficient is also a determining factor in the slope of the least squares line (which we give here just for the special case of a finite number of equiprobable points):

Theorem 6.4.5. *Least Squares Line.* *Let* (X, Y) *be a random point with* m *possible values* (x_i, y_i), *each having probability* $\frac{1}{m}$. *The line* $y = ax + b$, *such that the sum of the squared vertical distances*

$$Q(a, b) = \sum_{i=1}^{m} (ax_i + b - y_i)^2 \tag{6.161}$$

from the points to it is minimum, is given by the equation

$$y = \mu_Y + \rho \frac{\sigma_Y}{\sigma_X} (x - \mu_X), \tag{6.162}$$

or, equivalently, in standardized form by

$$\frac{y - \mu_Y}{\sigma_Y} = \rho \frac{x - \mu_X}{\sigma_X}. \tag{6.163}$$

The proof is left as Exercise 6.4.6.

Exercises

Exercise 6.4.1.

Prove that

$$Var(X + Y) = Var(X) + 2Cov(X, Y) + Var(Y) \tag{6.164}$$

whenever each term exists.

Exercise 6.4.2.

Let (X, Y) be uniform on the triangle $D = \{(x, y) : 0 < x, 0 < y, x + y < 1\}$. Compute $Cov(X, Y)$ and $\rho(X, Y)$.

Exercise 6.4.3.

Let X and Y have the same distribution and let $U = X + Y$ and $V = X - Y$:

1. Show that $Cov(U, V) = 0$, assuming that each variance and covariance exists.
2. Show that if X and Y denote the outcomes of throwing two dice, then U and V are not independent, although $Cov(U, V) = 0$ by part 1.

Exercise 6.4.4.

Let (X, Y) be uniform on the half disk $D = \{(x, y) : 0 < y, x^2 + y^2 < 1\}$. Compute $Cov(X, Y)$ and $\rho(X, Y)$.

Exercise 6.4.5.

Let X and Y be discrete random variables with joint probabilities $P(X = x_i, Y = y_j) = p_{ij}$ for $i = 1, 2, \ldots, m$ and $j = 1, 2, \ldots, n$. Using also p_i for $P(X = x_i)$ and q_j for $P(Y = y_j)$, write a formula for

1. $Cov(X, Y)$
2. $\rho(X, Y)$.

Exercise 6.4.6.

Prove Theorem 6.4.5. (Hint: Set the partial derivatives of $Q(a, b)$ in Equation 6.161 equal to zero to obtain the so-called *normal equations* of the least squares problem, and solve for a and b.)

Exercise 6.4.7.

Let X and Y be random variables such that $\rho(X, Y)$ exists, and let $U = aX + b$ and $V = cY + d$, with $a \neq 0, b, c \neq 0, d$ constants. Show that $\rho(U, V) = \text{sign}(ac)\rho(X, Y)$.

Exercise 6.4.8.

Let X and Y be random variables such that $Var(X), Var(Y)$, and $Cov(X, Y)$ exist, and let $U = aX + bY$ and $V = cX + dY$, with a, b, c, d constants. Find an expression for $Cov(U, V)$ in terms of $a, b, c, d, Var(X)$, $Var(Y)$, and $Cov(X, Y)$.

Exercise 6.4.9.

Let X and Y be random variables such that $Var(X) = 4, Var(Y) = 1$, and $\rho(X, Y) = \frac{1}{2}$. Find $Var(X - 3Y)$.

Exercise 6.4.10.

Prove that if X_i and Y_j are random variables such that $Cov(X_i, Y_j)$ exists for all i, j, and a_i, b_j are arbitrary constants, then

$$Cov\left(\sum_{i=1}^{m} a_i X_i, \sum_{j=1}^{n} b_j Y_j\right) = \sum_{i=1}^{m}\sum_{j=1}^{n} a_i b_j Cov(X_i, Y_j). \tag{6.165}$$

Exercise 6.4.11.

Suppose in Example 6.4.4, the first exam score of student E is changed from 80 to 90:

1. Recompute $\rho(X, Y)$ with this change.
2. Find the equation of the new least squares line.
3. Draw the scatter plot, together with the new line.

6.5 Conditional Expectation

In many applications, we need to consider expected values under given conditions. We define such expected values much as we defined unconditional ones; we just replace the unconditional distributions in the earlier definitions with conditional distributions:

Definition 6.5.1. *Conditional Expectation. Let A be any event with $P(A) \neq 0$ and X any discrete random variable. Then we define the conditional expectation of X under the condition A by*

$$E_A(X) = \sum_{x: f_{X|A}(x)>0} x f_{X|A}(x). \tag{6.166}$$

Let A be any event with $P(A) \neq 0$ and X any continuous random variable such that $f_{X|A}$ exists. Then we define the conditional expectation of X under the condition A by

$$E_A(X) = \int_{-\infty}^{\infty} x f_{X|A}(x) dx. \tag{6.167}$$

If X is discrete and Y any random variable such that $f_{X|Y}$ exists, then the conditional expectation of X given $Y = y$ is defined by

$$E_y(X) = \sum_{x: f_{X|Y}(x,y)>0} x f_{X|Y}(x,y). \tag{6.168}$$

If X is continuous and Y any random variable such that $f_{X|Y}$ exists, then the conditional expectation of X given $Y = y$ is defined by

$$E_y(X) = \int_{-\infty}^{\infty} x f_{X|Y}(x,y) dx. \tag{6.169}$$

All the theorems for unconditional expectations remain valid for conditional expectations as well, because the definitions are essentially the same, just that unconditional f's are replaced by conditional ones. The latter are still probability functions or densities, and so this change does not affect the proofs. In particular, conditional expectations of functions $g(X)$ can be computed for discrete X as

$$E_y(g(X)) = \sum_{x: f_{X|Y}(x,y)>0} g(x) f_{X|Y}(x,y), \tag{6.170}$$

and for continuous X as

$$E_y(g(X)) = \int_{-\infty}^{\infty} g(x) f_{X|Y}(x,y) dx. \tag{6.171}$$

Also,

$$E_y\left(aX + b\right) = aE_y\left(X\right) + b \tag{6.172}$$

and

$$E_y\left(X_1 + X_2\right) = E_y\left(X_1\right) + E_y\left(X_2\right). \tag{6.173}$$

Note that, whether X is discrete or continuous, $E_y\left(X\right)$ is a function of y, say $E_y\left(X\right) = g\left(y\right)$. If we replace y here by the random variable Y, we get a new random variable $g\left(Y\right) = E_Y\left(X\right)$. The next theorem says that the expected value of this new random variable is $E\left(X\right)$. In other words, we can obtain the expected value of X in two steps: first, averaging X under some given conditions and then averaging over the conditions with the appropriate weights. This procedure is analogous to the one in the theorem of total probability, in which we computed the probability (rather than the average) of an event A under certain conditions and then averaged over the conditions with the appropriate weights.

Theorem 6.5.1. ***Theorem of Total Expectation.*** *If all expectations below exist, then*

$$E\left(E_Y\left(X\right)\right) = E\left(X\right). \tag{6.174}$$

Proof. We give the proof for the continuous case only.
By Definition 6.5.1,

$$E_y\left(X\right) = \int_{-\infty}^{\infty} x f_{X|Y}\left(x, y\right) dx = \int_{-\infty}^{\infty} x \frac{f\left(x, y\right)}{f_Y\left(y\right)} dx. \tag{6.175}$$

Also, by Theorem 6.1.3,

$$E\left(E_Y\left(X\right)\right) = \int_{-\infty}^{\infty} E_y\left(X\right) f_Y\left(y\right) dy. \tag{6.176}$$

Thus,

$$\begin{aligned} E\left(E_Y\left(X\right)\right) &= \int_{-\infty}^{\infty} \left(\int_{-\infty}^{\infty} x \frac{f\left(x, y\right)}{f_Y\left(y\right)} dx\right) f_Y\left(y\right) dy \\ &= \int_{-\infty}^{\infty} x \left(\int_{-\infty}^{\infty} f\left(x, y\right) dy\right) dx \\ &= \int_{-\infty}^{\infty} x f_X\left(x\right) dx = E\left(X\right). \end{aligned} \tag{6.177}$$

∎

Example 6.5.1. Sum and Absolute Difference of Two Dice.

In Table 5.8 we displayed $f_{U|V}(u|v)$ for the random variables $U = X + Y$ and $V = |X - Y|$, where X and Y were the numbers obtained with rolling two dice. Hence, for $v = 0$ we get

$$E_v(U) = 2 \cdot \frac{1}{6} + 4 \cdot \frac{1}{6} + 6 \cdot \frac{1}{6} + 8 \cdot \frac{1}{6} + 10 \cdot \frac{1}{6} + 12 \cdot \frac{1}{6} = 7. \qquad (6.178)$$

Similarly, $E_v(U) = 7$ for all other values of v as well, and so, using the marginal probabilities $f_V(v)$, we obtain

$$E(U) = E(E_V(U)) = 7 \cdot \frac{6}{36} + 7 \cdot \frac{10}{36} + 7 \cdot \frac{8}{36} + 7 \cdot \frac{6}{36} + 7 \cdot \frac{4}{36} + 7 \cdot \frac{2}{36} = 7. \quad (6.179)$$

This is indeed the same value that we would obtain directly from the marginal probabilities $f_U(u)$ or from $E(U) = E(X) + E(Y) = 2 \cdot 3.5$.

Going the other way, from Table 5.9, we have, for instance, for $u = 4$

$$E_u(V) = 0 \cdot \frac{1}{3} + 2 \cdot \frac{2}{3} = \frac{4}{3}. \qquad (6.180)$$

The whole function $E_u(V)$ is given by the table:

u	2	3	4	5	6	7	8	9	10	11	12
$E_u(V)$	0	1	4/3	2	12/5	3	12/5	2	4/3	1	0

Thus

$$E(V) = E(E_U(V)) = 0 \cdot \frac{1}{36} + 1 \cdot \frac{2}{36} + \frac{4}{3} \cdot \frac{3}{36} + 2 \cdot \frac{4}{36} + \frac{12}{5} \cdot \frac{5}{36}$$
$$+ 3 \cdot \frac{6}{36} + \frac{12}{5} \cdot \frac{5}{36} + 2 \cdot \frac{4}{36} + \frac{4}{3} \cdot \frac{3}{36} + 1 \cdot \frac{2}{36} + 0 \cdot \frac{1}{36} = \frac{70}{36}. \quad (6.181)$$

As required by the theorem, the direct computation of $E(V)$ from $f_V(v)$ gives the same result:

$$E(V) = 0 \cdot \frac{6}{36} + 1 \cdot \frac{10}{36} + 2 \cdot \frac{8}{36} + 3 \cdot \frac{6}{36} + 4 \cdot \frac{4}{36} + 5 \cdot \frac{2}{36} = \frac{70}{36}. \quad (6.182)$$

♦

Example 6.5.2. Conditional Expectation $E_y(X)$ for (X, Y) Uniform on Unit Disk.

Let (X, Y) be uniform on the unit disk $D = \{(x, y) : x^2 + y^2 < 1\}$ as in Example 5.6.2. Then

$$E_y(X) = \int_{-\infty}^{\infty} x f_{X|Y}(x, y) \, dx$$

$$= \int_{-\sqrt{1-y^2}}^{\sqrt{1-y^2}} \frac{x}{2\sqrt{1-y^2}} \, dx = 0, \text{ for } y \in (-1, 1), \qquad (6.183)$$

just as we would expect by symmetry.

♦

Let us modify the last example so as to avoid the trivial outcome:

Example 6.5.3. Conditional Expectation $E_y(X)$ for (X,Y) Uniform on Half Disk.

Let (X,Y) be uniform on the right half disk $D = \{(x,y) : x^2 + y^2 < 1,\ 0 < x\}$. Then

$$
E_y(X) = \int_{-\infty}^{\infty} x f_{X|Y}(x,y)\, dx
$$

$$
= \int_0^{\sqrt{1-y^2}} \frac{x}{\sqrt{1-y^2}} dx = \frac{\sqrt{1-y^2}}{2}, \text{ for } y \in (-1,1), \qquad (6.184)
$$

and

$$
E(X) = E(E_Y(X)) = \int_{-\infty}^{\infty} E_y(X) f_Y(y)\, dy
$$

$$
= \int_{-1}^{1} \frac{\sqrt{1-y^2}}{2} \cdot \frac{2}{\pi} \sqrt{1-y^2} dy
$$

$$
= \int_{-1}^{1} \frac{1-y^2}{\pi} dy = \frac{4}{3\pi}. \qquad (6.185)
$$

◆

Before we state the next theorem, we present a lemma:

Lemma 6.5.1. *For any random variables X and Y and any functions $g(X)$ and $h(Y)$ such that $E_Y(g(X))$ exists,*

$$
E_Y(g(X) h(Y)) = h(Y) E_Y(g(X)). \qquad (6.186)
$$

Proof. We give the proof for the continuous case only.
By definition,

$$
E_y(g(X) h(Y)) = \int_{-\infty}^{\infty} g(x) h(y) f_{X|Y}(x,y)\, dx
$$

$$
= h(y) \int_{-\infty}^{\infty} g(x) f_{X|Y}(x,y)\, dx = h(y) E_y(g(X)). \qquad (6.187)
$$

If we replace y by Y here, we get the statement of the lemma.   ∎

The next theorem answers the following question: Suppose that, for given random variables X and Y, we want to find a function $p(Y)$ that is as close as possible to X. If we observe $Y = y$, then $p(y)$ may be considered to be a prediction of the corresponding value x of X. Thus we ask: What is the best prediction $p(Y)$ of X, given Y? "Best" is defined in the least squares sense, that is, in terms of minimizing the expected value of the squared difference of X and $p(Y)$. The answer is a generalization of the result of Theorem 6.2.5, that the mean of squared deviations is minimum if the deviations are taken from the mean, that is, that $E(X)$ is the best prediction in the least squares sense for X (e.g., if we toss a coin a hundred times, the best prediction for the number of heads is 50. On the other hand, if we toss only once, then $E(X) = \frac{1}{2}$ is not much of a prediction, but, still, that's the best we can do).

Theorem 6.5.2. *The Best Prediction of* X, *Given* Y. *For given random variables X and Y and all functions $p(Y)$, the mean squared difference $E\left([X - p(Y)]^2\right)$, if it exists, is minimized by the function $p(Y) = E_Y(X)$.*

Proof.

$$
\begin{aligned}
E\left([X - p(Y)]^2\right) &= E\left([X - E_Y(X) + E_Y(X) - p(Y)]^2\right) \\
&= E\left([X - E_Y(X)]^2\right) + E\left([E_Y(X) - p(Y)]^2\right) \\
&\quad + 2E\left[(X - E_Y(X))(E_Y(X) - p(Y))\right]. \quad (6.188)
\end{aligned}
$$

By Theorem 6.5.1, the last term can be reformulated as

$$
\begin{aligned}
2E\,&[(X - E_Y(X))(E_Y(X) - p(Y))] \\
&= 2E\left(E_Y\left[(X - E_Y(X))(E_Y(X) - p(Y))\right]\right). \quad (6.189)
\end{aligned}
$$

On the right here, we can apply the lemma, with $X - E_Y(X) = g(X)$ and $E_Y(X) - p(Y) = h(Y)$. Thus,

$$
\begin{aligned}
E_Y\,&[(X - E_Y(X))(E_Y(X) - p(Y))] \\
&= (E_Y(X) - p(Y))\,E_Y\left[(X - E_Y(X))\right] \\
&= (E_Y(X) - p(Y))\left[E_Y(X) - E_Y(E_Y(X))\right] = 0, \quad (6.190)
\end{aligned}
$$

because $E_Y(E_Y(X)) = E_Y(X)$. (The proof of this identity is left as Exercise 6.5.8.)

Hence,

$$
E\left([X - p(Y)]^2\right) = E\left([X - E_Y(X)]^2\right) + E\left([E_Y(X) - p(Y)]^2\right), \quad (6.191)
$$

and, since both terms are nonnegative, the sum on the right is minimum if $p(Y) = E_Y(X)$. ∎

The notion of conditional expectation can be used to define conditional variance:

Definition 6.5.2. *Conditional Variance.* *For given random variables X and Y, the conditional variance $Var_y(X)$ is defined as*

$$
Var_y(X) = E_y\left([X - E_y(X)]^2\right). \quad (6.192)
$$

Clearly, $Var_y(X)$ is a function of y, and so $Var_Y(X)$ is a function of the random variable Y and, as such, another random variable. However, the theorem of total expectation does not extend to conditional variances. (We leave the explanation as Exercise 6.5.13.)

Exercises

Exercise 6.5.1.

Prove Theorem 6.5.1 for discrete X and Y.

Exercise 6.5.2.

Roll two dice as in Example 6.5.1. Let $U = \max(X,Y)$ and $V = \min(X,Y)$. Find $E_v(U)$ and $E_u(V)$ for each possible value of v and u, and verify the relations $E(E_V(U)) = E(U)$ and $E(E_U(V)) = E(V)$.

Exercise 6.5.3.

Define a random variable X as follows: Toss a coin and if we get H, then let X be uniform on the interval $[0, 2]$, and if we get T, then throw a die and let X be the number obtained. Find $E(X)$.

Exercise 6.5.4.

Suppose a plant has X offspring in a year with $P(X = x) = \frac{1}{4}$ for $X = 1, 2, 3, 4$ and, independently, each offspring has from one to four offspring in the next year with the same discrete uniform distribution. Let Y denote the total number of offspring in the second generation. Find the values of $E_X(Y)$ and compute $E(E_X(Y))$.

Exercise 6.5.5.

Let (X, Y) be uniform on the triangle $D = \{(x, y) : 0 < x, 0 < y, x + y < 1\}$. Compute $E_x(Y), E_y(X), E(X)$, and $E(Y)$.

Exercise 6.5.6.

Let (X, Y) be uniform on the triangle $D = \{(x, y) : 0 < x < y < 1\}$. Compute $E_x(Y), E_y(X), E(X)$, and $E(Y)$.

Exercise 6.5.7.

Let (X, Y) be uniform on the open unit square $D = \{(x, y) : 0 < x < 1, 0 < y < 1\}$ and $Z = X + Y$ as in Exercise 5.6.7. Find $E_z(X)$ and $E_x(Z)$.

Exercise 6.5.8.

Prove that $E_Y (E_Y (X)) = E_Y (X)$ if $E_Y (X)$ exists.

Exercise 6.5.9.

Let X and Y be continuous random variables with joint density $f(x, y)$, and let $g(x, y)$ be any integrable function. Prove that $E(E_Y (g(X, Y))) = E(g(X, Y))$ if $E(E_Y (g(X, Y)))$ exists.

Exercise 6.5.10.

Show that for arbitrary X and Y with nonzero variances, if $E_y (X) = c$ for all y, where c is a constant, then X and Y are uncorrelated.

Exercise 6.5.11.

Show that for continuous X and Y, if $E_y (X) = c$ for all y, where c is a constant, then $E(X) = c$ and $Var(X) = E(Var_Y (X))$ if all quantities exist.

Exercise 6.5.12.

Let X and Y be as in Exercise 6.5.4. Find the values of $Var_X (Y)$, and compute $E(Var_X (Y))$ and $Var(Y)$.

Exercise 6.5.13.

Explain why $Var(X) \neq E(Var_Y (X))$ in general.

Exercise 6.5.14.

Show that for continuous X and Y, $Var(X) = E(Var_Y (X)) + Var (E_Y (X))$ if all quantities exist.

6.6 Median and Quantiles

The expected value of a random variable was introduced to provide a numerical value for the center of its distribution. For some random variables, however, it is preferable to use another quantity for this purpose, either because $E(X)$ does not exist or because the distribution of X is very skewed and $E(X)$ does not represent the center very well. The latter case occurs, for instance, when X stands for the income of a randomly selected person from a set of ten people, with nine earning 20 thousand dollars and one of them earning 20 million dollars. Saying that the average income is $E(X) = \frac{1}{10}(9 \cdot 20,000 + 20,000,000) \approx 2,000,000$ dollars is worthless and misleading. In such cases we use the *median* to represent the center. Also, for some random variables, $E(X)$ does not exist, but a median always does.

We want to define the median so that half of the probability is below it and half above it. This aim, however, cannot always be achieved, and even if it can, the median may not be unique, as will be seen below. Thus, we relax the requirements somewhat and make the following definition:

Definition 6.6.1. Median. *For any random variable X, a median of X, or of its distribution, is a number m such that $P(X < m) \leq \frac{1}{2}$ and $P(X > m) \leq \frac{1}{2}$.*

Note that $P(X < m)$ or $P(X > m)$ can be less than $\frac{1}{2}$ only if $P(X = m) \neq 0$, or, in other words, we have the following theorem:

Theorem 6.6.1. A Condition for $P(X < m) = \frac{1}{2}$. *For m a median of a random variable X, $P(X = m) = 0$ implies $P(X < m) = \frac{1}{2}$ and $P(X > m) = \frac{1}{2}$.*

Proof. Since m is a median,

$$P(X < m) \leq \frac{1}{2} \tag{6.193}$$

and

$$P(X > m) \leq \frac{1}{2}. \tag{6.194}$$

Also, because $P(X = m) = 0$, we have

$$P(X < m) + P(X > m) = 1. \tag{6.195}$$

Now, if we had $P(X < m) < \frac{1}{2}$, then adding corresponding sides of this inequality and Inequality 6.194, we would get $P(X < m) + P(X > m) < 1$, in contradiction to Equation 6.195. Thus, we must have $P(X < m) = \frac{1}{2}$ and then also $P(X > m) = \frac{1}{2}$. ∎

Observe that for continuous random variables, the condition $P(X = m) = 0$ is always true and so is therefore the conclusion of Theorem 6.6.1, too.

Before considering specific examples, we are going to show that for the large class of symmetric distributions, the center of symmetry is a median as well as $E(X)$ (see Theorem 6.1.1).

Theorem 6.6.2. The Center of Symmetry is a Median. *If the distribution of a random variable is symmetric about a point α, that is, the p.f. or the p.d.f. satisfies $f(\alpha - x) = f(\alpha + x)$ for all x, then α is a median of X.*

Proof. We give the proof for continuous X only; for discrete X the proof is similar and is left as an exercise.

If the density of X satisfies $f(\alpha - x) = f(\alpha + x)$ for all x, then, by obvious changes of variables,

$$
\begin{aligned}
\mathrm{P}(X < \alpha) &= \int_{-\infty}^{\alpha} f(t)\, dt = -\int_{\infty}^{0} f(\alpha - x)\, dx \\
&= \int_{0}^{\infty} f(\alpha - x)\, dx = \int_{0}^{\infty} f(\alpha + x)\, dx \\
&= \int_{\alpha}^{\infty} f(u)\, du = \mathrm{P}(X > \alpha).
\end{aligned}
\tag{6.196}
$$

Since, for continuous X, also

$$
\mathrm{P}(X < \alpha) + \mathrm{P}(X > \alpha) = 1,
\tag{6.197}
$$

we obtain

$$
\mathrm{P}(X < \alpha) = \mathrm{P}(X > \alpha) = \frac{1}{2},
\tag{6.198}
$$

which shows that α is a median of X.   ∎

Example 6.6.1. Median of Uniform Distributions.

If X is uniform on the interval $[a, b]$, then, by Theorem 6.6.2, the center $m = \frac{a+b}{2}$ is a median. Furthermore, this median is unique, because, if $c < m$ is another point, then $\mathrm{P}(X > c) > \frac{1}{2}$, and if $c > m$, then $\mathrm{P}(X < c) > \frac{1}{2}$, and so c is not a median in either case according to Definition 6.6.1.   ♦

The next example shows that even if the distribution is symmetric, the median need not be unique.

Example 6.6.2. Median of a Distribution Uniform on Two Intervals.

Let X be uniform on the union $[0, 1] \cup [2, 3]$ of two intervals, that is, let

$$
f(x) = \begin{cases} \frac{1}{2} & \text{if } 0 \leq x \leq 1 \\ \frac{1}{2} & \text{if } 2 \leq x \leq 3 \\ 0 & \text{otherwise} \end{cases}.
$$

Then $f(x)$ is symmetric about $\alpha = \frac{3}{2}$, and so, by Theorem 6.6.2, $\frac{3}{2}$ is a median, but, clearly, any m in $[1, 2]$ is also a median.   ♦

In the next example, $\mathrm{P}(X < m) \neq \frac{1}{2}$.

Example 6.6.3. Median of a Binomial.

Let X be binomial with parameters $n = 4$ and $p = \frac{1}{2}$. Then, by symmetry, $m = 2$ is a median. But, since $P(X = 2) = \binom{4}{2}\left(\frac{1}{2}\right)^4 = \frac{3}{8}$, we have $P(X < 2) = P(X > 2) = \frac{1}{2}\left(1 - \frac{3}{8}\right) = \frac{5}{16}$, and so, 2 is the only median.   ♦

Example 6.6.4. Median of the Exponential Distribution.

Let T be exponential with parameter λ. Then $P(T < t) = F(t)$ is continuous and strictly increasing on $(0, \infty)$, and so we can solve $F(m) = \frac{1}{2}$, that is, by Definition 5.2.3, solve

$$1 - e^{-\lambda m} = \frac{1}{2}. \tag{6.199}$$

Hence,

$$m = \frac{\ln 2}{\lambda} \tag{6.200}$$

is the unique median.

In physics, such a T is used to represent the lifetime of a radioactive particle. In that case, m is called (somewhat misleadingly) *the half-life* of the particle, for it is the length of time in which the particle decays with probability $\frac{1}{2}$ or, equivalently, the length of time in which half of a very large number of such particles decay.   ♦

An interesting property of medians is that they minimize the "mean absolute deviations" just as the expected value minimizes mean squared deviations (Exercise 6.2.5 and Theorem 6.5.2):

Theorem 6.6.3. Medians Minimize Mean Absolute Deviations. *For any random variable X such that the expected values below exist,*

$$\min_{c} E\left(|X - c|\right) = E\left(|X - m|\right) \tag{6.201}$$

for any median m of X.

Proof. We give the proof only for continuous X with density $f(x)$.

Let m be any median of X and c any number such that $c > m$. (For $c < m$ the proof would require just minor modifications.) Then

$$E\left(|X - c|\right) - E\left(|X - m|\right)$$

$$= \int_{-\infty}^{\infty} \left(|x - c| - |x - m|\right) f(x)\, dx$$

$$= \int_{-\infty}^{m} \left((c - x) - (m - x)\right) f(x)\, dx + \int_{m}^{c} \left((c - x) - (x - m)\right) f(x)\, dx$$

$$+ \int_{c}^{\infty} \left((x - c) - (x - m)\right) f(x)\, dx$$

$$= \int_{-\infty}^{m} (c - m) f(x)\, dx + \int_{m}^{c} (c + m - 2x) f(x)\, dx$$

$$+ \int_{c}^{\infty} (m - c) f(x)\, dx \tag{6.202}$$

Now, between m and c, we have $2x \leq 2c$ and $-2x \geq -2c$. Adding $c + m$ to both sides, we get $c + m - 2x \geq c + m - 2c = m - c$. Thus,

$$E\left(|X - c|\right) - E\left(|X - m|\right)$$
$$\geq \int_{-\infty}^{m} (c - m) f(x) \, dx + \int_{m}^{c} (m - c) f(x) \, dx + \int_{c}^{\infty} (m - c) f(x) \, dx$$
$$= \int_{-\infty}^{m} (c - m) f(x) \, dx + \int_{m}^{\infty} (m - c) f(x) \, dx$$
$$= (c - m) \left[\mathrm{P}\left(X < m\right) - \mathrm{P}\left(X > m\right)\right] = (c - m) \left(\frac{1}{2} - \frac{1}{2}\right) = 0. \quad (6.203)$$

Hence

$$E\left(|X - c|\right) \geq E\left(|X - m|\right), \quad (6.204)$$

for any c, which shows that the minimum of $E\left(|X - c|\right)$ occurs for $c = m$. $\blacksquare$

A useful generalization of the notion of a median is obtained by prescribing an arbitrary number $p \in (0, 1)$ and asking for a number x_p such that $F(x_p) = \mathrm{P}\left(X \leq x_p\right) = p$, instead of $\frac{1}{2}$. Unfortunately, for some distributions and certain values of p, this equation cannot be solved, or the solution is not unique, and for those the definition below is somewhat more complicated.

Definition 6.6.2. Quantiles. *Let X be a continuous random variable with $F(x)$ continuous and strictly increasing from 0 to 1 on some finite or infinite interval I. Then, for any $p \in (0, 1)$, the solution x_p of $F(x_p) = p$ or, in other words, $x_p = F^{-1}(p)$ is called the p quantile or the $100p$ percentile and the function F^{-1} the quantile function of X or of the distribution of X. For general X the p quantile is defined as $x_p = \min\left\{x : F(x) \geq p\right\}$, and we define the quantile function F^{-1} by $F^{-1}(p) = x_p$, for all $p \in (0, 1)$.*

Quantiles or percentiles are often used to describe statistical data such as exam scores, home prices, incomes, etc. For example, a student's score of, say, 650 on the math SAT is much better understood if it is also stated that this number is at the 78th percentile, meaning that 78% of the students who took the test scored 650 or less or, in other words, a randomly selected student's score is 650 or less with probability 0.78. Also, some distributions in statistics, as will be seen later, are usually described in terms of their quantile function F^{-1} rather than in terms of F or f.

Clearly, the 50th percentile is also a median. Furthermore, the 25th percentile is also called the *first quartile*, the 50th percentile the *second quartile*, and the 75th percentile *the third quartile*.

Example 6.6.5. **Quantiles of the Uniform Distribution.**

If X is uniform on the interval $[a, b]$, then

$$F(x) = \begin{cases} 0 & \text{if} \quad x < a \\ \dfrac{x - a}{b - a} & \text{if } a \leq x < b \\ 1 & \text{if} \quad x \geq b \end{cases} \tag{6.205}$$

is continuous and strictly increasing from 0 to 1 on (a, b), and so we can solve $F(x_p) = p$ for any $p \in (0, 1)$, that is, solve

$$\frac{x_p - a}{b - a} = p. \tag{6.206}$$

Hence,

$$x_p = a + p(b - a) \tag{6.207}$$

is the p quantile for any $p \in (0, 1)$.   ♦

Example 6.6.6. **Quantiles of the Exponential Distribution.**

Let T be exponential with parameter λ. Then $P(T < t) = F(t)$ is continuous and strictly increasing from 0 to 1 on $(0, \infty)$, and so we can solve $F(x_p) = p$ for any $p \in (0, 1)$, that is, solve

$$1 - e^{-\lambda x_p} = p. \tag{6.208}$$

Hence,

$$x_p = -\frac{\ln(1 - p)}{\lambda} \tag{6.209}$$

is the p quantile for any $p \in (0, 1)$.   ♦

Example 6.6.7. **Quantiles of a Binomial.**

Let X be binomial with parameters $n = 3$ and $p = \frac{1}{2}$. Then

$$F(x) = \begin{cases} 0 & \text{if } x < 0 \\ 1/8 & \text{if } 0 \leq x < 1 \\ 1/2 & \text{if } 1 \leq x < 2 \\ 7/8 & \text{if } 2 \leq x < 3 \\ 1 & \text{if } x \geq 3 \end{cases}. \tag{6.210}$$

In this case we have to use the formula $x_p = \min\{x : F(x) \geq p\}$ to find the quantiles. For example, if $p = \frac{1}{4}$, then the $\frac{1}{4}$ quantile $x_{0.25}$ is the lowest x-value such that $F(x) \geq \frac{1}{4}$. As seen from Equation 6.210, $x_{0.25} = 1$, because $F(1) = \frac{1}{2}$, and for $x < 1$ we have $F(x) = 0$ or $\frac{1}{8}$. So $x = 1$ is the lowest value where $F(x)$ jumps above $\frac{1}{4}$. Similarly, $x_p = 1$ for any $p \in \left(\frac{1}{8}, \frac{1}{2}\right]$, and working out the x_p values for all $p \in (0, 1]$, we obtain

$$F^{-1}(p) = x_p = \begin{cases} 0 \text{ if } 0 < p \leq 1/8 \\ 1 \text{ if } 1/8 < p \leq 1/2 \\ 2 \text{ if } 1/2 < p \leq 7/8 \\ 3 \text{ if } 7/8 < p \leq 1 \end{cases}. \tag{6.211}$$

♦

Exercises

Exercise 6.6.1.

Find all medians of the discrete uniform X on the set of increasingly numbered values $x_1, x_2, \ldots, x_n$:

1. For odd n,
2. For even n.

Exercise 6.6.2.

Prove Theorem 6.6.2 for discrete X.

Exercise 6.6.3.

Is the converse of Theorem 6.6.1 true? Prove your answer.

Exercise 6.6.4.

Prove that, for any X, a number m is a median if and only if $P(X \geq m) \geq \frac{1}{2}$ and $P(X \leq m) \geq \frac{1}{2}$.

Exercise 6.6.5.

Prove by differentiation that, for continuous X with continuous density $f(x) > 0$ and such that the expected values below exist, with m the median and c not a median, $E(|X - c|) > E(|X - m|)$, that is, $\min_c E(|X - c|)$ occurs only at the median.

Exercise 6.6.6.

Let X be uniform on the interval $(0, 1)$. Find the median of $\frac{1}{X}$.

Exercise 6.6.7.

Prove that for any X the 50th percentile is a median.

Exercise 6.6.8.

Find the quartiles of the first and second grades X and Y of a randomly selected student in Example 6.4.4.

Exercise 6.6.9.

Find and plot the quantile function for an X with density

$$f(x) = \begin{cases} \frac{x+1}{2} & \text{if } -1 < x < 1 \\ 0 & \text{otherwise} \end{cases}. \tag{6.212}$$

Exercise 6.6.10.

Find and plot the quantile function for an X uniform on the union $[0, 1] \cup [2, 3]$ as in Example 6.6.2.

Exercise 6.6.11.

Find and plot the quantile function for the X of Example 5.2.4.

Exercise 6.6.12.

Find and plot the quantile function for a binomial X with $n = 4$ and $p = .3$.

7. Some Special Distributions

7.1 Poisson Random Variables

Poisson random variables[1] are used to model the number of occurrences of certain events that come from a large number of independent sources, such as the number of calls to an office telephone during business hours, the number of atoms decaying in a sample of some radioactive substance, the number of visits to a website, or the number of customers entering a store.

Definition 7.1.1. *Poisson Distribution. A random variable X is Poisson with parameter $\lambda > 0$, if it is discrete with p.f. given by*

$$P(X = k) = \frac{\lambda^k e^{-\lambda}}{k!} \text{ for } k = 0, 1, \ldots . \tag{7.1}$$

The distribution of such an X is called the Poisson distribution with parameter λ.

We can easily check that the probabilities in Equation 7.1 form a distribution:

$$\sum_{k=0}^{\infty} \frac{\lambda^k e^{-\lambda}}{k!} = e^{-\lambda} \sum_{k=0}^{\infty} \frac{\lambda^k}{k!} = e^{-\lambda} e^{\lambda} = 1. \tag{7.2}$$

The histogram of a typical Poisson p.f. is shown in Figure 7.1.

Now where does the Formula 7.1 come from? It arises as the limit of the binomial distribution as $n \to \infty$, while $\lambda = np$ is kept constant, as will be shown below. This fact is the reason why the Poisson distribution is a good model for the kind of phenomena mentioned above. For instance, the number of people who may call an office, say between 1 and 2 PM, is generally a very

---

[1] Named after Simeon D. Poisson (1781–1840), who in 1837 introduced them to model the votes of jurors, although Abraham de Moivre had already considered them about a hundred years earlier.

© Springer International Publishing Switzerland 2016
G. Schay, *Introduction to Probability with Statistical Applications*,
DOI 10.1007/978-3-319-30620-9_7

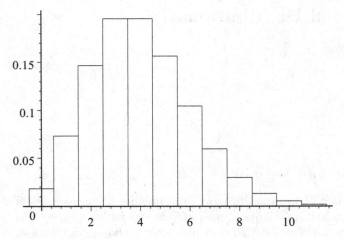

Fig. 7.1. Poisson p.f. for $\lambda = 4$.

large number n, but each one calls with only a very small probability p. If we assume that the calls are independent of each other, the probability that there will be k calls is then given by the binomial distribution. However, we generally do not know n and p, but we can establish the mean number np of calls by observing the phone over several days. (We assume that there is no change in the calling habits of the customers.) Now, when n is large (>100), p is small ($<.01$), and $\lambda = np$ is known, then the binomial probabilities will be very close to their limit as $n \to \infty$, the Poisson distribution. So, here is the theorem:

Theorem 7.1.1. *The Poisson Distribution as the Limit of the Binomial.* *If $n \to \infty$ and $p \to 0$ such that $np = \lambda$ is constant, then*

$$\binom{n}{k} p^k (1-p)^{n-k} \to \frac{\lambda^k e^{-\lambda}}{k!} \text{ for } k = 0, 1, \ldots \ . \tag{7.3}$$

Proof.

$$\binom{n}{k} p^k (1-p)^{n-k} = \frac{n(n-1)\cdots(n-k+1)}{k!} p^k (1-p)^{n-k}$$

$$= \frac{n(n-1)\cdots(n-k+1)}{k! n^k} n^k p^k (1-p)^{n-k}$$

$$= \frac{1}{k!} \frac{n-1}{n} \cdots \frac{n-k+1}{n} (np)^k \left(1 - \frac{np}{n}\right)^{n-k}$$

$$= \frac{1}{k!} \left(1 - \frac{1}{n}\right) \cdots \left(1 - \frac{k-1}{n}\right) \lambda^k \left(1 - \frac{\lambda}{n}\right)^n \left(1 - \frac{\lambda}{n}\right)^{-k}$$

$$\to \frac{1}{k!} \cdot 1 \cdots 1 \cdot \lambda^k \cdot e^{-\lambda} \cdot 1 = \frac{\lambda^k e^{-\lambda}}{k!}. \tag{7.4}$$

Since np is the expected value of the binomial distribution, we expect λ to be the expected value of the Poisson distribution. Similarly, since the variance of the binomial distribution is $npq = np\,(1-p) = np - np^2 = \lambda - \lambda p$, and $p \to 0$ in the proof above, we expect λ to equal the Poisson distribution's variance, too. Indeed:

Theorem 7.1.2. *Expectation and Variance of the Poisson Distribution. If X is Poisson with parameter λ, then*

$$E(X) = Var(X) = \lambda. \tag{7.5}$$

Proof.

$$E(X) = \sum_{k=0}^{\infty} k\frac{\lambda^k e^{-\lambda}}{k!} = \lambda \sum_{k=1}^{\infty} \frac{\lambda^{k-1} e^{-\lambda}}{(k-1)!}. \tag{7.6}$$

If we change from the variable k to $i = k - 1$, then the expression on the right becomes

$$E(X) = \lambda \sum_{i=0}^{\infty} \frac{\lambda^i e^{-\lambda}}{i!} = \lambda \cdot 1 = \lambda. \tag{7.7}$$

To obtain the variance, we first compute $E(X(X-1))$:

$$E(X(X-1)) = \sum_{k=0}^{\infty} k\,(k-1)\,\frac{\lambda^k e^{-\lambda}}{k!} = \lambda^2 \sum_{k=2}^{\infty} \frac{\lambda^{k-2} e^{-\lambda}}{(k-2)!} = \lambda^2 \sum_{i=0}^{\infty} \frac{\lambda^i e^{-\lambda}}{i!} = \lambda^2. \tag{7.8}$$

Hence

$$E(X(X-1)) = E(X^2) - E(X) = E(X^2) - \lambda = \lambda^2, \tag{7.9}$$

and so

$$E(X^2) = \lambda^2 + \lambda. \tag{7.10}$$

Thus

$$Var(X) = E(X^2) - (E(X))^2 = \lambda^2 + \lambda - \lambda^2 = \lambda. \tag{7.11}$$

∎

Theorem 7.1.3. *Moment Generating Function of the Poisson Distribution. If X is Poisson with parameter λ, then*

$$\psi(t) = \exp\left\{\lambda\left(e^t - 1\right)\right\}. \tag{7.12}$$

Proof.

$$E(e^{tX}) = \sum_{k=0}^{\infty} e^{kt} \frac{\lambda^k e^{-\lambda}}{k!} = e^{-\lambda} \sum_{k=1}^{\infty} \frac{(\lambda e^t)^k}{k!} = \exp\left\{\lambda\left(e^t - 1\right)\right\}. \qquad (7.13)$$

∎

Example 7.1.1. Misprints on a Page.

Suppose a page of a book contains $n = 1000$ characters, each of which is misprinted, independently of the others, with probability $p = 10^{-4}$. Find the probabilities of having a) no misprint, b) exactly one, and c) at least one misprint on the page, both by the binomial formula, exactly, and by the Poisson formula, approximately.

Let X denote the number of misprints. Then

a) by the binomial formula,

$$P(X = 0) = \binom{1000}{0} \left(10^{-4}\right)^0 \left(1 - 10^{-4}\right)^{1000-0} \approx 0.904\,833, \qquad (7.14)$$

and by the Poisson approximation with $\lambda = 1000 \cdot 10^{-4} = 0.1$,

$$P(X = 0) = \frac{0.1^0 e^{-0.1}}{0!} \approx 0.904\,837. \qquad (7.15)$$

b) By the binomial formula,

$$P(X = 1) = \binom{1000}{1} \left(10^{-4}\right)^1 \left(1 - 10^{-4}\right)^{1000-1} \approx .0904\,923, \qquad (7.16)$$

and by the Poisson approximation,

$$P(X = 1) = \frac{0.1^1 e^{-0.1}}{1!} \approx 0.0904\,837. \qquad (7.17)$$

c) By the binomial formula,

$$P(X \geq 1) = 1 - P(X = 0) \approx 1 - 0.904\,833 = .095\,167, \qquad (7.18)$$

and by the Poisson approximation,

$$P(X \geq 1) = 1 - P(X = 0) \approx 1 - 0.904\,837 = .095\,163. \qquad (7.19)$$

While the above approximations are interesting, they are not really necessary. Actually, even the binomial model is only an approximation, because misprints sometimes occur in clumps and may not be quite independent, their probabilities may vary, and not all pages have exactly 1000 characters. Also, it is difficult to measure the probability of a character being misprinted but relatively easy to establish the mean number of misprints per page. If we do not know n and p separately, but only the mean $\lambda = np$, then we cannot use the binomial distribution, but the Poisson distribution is still applicable. ♦

Example 7.1.2. Diners at a Restaurant.

Suppose that a restaurant has on the average 50 diners per night. What is the probability that on a certain night 40 or fewer will show up?

Suppose that the diners come from a large pool of potential customers, who show up independently with the same small probability for each. Then their number X may be taken to be Poisson, and so

$$P\left(X \leq 40\right) = \sum_{k=0}^{40} \frac{50^k e^{-50}}{k!} \approx 0.086. \tag{7.20}$$

On the other hand, if we assume that the customers come in independent pairs rather than singly, and denote the number of pairs by Y, then the corresponding probability is

$$P\left(Y \leq 20\right) = \sum_{k=0}^{20} \frac{25^k e^{-25}}{k!} \approx 0.185. \tag{7.21}$$

These numbers show that, in order to estimate the probability of a slow night, it is not enough to know how many people show up on average, but we need to know the sizes of the groups that decide independently from one another, whether to come or not.   ◆

. An important property of Poisson r.v.'s is contained in the following theorem:

Theorem 7.1.4. *The Sum of Independent Poisson Variables is Poisson.* *If X_1 and X_2 are independent Poisson r.v.'s with parameters λ_1 and λ_2, respectively, then $X_1 + X_2$ is Poisson with parameter $\lambda_1 + \lambda_2$.*

Proof. The joint distribution of X_1 and X_2 is given by

$$p_{ik} = P\left(X_1 = i, X_2 = k\right) = \frac{\lambda_1^i \lambda_2^k e^{-(\lambda_1+\lambda_2)}}{i!k!} \text{ for } i, k = 0, 1, \ldots, \tag{7.22}$$

and so

$$P\left(X_1 + X_2 = n\right) = \sum_{i=0}^{n} P\left(X_1 = i, X_2 = n - i\right)$$

$$= e^{-(\lambda_1+\lambda_2)} \sum_{i=0}^{n} \frac{\lambda_1^i \lambda_2^{n-i}}{i!\,(n-i)!} = \frac{e^{-(\lambda_1+\lambda_2)}}{n!} \sum_{i=0}^{n} \frac{n!}{i!\,(n-i)!} \lambda_1^i \lambda_2^{n-i}$$

$$= \frac{e^{-(\lambda_1+\lambda_2)}}{n!} \left(\lambda_1 + \lambda_2\right)^n \text{ for } n = 0, 1, \ldots. \tag{7.23}$$

∎

In most applications, we are interested not just in one Poisson r.v. but in a whole family of Poisson r.v.'s. For instance, in the foregoing examples, we may ask for the probabilities of the number of misprints on several pages and for the probabilities of the number of diners in a week or a month.

In general, a family of random variables $X(t)$ depending on a parameter t is called a *stochastic or random process.* . The parameter t is time in most applications, but not always, as in the generalization of Example 7.1.1, it would stand for the number of pages. Here we are concerned with the particular stochastic process called the Poisson process:

Definition 7.1.2. *Poisson Process.* *A family of random variables $X(t)$ depending on a parameter t is called a Poisson process with rate λ, for any $\lambda > 0$, if $X(t)$, the number of occurrences of some kind in any interval of length t, has a Poisson distribution with parameter λt for any $t > 0$, that is,*

$$P(X(t) = k) = \frac{(\lambda t)^k e^{-\lambda t}}{k!} \text{ for any } t > 0 \text{ and } k = 0, 1, \dots , \qquad (7.24)$$

and the numbers of occurrences in nonoverlapping time intervals are independent of each other.

Example 7.1.3. Misprints on Several Pages.

Suppose the pages of a book contain misprinted characters, independently of each other, with a rate of $\lambda = 0.1$ misprints per page. Assume that the numbers $X(t)$ of misprints on any t pages constitute a Poisson process. Find the probabilities of having a) no misprint on the first three pages, b) at least two misprints on the first two pages, and c) at least two misprints on the first two pages, if we know that there is at least one misprint on the first page.

a) In this case $t = 3$ and $\lambda t = 0.3$. Thus,

$$P(X(3) = 0) = \frac{0.3^0 e^{-0.3}}{0!} \approx 0.74. \qquad (7.25)$$

b) Now $t = 2$ and $\lambda t = 0.2$, and so

$$P(X(2) \geq 2) = 1 - [P(X(2) = 0) + P(X(2) = 1)]$$

$$= 1 - \left[\frac{0.2^0 e^{-0.2}}{0!} + \frac{0.2^1 e^{-0.2}}{1!}\right] \approx 0.0175. \qquad (7.26)$$

c) Let X_1 denote the number of misprints on the first page and X_2 the number of misprints on the second page. Then X_1 and X_2 are independent Poisson with parameter 0.1 both, and $X(2) = X_1 + X_2$. Hence,

$$P(X(2) \geq 2 | X_1 \geq 1) = \frac{P(X_1 \geq 1, X(2) \geq 2)}{P(X_1 \geq 1)}$$

$$= \frac{P(X_1 \geq 2) + P(X_1 = 1, X_2 \geq 1)}{P(X_1 \geq 1)} \qquad (7.27)$$

$$\approx \frac{[1 - e^{-0.1}(1 + 0.1)] + 0.1 e^{-0.1}[1 - e^{-0.1}]}{1 - e^{-0.1}} \approx 0.14.$$

♦

Poisson processes have three important properties given in the theorems below. The first of these is an immediate consequence of Definition 7.1.2 and Theorem 7.1.4 and says that the number of occurrences in an interval depends only on the length of the interval and not on where the interval begins. This property is called stationarity.

Theorem 7.1.5. *Poisson Processes are Stationary.* *For any $s, t > 0$,*

$$X(s+t) - X(s) = X(t). \qquad (7.28)$$

The next theorem expresses the independence assumption of Definition 7.1.2 with conditional probabilities. It says that the process is "memoryless," that is, the probability of k occurrences in an interval is the same regardless of how many went before.

Theorem 7.1.6. *Poisson Processes are Memoryless.* *For any $s, t > 0$ and $i, k = 0, 1, \dots$,*

$$P(X(s+t) = i + k | X(s) = i) = P(X(t) = k). \qquad (7.29)$$

Proof. For any $s, t > 0$ and $i, k = 0, 1, \dots$,

$$
\begin{aligned}
\mathrm{P}(X(s+t) = i + k | X(s) = i) &= \frac{\mathrm{P}(X(s+t) = i + k, X(s) = i)}{\mathrm{P}(X(s) = i)} \\
&= \frac{\mathrm{P}(X(s+t) - X(s) = k, X(s) = i)}{\mathrm{P}(X(s) = i)} \\
&= \frac{\mathrm{P}(X(t) = k, X(s) = i)}{\mathrm{P}(X(s) = i)} \\
&= \frac{\mathrm{P}(X(t) = k)\mathrm{P}(X(s) = i)}{\mathrm{P}(X(s) = i)} \\
&= \mathrm{P}(X(t) = k). \qquad (7.30)
\end{aligned}
$$

∎

The next theorem shows that in a Poisson process, the *"waiting time"* for an occurrence and the *"interarrival time,"* (the time between any two consecutive occurrences) both have the same exponential distribution with parameter λ. (In this context, it is customary to regard the parameter t to be time and the occurrences to be arrivals.)

Theorem 7.1.7. *Waiting Time and Interarrival Time in Poisson Processes.*

1. *Let $s \geq 0$ be any instant and let $T > 0$ denote the length of time we have to wait for the first arrival after s, that is, let this arrival occur at the instant $s + T$. Then T is an exponential random variable with parameter λ.*

2. *Assume that an arrival occurs at an instant $s \geq 0$ and let $\overline{T} \geq 0$ denote the time between this arrival and the next one, that is, let the next arrival occur at the instant $s + \overline{T}$. Then $\overline{T}$ is an exponential random variable with parameter λ.*

Proof. 1. Clearly, for any $t > 0$, the waiting time T is $\leq t$, if and only if there is at least one arrival in the time interval $(s, s + t]$. Thus,

$$P\left(T \leq t\right) = P\left(X\left(s + t\right) - X\left(s\right) > 0\right) = P\left(X\left(t\right) > 0\right) = 1 - e^{-\lambda t},$$

$$(7.31)$$

which, together with $P(T \leq t) = 0$ for $t \leq 0$, shows that T has the distribution function of an exponential random variable with parameter λ.

2. Instead of assuming that an arrival occurs at the instant s, we assume that it occurs in the time interval $[s - \Delta s, s]$ and let $\Delta s \to 0$. Then, similarly to the first part, for any $t > 0$,

$$P\left(\overline{T} \leq t\right) = \lim_{\Delta s \to 0} P\left(X\left(s + t\right) - X\left(s\right) > 0 | X\left(s\right) - X\left(s - \Delta s\right) = 1\right)$$

$$= P\left(X\left(s + t\right) - X\left(s\right) > 0\right) = P\left(X\left(t\right) > 0\right) = 1 - e^{-\lambda t},$$

$$(7.32)$$

and $P\left(\overline{T} \leq t\right) = 0$ for $t \leq 0$. Thus $\overline{T}$, too, has the distribution function of an exponential random variable with parameter λ. ∎

Theorem 7.1.7 has, by Example 6.1.5, the following corollary:

Corollary 7.1.1. *If in a Poisson process the arrival rate, that is, the mean number of arrivals per unit time, is λ, then the mean interarrival time is $\frac{1}{\lambda}$.*

The converse of Theorem 7.1.7 is also true, that is, if we have a stream of random arrivals such that the waiting time for the first one and the successive interarrival times are independent exponential random variables with parameter λ, then the number of arrivals $X(t)$, during time intervals of length t, form a Poisson process with rate λ. We omit the proof.

Exercises

In all the exercises below, assume a Poisson model.

Exercise 7.1.1.

Customers enter a store at a mean rate of 1 per minute. Find the probabilities that:

1. More than one will enter in the first minute,
2. More than two will enter in the first two minutes,
3. More than one will enter in each of the first two minutes,

4. Two will enter in the first minute and two in the second minute if four have entered in the first two minutes.

Exercise 7.1.2.

A textile plant turns out cloth that has on average 1 defect per 20 square yards. Assume that 2 square yards of this material is in a pair of pants and 3 square yards in a coat:

1. About what percentage of the pants will be defective?
2. About what percentage of the coats will be defective?
3. Explain the cause of the difference between the two preceding results.

Exercise 7.1.3.

In each gram of a certain radioactive substance, two atoms will decay on average per minute. Find the probabilities that:

1. In one gram, more than two atoms will decay in one minute,
2. In two grams, more than four atoms will decay in one minute,
3. In one gram, more than four atoms will decay in one minute,
4. In one gram, the time between two consecutive decays is more than a minute,
5. In two grams, the time between two consecutive decays is more than half a minute.

Exercise 7.1.4.

In a certain city, there are 12 murders on average per year. Assume that they are equally likely at any time and independent of each other. Approximate the length of each month as 1/12 of a year. Find the probabilities that:

1. There will be no murders in January and February,
2. There will be none in exactly two, not necessarily consecutive, months of the year,
3. There will be none in at most two, not necessarily consecutive, months of the year,
4. There will be none in February if there was none in January.

Exercise 7.1.5.

Show that in a Poisson process with rate λ, the probability of an even number of arrivals in any interval of length t is $\left(1 + e^{-2\lambda t}\right)/2$ and of an odd number of arrivals is $\left(1 - e^{-2\lambda t}\right)/2$. (Hint: First find P(even)− P(odd).)

Exercise 7.1.6.

Suppose that a Poisson stream $X(t)$ of arrivals with rate λ is split into two streams A and B, so that each arrival goes to stream A with probability p and to stream B with probability $q = 1 - p$ independently of one another. Prove that the new streams are also Poisson processes, with rates $p\lambda$ and $q\lambda$, respectively. (Hint: First find a formula for the joint probability $P(X_A(t) = m, X_B(t) = n)$ in terms of the original Poisson process $X(t)$ and the binomial p.f.)

Exercise 7.1.7.

Show that in a Poisson process, any two distinct interarrival times are independent of each other.

Exercise 7.1.8.

Show that for a Poisson r.v. X with parameter λ, $\max_k P(X = k)$ occurs exactly at $\lambda - 1$ and at λ if λ is an integer and only at $[\lambda]$ otherwise. (Here $[\lambda]$ denotes the greatest integer $\leq \lambda$.) *Hint*: First show that $P(X = k) = \frac{\lambda}{k}P(X = k - 1)$ for any $k > 0$.

7.2 Normal Random Variables

Definition 7.2.1. *Normal Distribution. A random variable X is normal or normally distributed with parameters μ and σ^2, (abbreviated $N(\mu, \sigma^2)$), if it is continuous with p.d.f.*

$$f(x) = \frac{1}{\sqrt{2\pi}\sigma}e^{-(x-\mu)^2/2\sigma^2} \quad for \ -\infty < x < \infty. \tag{7.33}$$

The distribution of such an X is called the normal distribution with parameters μ and σ^2 and the above p.d.f. the normal density function with parameters μ and σ^2.

The graph of such a function, for arbitrarily fixed μ and σ^2, is shown in Figure 7.2. It is symmetric about $x = \mu$, and its inflection points are at $x = \mu \pm \sigma$.

This distribution was discovered by Abraham de Moivre around 1730 as the limiting distribution of the (suitably scaled) binomial distribution as $n \to \infty$. Nevertheless it used to be referred to as the Gaussian distribution, because many people learned about it from the much later works of Gauss. The name "normal" comes from the fact that it occurs in so many applications that, with some exaggeration, it may seem abnormal if we encounter any other distribution. The reason for its frequent occurrence is the so-called Central Limit Theorem (Section 7.3), which says, roughly speaking that, under very general conditions, the sum and the average of n arbitrary independent random variables are asymptotically normal for large n. Thus, any physical

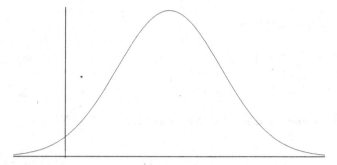

Fig. 7.2. The p.d.f. of a typical normal distribution.

quantity that arises as the sum of a large number of independent random influences will have an approximately normal distribution. For instance, the height and the weight of a more or less homogeneous population are approximately normally distributed. Other examples of normal random variables are the x-coordinates of shots aimed at a target, the repeated measurements of almost any kind of laboratory data, the blood pressure and the temperature of people, the grades on the SAT, and so on.

We are going to list several properties of the normal distribution as theorems, beginning with one that shows that Definition 7.2.1 does indeed define a probability density.

Theorem 7.2.1. *The Area Under the Normal p.d.f. is 1.*

$$\int_{-\infty}^{\infty} \frac{1}{\sqrt{2\pi}\sigma} e^{-(t-\mu)^2/2\sigma^2} dt = 1. \tag{7.34}$$

Proof. This p.d.f. is one of those functions whose indefinite integral cannot be expressed in terms of common elementary functions, but the definite integral above can be evaluated by a special trick: First, we substitute $x = (t - \mu)/\sigma$. Then the integral becomes

$$I = \frac{1}{\sqrt{2\pi}} \int_{-\infty}^{\infty} e^{-x^2/2} dx. \tag{7.35}$$

Now, we write y for the variable of integration and multiply the two forms of I, obtaining

$$I^2 = \frac{1}{\sqrt{2\pi}} \int_{-\infty}^{\infty} e^{-x^2/2} dx \frac{1}{\sqrt{2\pi}} \int_{-\infty}^{\infty} e^{-y^2/2} dy = \frac{1}{2\pi} \int_{-\infty}^{\infty} \int_{-\infty}^{\infty} e^{-(x^2+y^2)/2} dx dy. \tag{7.36}$$

Changing to polar coordinates, we get

$$I^2 = \frac{1}{2\pi} \int_0^{2\pi} \int_0^{\infty} e^{-r^2/2} r dr d\theta = \frac{1}{2\pi} \int_0^{2\pi} d\theta \int_0^{\infty} e^{-r^2/2} r dr = \int_0^{\infty} e^{-r^2/2} r dr. \tag{7.37}$$

Substituting $u = r^2/2, du = rdr$ yields

$$I^2 = \int_0^\infty e^{-u} du = -e^{-u}|_0^\infty = 1, \tag{7.38}$$

and so, since I is nonnegative (why?), $I = 1$. ∎

Theorem 7.2.2. _The Expectation of a Normal r.v._. If X is $N(\mu, \sigma^2)$, then

$$E(X) = \mu. \tag{7.39}$$

Proof. The p.d.f. in Definition 7.2.1 is symmetric about $x = \mu$, and so Theorem 6.1.1 yields Equation 7.39. ∎

Theorem 7.2.3. _The Variance of a Normal r.v._ If X is $N(\mu, \sigma^2)$, then

$$Var(X) = \sigma^2. \tag{7.40}$$

Proof. By definition,

$$Var(X) = E((X - \mu)^2) = \int_{-\infty}^\infty (x - \mu)^2 \frac{1}{\sqrt{2\pi}\sigma} e^{-(x-\mu)^2/2\sigma^2} dx. \tag{7.41}$$

Substituting $u = (x - \mu)/\sigma$ and integrating by parts, we get

$$Var(X) = \frac{\sigma^2}{\sqrt{2\pi}} \int_{-\infty}^\infty u^2 e^{-u^2/2} du = \frac{\sigma^2}{\sqrt{2\pi}} \int_{-\infty}^\infty u \cdot u e^{-u^2/2} du \tag{7.42}$$

$$= \frac{\sigma^2}{\sqrt{2\pi}} \left(-\left[u e^{-u^2/2} \right]_{-\infty}^\infty + \int_{-\infty}^\infty e^{-u^2/2} du \right) = \frac{\sigma^2}{\sqrt{2\pi}} \left(0 + \sqrt{2\pi} \right) = \sigma^2.$$

∎

Theorem 7.2.4. _A Linear Function of a Normal Random Variable is Normal._ If X is $N(\mu, \sigma^2)$, then, for any constants $a \neq 0$ and b, $Y = aX + b$ is normal with $E(Y) = a\mu + b$ and $Var(Y) = (a\sigma)^2$.

Proof. Assume $a > 0$. (The proof of the opposite case is left as an exercise.) Then the d.f. of Y can be computed as

$$F_Y(y) = P(Y \leq y) = P(aX + b \leq y) = P(X \leq \frac{y - b}{a})$$

$$= \frac{1}{\sqrt{2\pi}\sigma} \int_{-\infty}^{(y-b)/a} e^{-(x-\mu)^2/2\sigma^2} dx \tag{7.43}$$

and from here the chain rule and the fundamental theorem of calculus give its p.d.f. as

$$f_Y(y) = F_Y'(y) = \frac{1}{\sqrt{2\pi}\sigma}\frac{d}{dy}\left(\frac{y-b}{a}\right) \cdot e^{-(((y-b)/a)-\mu)^2/2\sigma^2}$$

$$= \frac{1}{\sqrt{2\pi}a\sigma}e^{-(y-(a\mu+b))^2/2(a\sigma)^2}. \tag{7.44}$$

■

A comparison with Definition 7.2.1 shows that this function is the p.d.f. of a normal r.v. with $a\mu + b$ in place of μ and $(a\sigma)^2$ in place of σ^2.

Corollary 7.2.1. *Standardization of a Normal Random Variable.* *If* X *is* $N(\mu, \sigma^2)$, *then* $Z = \frac{X-\mu}{\sigma}$ *is* $N(0,1)$.

Proof. Apply 7.2.4 with $a = \frac{1}{\sigma}$ and $b = \frac{-\mu}{\sigma}$.   ■

Definition 7.2.2. *Standard Normal Distribution.* *The distribution* $N(0,$ 1) *is called the standard normal distribution, and its p.d.f. and d.f. are denoted by* φ *and* Φ, *respectively, that is,*

$$\varphi(z) = \frac{1}{\sqrt{2\pi}}e^{-z^2/2} \quad for \quad -\infty < z < \infty, \tag{7.45}$$

and

$$\Phi(z) = \frac{1}{\sqrt{2\pi}}\int_{-\infty}^{z}e^{-t^2/2}dt \quad for \quad -\infty < z < \infty. \tag{7.46}$$

Corollary 7.2.2. *Arbitrary Normal d.f. in Terms of the Standard Normal d.f.* *If* X *is* $N(\mu, \sigma^2)$, *then* $F_X(x) = \Phi\left(\frac{x-\mu}{\sigma}\right)$.

Proof. $F_X(x) = P(X \leq x) = P\left(\frac{X-\mu}{\sigma} \leq \frac{x-\mu}{\sigma}\right) = P\left(Z \leq \frac{x-\mu}{\sigma}\right) = \Phi\left(\frac{x-\mu}{\sigma}\right).$
■

As mentioned before, the p.d.f. of a normal r.v. cannot be integrated in terms of the common elementary functions, and therefore the probabilities of X falling in various intervals are obtained from tables or by computer. Now, it would be overwhelming to construct tables for all μ and σ values required in applications, but Corollary 7.2.2 makes this unnecessary. It enables us to compute the probabilities for any $N(\mu, \sigma^2)$ r.v. X from the single table of the standard normal distribution function, which is given (with minor variations) in most probability or statistics books, including this one. The next examples illustrate the procedure.

Example 7.2.1. Height Distribution of Men.

Assume that the height X, in inches, of a randomly selected man in a certain population is normally distributed[2] with $\mu = 69$ and $\sigma = 2.6$. Find:

---

[2] Any such assumption is always just an approximation that is usually valid only within three or four standard deviations from the mean. But that is the range where almost all of the probability of the normal distribution falls, and although theoretically the tails of the normal distribution are infinite, $\varphi(z)$ is so small for $|z| > 4$, that as a practical matter we can ignore the fact that it gives nonzero probabilities to impossible events such as people having negative heights or heights over ten feet.

1. $P(X < 72)$,
2. $P(X > 72)$,
3. $P(X < 66)$,
4. $P(|X - \mu| < 3)$.

In each case, we transform the inequalities so that X will be standardized and use the Φ-table to find the required probabilities. However, the table gives $\Phi(z)$ only for $z \geq 0$, and for $z < 0$ we need to make use of the symmetry of the normal distribution, which implies that, for any z, $P(Z < -z) = P(Z > z)$,.that is, $\Phi(-z) = 1 - \Phi(z)$ (See Figure 7.3.) Thus,

1. $P(X < 72) = P(\frac{X-\mu}{\sigma} < \frac{72-69}{2.6}) \approx P(Z < 1.15) = \Phi(1.15) \approx .875$.
2. $P(X > 72) = P(\frac{X-\mu}{\sigma} > \frac{72-69}{2.6}) \approx P(Z > 1.15) = 1- P(Z \leq 1.15) = 1 - \Phi(1.15) \approx 1 - .875 = .125$.
3. $P(X < 66) = P(\frac{X-\mu}{\sigma} < \frac{66-69}{2.6}) \approx P(Z < -1.15) = P(Z > 1.15) = .125$.
4. $P(|X-\mu| < 3) = P(\left|\frac{X-\mu}{\sigma}\right| < \frac{3}{2.6}) \approx P(|Z| < 1.15) = 1-[P(Z < -1.15) + P(Z > 1.15)] = 2\Phi(1.15) - 1 \approx .75$.   ◆

Example 7.2.2. Percentiles of Normal Test Scores.

Assume that the math scores on the SAT at a certain school were normally distributed with $\mu = 560$ and $\sigma = 50$. Find the quartiles and the 90th percentile of this distribution.

For the third quartile, we have to find the score x for which $P(X < x) = .75$ or, equivalently, $P(\frac{X-\mu}{\sigma} < \frac{x-560}{50}) = .75$. The quantity $z = \frac{x-560}{50}$ is called the *z-score* or the value of x *in standard units*, and, by Corollary 7.2.2, we thus first need to find the z-score for which $\Phi(z) = .75$, or $z = \Phi^{-1}(.75)$. In the body of the Φ-table look for .75, and for the corresponding z-value find .675. Solving $z = \frac{x-560}{50}$ for x, we obtain $x = 50z+560 = 50\cdot.675+560 \approx 594$. Hence 75% of the SAT scores were under 594.

For the first quartile, we have to find the score x for which $P(X < x) = .25$ or, equivalently, $\Phi(z) = .25$. However, no $p = \Phi(z)$ value less than .5 is listed in the table. The corresponding z would be negative, and we use the

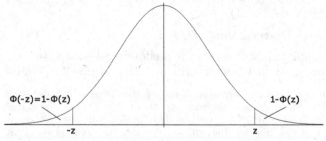

Fig. 7.3. The area of any left tail of φ equals the area of the corresponding right tail, that is, $\Phi(-z) = 1 - \Phi(z)$.

symmetry of φ to find instead the $|z|$ for the corresponding right tail that has area .25. Thus $\Phi(z) = .25$ is equivalent to $1 - \Phi(|z|) = .25$ or $\Phi(|z|) = .75$, and the table gives $|z| = .675$. Hence $z = -.675$ and $x = 50z + 560 = 50 \cdot (-.675) + 560 \approx 526$.

The 90th percentile can be computed from $P(X < x) = .90$ or, equivalently, from $\Phi(z) = .90$. The table shows $z \approx 1.282$, and so $x = 50z + 560 = 50 \cdot 1.282 + 560 \approx 624$.   ◆

Theorem 7.2.5. The Moment Generating Function of the Normal Distribution. *If X is $N(\mu, \sigma^2)$, then*

$$\psi(t) = e^{\mu t + \sigma^2 t^2/2} \quad for \quad -\infty < t < \infty. \tag{7.47}$$

Proof. First compute the moment generating function of a standard normal r.v. Z:
By definition,

$$\psi_Z(t) = E(e^{tZ}) = \frac{1}{\sqrt{2\pi}} \int_{-\infty}^{\infty} e^{tz - z^2/2} dz = \frac{1}{\sqrt{2\pi}} \int_{-\infty}^{\infty} e^{(t^2 - (z-t)^2)/2} dz$$

$$= e^{t^2/2} \frac{1}{\sqrt{2\pi}} \int_{-\infty}^{\infty} e^{-(z-t)^2/2} dz = e^{t^2/2} \quad for \quad -\infty < t < \infty. \tag{7.48}$$

Now $X = \sigma Z + \mu$ is $N(\mu, \sigma^2)$, and

$$\psi_X(t) = E(e^{t(\sigma Z + \mu)}) = e^{\mu t} E(e^{\sigma t Z}) = e^{\mu t} \psi_Z(\sigma t)$$

$$= e^{\mu t + \sigma^2 t^2/2} \quad for \quad -\infty < t < \infty. \tag{7.49}$$

∎

Theorem 7.2.6. Any Nonzero Linear Combination of Independent Normal Random Variables is Normal. *Let X_i be independent and $N(\mu_i, \sigma_i^2)$ random variables for $i = 1, \ldots, n$, and let $X = \sum a_i X_i$ with the a_i arbitrary constants, not all zero. Then X is $N(\mu, \sigma^2)$, with $\mu = \sum a_i \mu_i$, and $\sigma^2 = \sum (a_i \sigma_i)^2$.*

Proof. Let ψ_i denote the moment generating function of X_i. Then, by Theorem 6.3.2 and Equation 7.47,

$$\psi_X(t) = \prod \psi_i(a_i t) = \prod e^{\mu_i a_i t + \sigma_i^2 (a_i t)^2/2} = e^{\sum (\mu_i a_i t + \sigma_i^2 (a_i t)^2/2)}$$

$$= e^{(\sum a_i \mu_i) t + (\sum (a_i \sigma_i)^2) t^2/2}. \tag{7.50}$$

Comparing this expression with Equation 7.47 and using the uniqueness of the m.g.f., we obtain the result of the theorem.   ∎

Definition 7.2.3. Random Sample and Sample Mean. *n independent and identically distributed (abbreviated: i.i.d.) random variables $X_1, \ldots, X_n$ are said to form a random sample of size n from their common distribution and $\overline{X}_n = \frac{1}{n} \sum X_i$ is called the sample mean.*

Corollary 7.2.3. *Distribution of the Sample Mean.* Let X_i be i.i.d. $N(\mu, \sigma^2)$ *random variables for* $i = 1, \ldots, n$. *Then the sample mean is* $N(\mu, \sigma^2/n)$.

Proof. Set $a_i = \frac{1}{n}$, $\mu_i = \mu$ and $\sigma_i = \sigma$ in Theorem 7.2.6 for all i.   ∎

Example 7.2.3. *Heights of Men and Women.*

Assume that the height X, in inches, of a randomly selected woman in a certain population is normally distributed with $\mu_X = 66$ and $\sigma_X = 2.6$ and the height Y, in inches, of a randomly selected man is normally distributed with $\mu_Y = 69$ and $\sigma_Y = 2.6$. Find the probability that a randomly selected woman is taller than an independently randomly selected man.

The probability we want to find is $P(Y - X < 0)$. By Theorem 7.2.6, $Y - X$ is $N(3, 2 \cdot 2.6^2) = N(3, 13.52)$. Thus

$$P(Y - X < 0) = P\left(\frac{Y - X - 3}{\sqrt{13.52}} < \frac{0 - 3}{\sqrt{13.52}}\right) \approx \Phi(-0.816) = 1 - \Phi(0.816)$$

$$\approx 1 - 0.793 = 0.207. \tag{7.51}$$

◆

Exercises

Exercise 7.2.1.

For a standard normal r.v. Z, find:

1. $P(Z < 2)$
2. $P(Z > 2)$
3. $P(Z = 2)$
4. $P(Z < -2)$
5. $P(-2 < Z < 2)$
6. $P(|Z| > 2)$
7. $P(-2 < Z < 1)$
8. z such that $P(z < Z) = .05$
9. z such that $P(-z < Z < z) = .9$
10. z such that $P(-z < Z < z) = .8$

Exercise 7.2.2.

Let X be a normal r.v. with $\mu = 10$ and $\sigma = 2$. Find:

1. $P(X < 11)$
2. $P(X > 11)$
3. $P(X < 9)$
4. $P(9 < X < 11)$
5. $P(9 < X < 12)$

6. x such that $P(x < X) = .05$
7. x such that $P(10 - x < X < 10 + x) = .9$
8. x such that $P(10 - x < X < 10 + x) = .8$

Exercise 7.2.3.

1. Prove that the standard normal density φ has inflection points at $z = \pm 1$.
2. Prove that the general normal density given in Definition 7.2.1 has inflection points at $x = \mu \pm \sigma$.

Exercise 7.2.4.

Assume that the height X, in inches, of a randomly selected woman in a certain adult population is normally distributed with $\mu_X = 66$ and $\sigma_X = 2.6$ and the height Y, in inches, of a randomly selected man is normally distributed with $\mu_Y = 69$ and $\sigma_Y = 2.6$ and half the adult population is male and half female.

1. Find the probability density of the height H of a randomly selected adult from this population and sketch its graph.
2. Find $E(H)$ and $SD(H)$.
3. Find $P(66 < H < 69)$.

Exercise 7.2.5.

Prove Theorem 7.2.4 for $a < 0$

1. By modifying the proof given for $a > 0$,
2. By using the moment generating function.

Exercise 7.2.6.

Assume that the math scores on the SAT at a certain school were normally distributed with unknown μ and σ and two students got their reports back with the following results: 750 (95th percentile) and 500 (46th percentile). Find μ and σ. (Hint: Obtain and solve two simultaneous equations for the two unknowns μ and σ.)

Exercise 7.2.7.

The p.d.f. of a certain distribution is determined to be of the form $ce^{-(x+2)^2/24}$. Find μ, σ and c.

Exercise 7.2.8.

The p.d.f. of a certain distribution is determined to be of the form ce^{-x^2-4x}. Find μ, σ and c.

Exercise 7.2.9.

Assume that the weight X, in ounces, of a randomly selected can of coffee of a certain brand is normally distributed with $\mu = 16$ and $\sigma = .32$. Find the probability that the weights of two independently selected cans from this brand differ by more than $1/2$ oz.

Exercise 7.2.10.

Let Z_n denote the sample mean for a random sample of size n from the standard normal distribution. For $n = 1, 4$ and 16

1. Sketch the p.d.f. of each Z_n in the same coordinate system,
2. Compute the quartiles of each Z_n.

Exercise 7.2.11.

Prove that $\Phi^{-1}(1 - p) = -\Phi^{-1}(p)$ for $0 < p < 1$.

Exercise 7.2.12.

Prove that if X is $N(\mu, \sigma^2)$, then $F_X^{-1}(p) = \mu + \sigma\Phi^{-1}(p)$ for $0 < p < 1$.

7.3 The Central Limit Theorem

Earlier, we saw that the binomial distribution becomes Poisson, if $n \to \infty$ while $p \to 0$ such that $np = \lambda$ remains constant. About a hundred years before Poisson, de Moivre noticed a different approximation to the binomial distribution. He observed and proved that if n is large *with p fixed*, then the binomial probabilities are approximately on a normal curve. An illustration of this fact can be seen in Figure 7.4 and is stated more precisely in the subsequent theorem.[3]

Theorem 7.3.1. *De Moivre-Laplace Limit Theorem. The binomial probabilities $p_k = \binom{n}{k}p^k q^{n-k}$ can be approximated by the corresponding values of the $N(\mu, \sigma^2)$ distribution with matching parameters, that is, with $\mu = np$ and $\sigma^2 = npq$. More precisely,*

$$p_k \sim \frac{1}{\sqrt{2\pi}\sigma}e^{-\frac{(k-\mu)^2}{2\sigma^2}}, \tag{7.52}$$

where the symbol $\sim$ means that the ratio of the two sides tends to 1 as $n \to \infty$.

---

[3] De Moivre discovered the normal curve and proved this theorem only for $p = \frac{1}{2}$. For $p \neq \frac{1}{2}$ he only sketched the result. It was Pierre-Simon de Laplace who around 1812 gave the details for arbitrary p, and outlined a further generalization, the Central Limit Theorem, which we will discuss shortly.

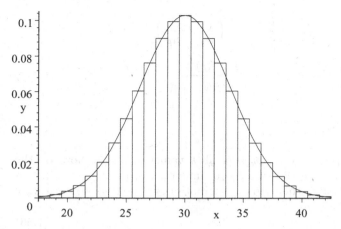

Fig. 7.4. Histogram of the binomial p.f. for $n = 60$ and $p = \frac{1}{2}$, with the approximating normal p.d.f. superimposed.

Proof. We just give an outline, with various technical details omitted.

The proof rests on the linearization $\ln(1 + x) \sim x$ as $x \to 0$, known from calculus.

We want to express p_k for k values near the mean np. Thus, writing $m = [np]$ we have, for $k > m$ (for $k < m$ the argument would be similar; we omit it),

$$\frac{p_k}{p_m} = \frac{n!}{k!\,(n-k)!} \cdot \frac{m!\,(n-m)!}{n!} \cdot \frac{p^k q^{n-k}}{p^m q^{n-m}}$$

$$= \frac{m!}{k!} \cdot \frac{(n-m)!}{(n-k)!} \cdot \left(\frac{p}{q}\right)^{k-m} = \prod_{i=m+1}^{k} \left(\frac{n-i+1}{i} \cdot \frac{p}{q}\right). \tag{7.53}$$

Next, we take logarithms and replace i with $j = i - m$. Then[4]

$$\ln\left(\frac{p_k}{p_m}\right) \sim \sum_{j=1}^{k-m} \ln\left(\frac{n-(m+j)}{m+j} \cdot \frac{p}{q}\right)$$

$$\sim \sum_{j=1}^{k-m} \ln\left(\frac{n-(np+j)}{np+j} \cdot \frac{p}{q}\right)$$

$$\sim \sum_{j=1}^{k-m} \ln\left(\frac{nq-j}{np+j} \cdot \frac{p}{q}\right) = \sum_{j=1}^{k-m} \ln\left(\frac{npq-pj}{npq+qj}\right)$$

$$\sim \sum_{j=1}^{k-m} \ln\left(\frac{1-\frac{j}{nq}}{1+\frac{j}{np}}\right) = \sum_{j=1}^{k-m} \left(\ln\left(1-\frac{j}{nq}\right) - \ln\left(1+\frac{j}{np}\right)\right)$$

---

[4] If $n \to \infty$ with p and k fixed, then $n - i + 1 \sim n - i$ for all $i \leq k$ and $np \sim [np]$, as well.

$$\sim \sum_{j=1}^{k-m} \left(-\frac{j}{nq} - \frac{j}{np} \right) = -\left(\frac{1}{nq} + \frac{1}{np} \right) \sum_{j=1}^{k-m} j$$

$$\sim -\frac{1}{npq} \frac{(k-m)^2}{2} \sim -\frac{(k-np)^2}{2npq}. \tag{7.54}$$

Hence

$$p_k \sim p_m e^{-(k-np)^2/2npq}. \tag{7.55}$$

The sum of the probabilities p_k over all k values equals 1. Now, on the right, we can approximate the sum by an integral (again, if $n \to \infty$), and then from the definition of the normal distribution, we see that we must have

$$p_m \sim \frac{1}{\sqrt{2\pi}\sigma} \tag{7.56}$$

with $\sigma = \sqrt{npq}$, and

$$p_k \sim \frac{1}{\sqrt{2\pi npq}} e^{-(k-np)^2/2npq}. \tag{7.57}$$

∎

If S_n is a binomial r.v. with parameters n and p, then, for integers a and b with $a \leq b$,

$$P(a \leq S_n \leq b) = \sum_{k=a}^{b} p_k. \tag{7.58}$$

Now the sum on the right equals the sum of the areas of the rectangles of the histogram of the p_k values, and it can be approximated by the area under the corresponding normal curve, that is, by the integral of the p.d.f. on the right of Equation 7.57, from the beginning $a - \frac{1}{2}$ of the first rectangle to the end $b + \frac{1}{2}$ of the last rectangle:

$$P(a \leq S_n \leq b) \approx \frac{1}{\sqrt{2\pi}\sigma} \int_{a-\frac{1}{2}}^{b+\frac{1}{2}} e^{-(x-\mu)^2/2\sigma^2} dx, \tag{7.59}$$

where $\mu = np$ and $\sigma = \sqrt{npq}$. Changing variables from x to $z = \frac{x-\mu}{\sigma}$, we can write the expression on the right in terms of the standard normal d.f. and we obtain

Corollary 7.3.1. *Normal Approximation with Continuity Correction.* *For large values of n*

$$P(a \leq S_n \leq b) \approx \Phi\left(\frac{b + \frac{1}{2} - \mu}{\sigma} \right) - \Phi\left(\frac{a - \frac{1}{2} - \mu}{\sigma} \right). \tag{7.60}$$

Remarks.

1. The term $\frac{1}{2}$ in the arguments on the right is called the correction for continuity. It may be ignored for large values of σ; say, for $\sigma \geq 10$, unless $b - a$ is small; say, $b - a \leq 10$.
2. The closer p is to $\frac{1}{2}$, the better the approximation is. For p between .4 and .6, it can be used for $n \geq 25$, but for $p \approx .1$ or .9 it is good for $n \geq 50$ only.

Example 7.3.1. Coin Tossing.

We toss a fair coin $n = 100$ times. Letting S_n denote the number of heads obtained, find the normal approximation to $P(45 \leq S_n \leq 55)$.

Here $p = \frac{1}{2}, \mu = np = 50$ and $\sigma = \sqrt{npq} = 5$. By Equation 7.60,

$$P\left(a \leq S_n \leq b\right) \approx \Phi\left(\frac{55 + \frac{1}{2} - 50}{5}\right) - \Phi\left(\frac{45 - \frac{1}{2} - 50}{5}\right)$$

$$= \Phi(1.1) - \Phi(-1.1) = 2 \cdot \Phi(1.1) - 1 \approx .72867. \qquad (7.61)$$

This result is an excellent approximation to the exact value

$$\sum_{k=45}^{55} \binom{100}{k} \left(\frac{1}{2}\right)^{100} = 0.72875\ldots. \qquad (7.62)$$

It is also interesting to compare the approximation (Equation 7.57) with the binomial value. For instance, for $k = 50$ we have

$$p_{50} = \binom{100}{50} \left(\frac{1}{2}\right)^{100} \approx 0.079589, \qquad (7.63)$$

and from Formula 7.57 we get

$$p_{50} \approx \frac{1}{\sqrt{2\pi \cdot 25}} e^{-0} \approx 0.079788. \qquad (7.64)$$

We can also use Formula 7.60 with $a = b = 50$ to approximate p_{50}. This method yields

$$p_{50} \approx \Phi\left(\frac{50 + \frac{1}{2} - 50}{5}\right) - \Phi\left(\frac{50 - \frac{1}{2} - 50}{5}\right)$$

$$= \Phi(0.1) - \Phi(-0.1) = 2 \cdot \Phi(0.1) - 1 \approx 0.079656, \qquad (7.65)$$

a slightly better approximation then the preceding one.   ♦

Example 7.3.2. Difference of Two Polls.

Suppose that two polling organizations each take a random sample of 200 voters in a state about their preference for a certain candidate, with a yes or no answer. Find an approximate upper bound for the probability that the proportions of yes answers in the two polls differ by more than 4%.

Denoting the proportions of yes answers in the two polls by $\overline{X}$ and $\overline{Y}$, we are interested in the probability $P\left(\left|\overline{X} - \overline{Y}\right| > 0.04\right)$, which can be written as $P\left(\left|200\overline{X} - 200\overline{Y}\right| > 8\right)$, where $200\overline{X}$ and $200\overline{Y}$ are i.i.d. binomial random variables with parameters $n = 200$ and an unknown p. The mean of the difference is 0 and the variance is $400p\left(1 - p\right)$. Thus, we can standardize the desired probability and apply Theorem 7.3.1:[5]

$$P\left(\left|\overline{X} - \overline{Y}\right| > 0.04\right) = P\left(\frac{\left|200\overline{X} - 200\overline{Y}\right|}{20\sqrt{p\left(1 - p\right)}} > \frac{8}{20\sqrt{p\left(1 - p\right)}}\right)$$

$$\approx P\left(\left|Z\right| > \frac{2}{5\sqrt{p\left(1 - p\right)}}\right). \tag{7.66}$$

Now, it is easy to show that $p\left(1 - p\right)$ is maximum when $p = \frac{1}{2}$, and so $\frac{2}{5\sqrt{p(1-p)}}$ is minimum then and $\frac{2}{5\sqrt{p(1-p)}} = \frac{4}{5}$. Hence

$$P\left(\left|\overline{X} - \overline{Y}\right| > 0.04\right) \lesssim P\left(\left|Z\right| > \frac{4}{5}\right) = 2\left(1 - \Phi\left(0.8\right)\right) \approx 0.424. \tag{7.67}$$

Thus, there is a rather substantial chance that the two polls will differ by more than 4 percentage points.   ◆

As mentioned in Footnote 3 on page 246, Laplace discovered a very important generalization of Theorem 7.3.1, the first version of which was, however, proved only in 1901 by A. Liapounov. This generalization is based on the decomposition of a binomial random variable into a sum of i.i.d. Bernoulli random variables as $S_n = \sum_{i=1}^{n} X_i$ where $X_i = 1$ if the ith trial results in success and 0 otherwise. Now, when we standardize S_n, then we divide by $\sqrt{npq}$ and so, as $n \to \infty$, we have an increasing number of smaller and smaller terms. What Laplace noticed was that the limiting distribution is still normal in many cases even if the X_i are other than Bernoulli random variables. This fact has been proved under various conditions on the X_i (they don't even have to be i.i.d.) and is known as the Central Limit Theorem (CLT). We present a version from 1922, due to J. W. Lindeberg.

Theorem 7.3.2. The Central Limit Theorem. *For any positive integer n, let $X_1, X_2, \ldots, X_n$ be i.i.d. random variables with mean μ and standard deviation σ and let S_n^* denote the standardization of their sum, that is, let*

$$S_n^* = \frac{1}{\sqrt{n}\sigma}\left(\sum_{i=1}^{n} X_i - n\mu\right). \tag{7.68}$$

---

[5] By 7.3.1 $200\overline{X}$ and $200\overline{Y}$ are both approximately normal and they are also independent, hence their difference is also approximately normal.

Then, for any real x,

$$\lim_{n \to \infty} P\left(S_n^* < x\right) = \Phi\left(x\right). \tag{7.69}$$

Proof. Again, we just give an outline of the proof and omit some difficult technical details.

We are going to use moment generating functions to deal with the distribution of the sum in the theorem because, as we know, the m.g.f. of a sum of independent r.v.'s is simply the product of the m.g.f.'s of the terms. Now the assumption of the existence of μ and σ does not guarantee the existence of the m.g.f. of X_i and, in general, this problem is handled by truncating the X_i, but we skip this step and assume the existence of the m.g.f. of X_i or, equivalently, the existence of the m.g.f. ψ of the standardization $X_i^* = (X_i - \mu)/\sigma$.

Then

$$S_n^* = \frac{1}{\sqrt{n}\sigma} \sum_{i=1}^{n} (X_i - \mu) = \frac{1}{\sqrt{n}} \sum_{i=1}^{n} X_i^*, \tag{7.70}$$

and so the m.g.f. ψ_n of S_n^* is given by

$$\psi_n\left(t\right) = \left[\psi\left(\frac{t}{\sqrt{n}}\right)\right]^n. \tag{7.71}$$

Now, $\psi(0) = 1$, $\psi'(0) = E(X_i^*) = 0$ and $\psi''(0) = E\left((X_i^*)^2\right) = 1$. Hence Taylor's Formula gives

$$\psi\left(t\right) = 1 + \frac{1}{2}t^2 + \frac{\psi'''\left(c\right)}{3!}t^3, \tag{7.72}$$

where c is some number between 0 and t. From here,

$$\psi_n\left(t\right) = \left[1 + \frac{1}{2}\frac{t^2}{n} + \frac{\psi'''\left(c\right)}{3!}\left(\frac{t}{\sqrt{n}}\right)^3\right]^n, \tag{7.73}$$

and with some calculus it can be shown that

$$\lim_{n \to \infty} \psi_n\left(t\right) = e^{t^2/2}. \tag{7.74}$$

The expression on the right is the m.g.f. of the standard normal distribution, and so the limiting distribution of S_n^*, as $n \to \infty$, is the standard normal distribution. ∎

In statistical applications, it is often the mean of a random sample (see Definition 7.2.3) rather than the sum that we need, and fortunately, the distribution of the sample mean also approaches a normal distribution:

Corollary 7.3.2. Central Limit Theorem for the Sample Mean. *Let* $X_1, X_2, \ldots, X_n$ *be as in the theorem above, let* $\overline{X}_n = \frac{1}{n} \sum_{i=1}^{n} X_i$ *denote their average and let*

$$\overline{X}_n^* = \frac{\overline{X}_n - \mu}{\sigma / \sqrt{n}} \tag{7.75}$$

be the standardization of $\overline{X}_n$*. Then, for any real* x*,*

$$\lim_{n \to \infty} P\left(\overline{X}_n^* < x\right) = \Phi(x). \tag{7.76}$$

$$\overline{X}_n^* = \frac{\sqrt{n}}{\sigma}\left(\frac{1}{n}\sum_{i=1}^{n} X_i - \mu\right) = \frac{\sqrt{n}}{\sigma n}\left(\sum_{i=1}^{n} X_i - n\mu\right) = S_n^*. \tag{7.77}$$

Example 7.3.3. Total Weight of People in a Sample.

Assume that the weight of the adults in a population has mean $\mu = 150$ pounds and s.d. $\sigma = 30$ pounds. Find (approximately) the probability that the total weight of a random sample of 36 such people exceeds 5700 pounds.

The weight of any individual is not a normal r.v. but a mixture (that is, the p.d.f. is a weighted average) of two, approximately normal, random variables: one for the women and one for the men. This fact is, however, immaterial because, by the CLT, the total weight W is approximately normal with mean $n\mu = 36 \cdot 150 = 5400$ and $SD = \sqrt{n}\sigma = \sqrt{36} \cdot 30 = 180$. Thus,

$$P(W > 5700) = P\left(\frac{W - 5400}{180} > \frac{5700 - 5400}{180}\right) \tag{7.78}$$

$$\approx P(Z > 1.667) = 1 - \Phi(1.667) \approx .048. \tag{7.79}$$

◆

The law of large numbers is a straightforward consequence of the CLT whenever the latter holds:

Corollary 7.3.3. Law of Large Numbers. *For any positive integer* n*, let* $X_1, X_2, \ldots, X_n$ *be i.i.d. random variables with mean* μ *and standard deviation* σ*. Then, for any* $\varepsilon > 0$*, their mean* $\overline{X}_n$ *satisfies the relation*

$$\lim_{n \to \infty} P\left(|\overline{X}_n - \mu| < \varepsilon\right) = 1. \tag{7.80}$$

Proof. By Equation 7.75

$$\overline{X}_n - \mu = \frac{\sigma}{\sqrt{n}}\overline{X}_n^*, \tag{7.81}$$

and so

$$\lim_{n \to \infty} P\left(\left|\overline{X}_n - \mu\right| < \varepsilon\right) = \lim_{n \to \infty} P\left(\left|\frac{\sigma}{\sqrt{n}}\overline{X}_n^*\right| < \varepsilon\right)$$

$$= \lim_{n \to \infty} P\left(-\frac{\sqrt{n}}{\sigma}\varepsilon < \overline{X}_n^* < \frac{\sqrt{n}}{\sigma}\varepsilon\right)$$

$$= \Phi(\infty) - \Phi(-\infty) = 1. \tag{7.82}$$

■

Example 7.3.4. Determining Sample Size.

Suppose that in a public opinion poll, the proportion p of voters who favor a certain proposition is to be determined. In other words, we want to estimate the unknown probability p of a randomly selected voter being in favor of the proposition. We take a random sample, with the responses being i.i.d. Bernoulli random variables X_i with parameter p and use $\overline{X}_n$ to estimate p. Approximately how large a random sample must be taken to ensure that

$$P\left(\left|\overline{X}_n - p\right| < 0.1\right) \geq 0.95? \tag{7.83}$$

By the CLT

$$P\left(\left|\overline{X}_n - p\right| < 0.1\right) = P\left(\frac{\left|\overline{X}_n - p\right|}{\sqrt{p(1-p)/n}} < \frac{0.1}{\sqrt{p(1-p)/n}}\right)$$

$$\approx P\left(|Z| < \frac{0.1}{\sqrt{p(1-p)/n}}\right)$$

$$= 2\Phi\left(\frac{0.1}{\sqrt{p(1-p)/n}}\right) - 1. \tag{7.84}$$

Now, this quantity is $\geq .95$ if

$$\Phi\left(\frac{0.1}{\sqrt{p(1-p)/n}}\right) \geq 0.975 \tag{7.85}$$

or, equivalently, if

$$\frac{0.1}{\sqrt{p(1-p)/n}} \geq \Phi^{-1}(.975) \approx 1.96 \tag{7.86}$$

or

$$\sqrt{n} \gtrsim \frac{1.96}{0.1}\sqrt{p(1-p)} \tag{7.87}$$

or

$$n \gtrsim 384.16p\,(1-p)\,. \tag{7.88}$$

Here $p\,(1-p)$ has its maximum at $p = 1/2$, and then $p\,(1-p) = 1/4$. Thus

$$n \gtrsim 97 \tag{7.89}$$

ensures that $\mathrm{P}\big(|\overline{X}_n - p| < 0.1\big) \geq .95$ for any value of p.

The lower bound for n obtained above by the normal approximation is actually the same as the precise value given by the binomial distribution. Indeed, for $n = 97$ and $p = 1/2$ a computer evaluation gives $\mathrm{P}\big(|\overline{X}_n - p| < 0.1\big) = 0.958\dots$ and for $n = 96$ and $p = 1/2$ it gives $\mathrm{P}\big(|\overline{X}_n - p| < 0.1\big) = 0.948\dots$.

If we know in advance an approximate value for p, which is far from $1/2$, then Formula 7.88 can be used to obtain a lower value for the required sample size.   ◆

Exercises

Exercise 7.3.1.

A die is rolled 20 times. Find the probability of obtaining 3 sixes, both by the binomial p.f. and by the normal approximation with continuity correction (Equation 7.60.).

Exercise 7.3.2.

A die is rolled 20 times. Find the probability of obtaining 3, 4, or 5 sixes, both by the binomial p.f. and by the normal approximation with continuity correction (Equation 7.60.).

Exercise 7.3.3.

Choose 100 independent random numbers, uniformly distributed on the interval $[0, 1]$. What is the approximate probability of their average falling in the interval $[0.49, 0.51]$?

Exercise 7.3.4.

The heights of 100 persons are measured to the nearest inch. What is the approximate probability that the average of these rounded numbers differs from the true average by less than 1%?

Exercise 7.3.5.

A scale is calibrated by repeatedly measuring a standard weight of 10 grams and taking the average $\overline{X}$ of these measurements. Due to unpredictable causes, such as changes in temperature, air pressure, and friction, the individual measurements vary slightly. They are taken to be independent random variables with $\sigma = 6\mu g$ each[6]:

---

[6] $1\mu g = 1$ microgram $= 10^{-6}g$

1. How many weighings are needed to make $\sigma_{\overline{X}} \leq 0.5\mu g$?
2. How many weighings are needed to make $P\left(\left|\overline{X} - 10g\right| < 0.5\mu g\right) \geq .9$?

Exercise 7.3.6.

In Example 7.1.2, we obtained the exact answer to the question of finding the probability that on a certain night 40 or fewer diners will show up at a restaurant if the number of diners is Poisson with $\lambda = 50$. Answer the same question approximately, by using the CLT and the fact that a Poisson r.v. with $\lambda = 50$ is the sum of 50 independent Poisson r.v.'s with $\lambda = 1$.

7.4 Negative Binomial, Gamma and Beta Random Variables

In this section, we shall discuss three other named families of random variables that occur in various applications.

The negative binomial distribution is a generalization of the geometric distribution: in a sequence of i.i.d. Bernoulli trials, we wait for the rth success, rather than just the first one. The probability that the rth success occurs on the kth trial equals the probability that in the first $k-1$ trials, we have exactly $r - 1$ successes and the rth trial is a success, that is, $\binom{k-1}{r-1}p^{r-1}q^{(k-1)-(r-1)}$ times p. Thus, we make the following definition:

Definition 7.4.1. *Negative Binomial Random Variables. Suppose we perform i.i.d. Bernoulli trials with parameter p, until we obtain r successes, for a fixed positive integer r. The number X_r of such trials up to and including the rth success is called a negative binomial random variable[7] with parameters p and r. It has the probability function*

$$f(k) = P(X_r = k) = \binom{k - 1}{r - 1}p^r q^{k-r} \quad for \quad k = r, r+1, r+2, \ldots . \quad (7.90)$$

The distribution of X_r is called negative binomial, too.

The reason for the name "negative binomial" is that $f(k)$ can be written as

$$f(k) = \binom{-r}{k - r}p^r \left(-q\right)^{k-r} \quad for \quad k = r, r+1, r+2, \ldots , \quad (7.91)$$

with the definition of binomial coefficients extended for negative numbers on top as

---

[7] Some authors define a negative binomial r.v. as the number of *failures* before the rth success, rather than the total number of trials.

$$\binom{-r}{i} = \frac{(-r)(-r-1)\cdots(-r-i+1)}{i!} = (-1)^i \binom{r+i-1}{i} \qquad (7.92)$$

for nonnegative integers r and i, and the binomial theorem can also be extended[8] for negative exponents as

$$(1+x)^{-r} = \sum_{i=0}^{\infty} \binom{-r}{i} x^i. \qquad (7.93)$$

From Equations 7.91 and 7.93,

$$\sum_{k=r}^{\infty} \binom{-r}{k-r} p^r (-q)^{k-r} = p^r \sum_{i=0}^{\infty} \binom{-r}{i} (-q)^i = p^r (1-q)^{-r} = 1, \quad (7.94)$$

and so the probabilities $f(k)$ do, indeed, add up to 1 and are p^r times the terms of a series for a binomial expression with a negative exponent.

Clearly, the geometric distribution is a special case of the negative binomial, with $r = 1$. Also, X_r is the sum of r i.i.d. geometric random variables $Z_1, Z_2, \ldots, Z_r$ with parameter p, because to get r successes, we first have to wait Z_1 trials for the first success, then Z_2 trials for the second success, independently of what happened before, and so on. Thus, we can easily compute $E(X_r)$ and $Var(X_r)$ as r times $E(Z_i)$ and $Var(Z_i)$, and so, by Example 6.1.12 and Exercise 6.3.4,

$$E(X_r) = \frac{r}{p} \qquad (7.95)$$

and

$$Var(X_r) = \frac{rq}{p^2}. \qquad (7.96)$$

Similarly, from Example 6.3.2, the m.g.f. of X_r is the rth power of the m.g.f. of Z_i:

$$\psi(t) = \left(\frac{pe^t}{1 - qe^t} \right)^r. \qquad (7.97)$$

Example 7.4.1. Number of Children.

A couple wants to have two boys. Find the distribution of the number of children they must have to achieve this goal. Assume that the children are boys or girls independently of each other and $P(\text{boy}) = \frac{1}{2}$.

Clearly, the number of children is a negative binomial random variable with parameters $p = \frac{1}{2}$ and $r = 2$. Thus, with $f(k)$ denoting the probability of needing k children, we have

---

[8] Expand $(1+x)^{-r}$ in a Taylor series.

$$f(k) = (k-1) \left(\frac{1}{2}\right)^k \quad \text{for} \quad k = 2, 3, 4, \ldots \, . \tag{7.98}$$

Furthermore, the expected number of children they need in order to have two boys is $\frac{r}{p} = 4$. ◆

The next type of random variable we want to consider is called a gamma random variable because its density contains the so-called *gamma function*,

$$\Gamma(t) = \int_0^\infty x^{t-1} e^{-x} dx \quad \text{for } t > 0. \tag{7.99}$$

This integral cannot be evaluated in terms of elementary functions but only by approximate methods, except for some specific values of t, which include the positive integers.

Integration by parts yields the reduction formula

$$\Gamma(t+1) = t\Gamma(t) \quad \text{for } t > 0, \tag{7.100}$$

and from here, using the straightforward evaluation $\Gamma(1) = 1$, we obtain

$$\Gamma(r) = (r-1)! \quad \text{for } r = 1, 2, 3, \ldots \, . \tag{7.101}$$

Thus, the gamma function is a generalization of the factorial function from integer to positive real arguments.

Definition 7.4.2. *Gamma Random Variables.* *A continuous random variable with density function*

$$f(x) = \begin{cases} 0 & \text{if } x \leq 0 \\ \frac{\lambda^\alpha}{\Gamma(\alpha)} x^{\alpha-1} e^{-\lambda x} & \text{if } x > 0 \end{cases} \tag{7.102}$$

is called a gamma random variable and $f(x)$ the gamma density with parameters α and λ, for any real $\alpha > 0$ and $\lambda > 0$.

The essential part of this definition is the fact that $f(x)$ is proportional to $x^{\alpha-1} e^{-\lambda x}$ for $x > 0$, and the coefficient $\frac{\lambda^\alpha}{\Gamma(\alpha)}$ just normalizes this expression. Indeed, with the change of variable $u = \lambda x$, we get

$$\int_0^\infty \frac{\lambda^\alpha}{\Gamma(\alpha)} x^{\alpha-1} e^{-\lambda x} dx = \frac{\lambda^\alpha}{\Gamma(\alpha)} \int_0^\infty \left(\frac{u}{\lambda}\right)^{\alpha-1} e^{-u} \frac{1}{\lambda} du$$

$$= \frac{1}{\Gamma(\alpha)} \int_0^\infty u^{\alpha-1} e^{-u} du = 1. \tag{7.103}$$

Clearly, for $\alpha = 1$ the gamma density becomes the exponential density with parameter λ. More generally, for $\alpha = n$, a positive integer, the gamma density turns out to be the density of the sum of n i.i.d. exponential r.v.'s with parameter λ, as shown below. Hence, for $\alpha = n$, a gamma r.v. is a continuous analog of the negative binomial: it is the waiting time for the occurrence of the nth arrival in a Poisson process with parameter λ.

Theorem 7.4.1. *Gamma as the Sum of Exponentials.* *For any positive integer n, let $T_1, T_2, \ldots, T_n$ be i.i.d. exponential random variables with density*

$$f(t) = \begin{cases} 0 & if\ t \leq 0 \\ \lambda e^{-\lambda t} & if\ t > 0. \end{cases} \tag{7.104}$$

Then $S_n = T_1 + T_2 + \ldots + T_n$ is a gamma random variable with parameters $\alpha = n$ and λ, that is, its density is

$$f_n(t) = \begin{cases} 0 & if\ t \leq 0 \\ \frac{\lambda^n}{(n-1)!} t^{n-1} e^{-\lambda t} & if\ t > 0. \end{cases} \tag{7.105}$$

Proof. We use induction.

For $n = 1$ Equation 7.105 reduces to Equation 7.104, which is the density of $S_1 = T_1$, and so the statement is true in this case.

Now, assume that Equation 7.105 is true for arbitrary n. Then the convolution formula (Equation 5.128) gives

$$f_{n+1}(t) = \int_0^t f_n(x) f_1(t - x) dx = \int_0^t \frac{\lambda^n}{(n-1)!} x^{n-1} e^{-\lambda x} \lambda e^{-\lambda(t-x)} dx$$

$$= \frac{\lambda^{n+1}}{(n-1)!} e^{-\lambda t} \int_0^t x^{n-1} dx = \frac{\lambda^{n+1}}{n!} t^n e^{-\lambda t} \quad \text{for } t > 0. \tag{7.106}$$

The expression on the right is the same as the one in Equation 7.105 with $n + 1$ in place of n. Thus, if Equation 7.105 gives the density of S_n, for any n, then it gives the density of S_{n+1}, with $n + 1$ in place of n, too, and so it gives the density of S_n for every n. ∎

The preceding theorem implies that the sum of two independent gamma random variables, with integer α values, say m and n, and a common λ, is gamma with $\alpha = m + n$ and the same λ, because it is the sum of $m + n$ i.i.d. exponential random variables with parameter λ. The sum is still gamma even if the parameters are not integers:

Theorem 7.4.2. *Sum of Independent Gamma Variables.* *For any $r, s > 0$, let R and S be two independent gamma random variables with parameters $\alpha = r$ and $\alpha = s$, respectively, and a common λ. Then $T = R + S$ is gamma with parameters $r + s$ and λ.*

Proof. By the convolution formula (Equation 5.128), for $t > 0$,

$$f_T(t) = \int_0^t f_R(x) f_S(t - x) dx$$

$$= \int_0^t \frac{\lambda^r}{\Gamma(r)} x^{r-1} e^{-\lambda x} \frac{\lambda^s}{\Gamma(s)} (t - x)^{s-1} e^{-\lambda(t-x)} dx$$

$$= \frac{\lambda^{r+s}}{\Gamma(r) \Gamma(s)} e^{-\lambda t} \int_0^t x^{r-1} (t - x)^{s-1} dx. \tag{7.107}$$

In the last integral we change the variable x to u by substituting $x = tu$ and $dx = tdu$. Then we get

$$f_T(t) = \frac{\lambda^{r+s}}{\Gamma(r)\Gamma(s)}e^{-\lambda t}\int_0^1 (tu)^{r-1}(t-tu)^{s-1}t\,du$$

$$= \frac{\lambda^{r+s}}{\Gamma(r+s)}t^{r+s-1}e^{-\lambda t}\int_0^1 \frac{\Gamma(r+s)}{\Gamma(r)\Gamma(s)}u^{r-1}(1-u)^{s-1}\,du. \quad (7.108)$$

Here the function $\frac{\lambda^{r+s}}{\Gamma(r+s)}t^{r+s-1}e^{-\lambda t}$ is the gamma density with parameters $r+s$ and λ. Since the whole expression on the right is a density as well, we must have

$$\int_0^1 \frac{\Gamma(r+s)}{\Gamma(r)\Gamma(s)}u^{r-1}(1-u)^{s-1}\,du = 1 \quad (7.109)$$

and

$$f_T(t) = \frac{\lambda^{r+s}}{\Gamma(r+s)}t^{r+s-1}e^{-\lambda t}. \quad (7.110)$$

∎

We have an important by-product of the above proof:

Corollary 7.4.1.

$$\int_0^1 u^{r-1}(1-u)^{s-1}\,du = \frac{\Gamma(r)\Gamma(s)}{\Gamma(r+s)}. \quad (7.111)$$

This integral is called the *beta integral* and its value

$$B(r,s) = \frac{\Gamma(r)\Gamma(s)}{\Gamma(r+s)} \quad (7.112)$$

the *beta function* of r and s. It will show up again shortly in the density of beta random variables.

Example 7.4.2. Expectation and Variance of Gamma Variables.

For the X_i in 7.4.1, $E(X_i) = \frac{1}{\lambda}$ and $Var(X_i) = \frac{1}{\lambda^2}$, and so if T is gamma with parameters $\alpha = n$ and arbitrary λ, then $E(T) = \frac{n}{\lambda}$ and $Var(T) = \frac{n}{\lambda^2}$. These expressions remain valid for arbitrary α in place of n: For a gamma random variable T with arbitrary α and λ, we have, with the change of variable $u = \lambda x$,

$$E(T) = \int_0^\infty \frac{\lambda^\alpha}{\Gamma(\alpha)}x^\alpha e^{-\lambda x}dx = \frac{\lambda^\alpha}{\Gamma(\alpha)}\int_0^\infty \left(\frac{u}{\lambda}\right)^\alpha e^{-u}\frac{1}{\lambda}du$$

$$= \frac{1}{\lambda\Gamma(\alpha)}\int_0^\infty u^\alpha e^{-u}du = \frac{\Gamma(\alpha+1)}{\lambda\Gamma(\alpha)} = \frac{\alpha}{\lambda}. \quad (7.113)$$

Similarly,

$$Var\,(T) = \frac{\alpha}{\lambda^2}. \tag{7.114}$$

(The proof is left as an exercise.)   ◆

Another important case in which we obtain a gamma random variable is described in the following theorem.

Theorem 7.4.3. *Square of a Normal Random Variable.* Let X be an $N\left(0, \sigma^2\right)$ *random variable. Then* $Y = X^2$ *is gamma with* $\alpha = \frac{1}{2}$ *and* $\lambda = \frac{1}{2\sigma^2}$.

Proof. By Equation 5.57

$$f_Y(y) = \begin{cases} \frac{1}{2\sqrt{y}}\left[f_X\left(\sqrt{y}\right) + f_X\left(-\sqrt{y}\right)\right] & \text{if } y > 0 \\ 0 & \text{if } y \le 0 \end{cases}. \tag{7.115}$$

In the present case,

$$f_X(x) = \frac{1}{\sqrt{2\pi}\sigma} e^{-x^2/2\sigma^2} \quad \text{for} \quad -\infty < x < \infty, \tag{7.116}$$

and so

$$f_Y(y) = \begin{cases} \frac{1}{\sigma\sqrt{2\pi y}} e^{-y/2\sigma^2} & \text{if } y > 0 \\ 0 & \text{if } y \le 0 \end{cases}. \tag{7.117}$$

For $y > 0$, this density is proportional to $y^{-1/2} e^{-y/2\sigma^2}$ and is therefore gamma with $\alpha = \frac{1}{2}$ and $\lambda = \frac{1}{2\sigma^2}$.   ∎

Since the coefficient in Definition 7.4.2 with $\alpha = \frac{1}{2}$ and $\lambda = \frac{1}{2\sigma^2}$ and the coefficient in Equation 7.117 normalize the same function, they must be equal, that is,

$$\frac{\left[1/\left(2\sigma^2\right)\right]^{1/2}}{\Gamma\left(1/2\right)} = \frac{1}{\sigma\sqrt{2\pi}} \tag{7.118}$$

must hold. Hence, we must have

$$\Gamma\left(\frac{1}{2}\right) = \sqrt{\pi}. \tag{7.119}$$

Using Equation 7.119 and the reduction formula, Equation 7.100, we obtain (Exercise 7.4.11) the values of the gamma function for positive half-integer arguments:

$$\Gamma\left(\frac{2k+1}{2}\right) = \frac{\sqrt{\pi}\,(2k)!}{2^{2k}k!} \quad \text{for } k = 0, 1, 2, \dots. \tag{7.120}$$

In various statistical applications, we encounter the sum of the squares of independent standard normal random variables. As a consequence of Theorems 7.4.3 and 7.4.2, such sums have gamma distributions, but we have a special name associated with them and their densities are specially tabulated. Thus, we have the following definition and theorem:

Definition 7.4.3. *Chi-Square Random Variables.* *For independent standard normal random variables $Z_1, Z_2, \ldots, Z_n$, the sum $\chi_n^2 = Z_1^2 + Z_2^2 + \ldots + Z_n^2$ is called a chi-square random variable with n degrees of freedom.*

Theorem 7.4.4. *Chi-Square is Gamma.* *The distribution of χ_n^2 is gamma with parameters $\alpha = \frac{n}{2}$ and $\lambda = \frac{1}{2}$, and its density is*

$$f_{\chi_n^2}(x) = \begin{cases} 0 & if\ x \leq 0 \\ \dfrac{1}{2^{n/2}\Gamma(n/2)}x^{\frac{n}{2}-1}e^{-\frac{x}{2}} & if\ x > 0. \end{cases} \tag{7.121}$$

Proof. By Theorem 7.4.3, Z_i^2 is gamma with parameters $\alpha = \frac{1}{2}$ and $\lambda = \frac{1}{2}$, for all i, and so, by repeated application of the result of Theorem 7.4.2, the statement follows. ∎

Corollary 7.4.2. *Expectation and Variance of Chi-Square.* $E\left(\chi_n^2\right) = n$ and $Var\left(\chi_n^2\right) = 2n$.

Proof. These values follow at once from Example 7.4.2 and Equation 7.121. ∎

Corollary 7.4.3. *Density of* χ_n. *The density of $\chi_n = \sqrt{\chi_n^2}$ is*

$$f_{\chi_n}(x) = \begin{cases} 0 & if\ x \leq 0 \\ \dfrac{2}{2^{n/2}\Gamma(n/2)}x^{n-1}e^{-\frac{x^2}{2}} & if\ x > 0. \end{cases} \tag{7.122}$$

We leave the proof as Exercise 7.4.10.

Example 7.4.3. *Moment Generating Function of Chi-Square.*

For independent standard normal random variables $Z_1, Z_2, \ldots, Z_n$, the m.g.f. of $\chi_n^2 = Z_1^2 + Z_2^2 + \ldots + Z_n^2$ is given by

$$\psi_n(t) = E\left(e^{\sum_{i=1}^n Z_i^2 t}\right) = \prod_{i=1}^n E\left(e^{Z_i^2 t}\right). \tag{7.123}$$

Here

$$E\left(e^{Z_i^2 t}\right) = \frac{1}{\sqrt{2\pi}}\int_{-\infty}^{\infty} e^{z^2 t}e^{-z^2/2}dz = \frac{1}{\sqrt{2\pi}}\int_{-\infty}^{\infty} e^{-z^2(1-2t)/2}dz \ \ \text{for} \ \ t < \frac{1}{2}. \tag{7.124}$$

Making the change of variable $u = z\sqrt{1-2t}$, we get

$$E\left(e^{Z_i^2 t}\right) = \frac{1}{\sqrt{2\pi}\sqrt{1-2t}} \int_{-\infty}^{\infty} e^{-u^2/2} du = \frac{1}{\sqrt{1-2t}} \quad \text{for } t < \frac{1}{2}. \quad (7.125)$$

Thus,

$$\psi_n(t) = \prod_{i=1}^{n} E\left(e^{Z_i^2 t}\right) = \left(\frac{1}{\sqrt{1-2t}}\right)^n = (1-2t)^{-n/2} \quad \text{for } t < \frac{1}{2}. \quad (7.126)$$

$\blacklozenge$

Gamma random variables, with values of the parameter α not just integers or half-integers, are often used to model continuous random variables with unknown or approximately known distribution on $(0, \infty)$. Similarly, continuous random variables with unknown distribution on $[0, 1]$ are often modelled by beta random variables, to be defined below. This is especially true in some statistical applications of Bayes' theorem, in which the prior probability P of an event is taken to be a random variable with such a distribution on $[0, 1]$, and then the posterior distribution turns out to be beta, too. (See Example 7.4.4 below.)

Definition 7.4.4. Beta Random Variables. *A continuous random variable with density function*[9]

$$f(x) = \begin{cases} \frac{1}{B(r,s)} x^{r-1} (1-x)^{s-1} & \text{if } 0 \le x \le 1 \\ 0 & \text{otherwise} \end{cases} \quad (7.127)$$

is called a beta random variable and $f(x)$ the beta density with parameters r and s, for any real $r > 0$ and $s > 0$. Here

$$B(r,s) = \frac{\Gamma(r)\,\Gamma(s)}{\Gamma(r+s)}. \quad (7.128)$$

Notice that the beta distribution with $r = s = 1$ is the uniform distribution on $[0, 1]$.

Example 7.4.4. Updating Unknown Probabilities by Bayes' Theorem.

Suppose the probability P of an event A is unknown and is taken to be a uniform random variable on $[0, 1]$, which is a (somewhat controversial) way of expressing the fact that we have no idea what the value of P is. Assume that we conduct $n \ge 1$ independent performances of the same experiment, and obtain k successes, that is, obtain A exactly k times. How should we revise the distribution of P in light of this result?

---

[9] We assume $0^0 = 1$ where necessary.

We have already treated this problem for $n = 1$ in Example 5.6.3. The computation for general $n > 1$ is similar; we just use a binomial distribution for the number X of successes instead of the Bernoulli distribution used there. Thus,

$$f_{X|P}(k,p) = \binom{n}{k} p^k (1-p)^{n-k} \text{ for } k = 0, 1, \ldots, n \qquad (7.129)$$

and

$$f_P(p) = \begin{cases} 1 \text{ for } p \in [0,1] \\ 0 \text{ otherwise} \end{cases} \qquad (7.130)$$

By Equation 5.171, (the $\binom{n}{k}$ in the numerator and denominator cancel)

$$f_{P|X}(p,k) = \begin{cases} \frac{p^k(1-p)^{n-k}}{\int_0^1 p^k(1-p)^{n-k}\,dp} \text{ for } p \in [0,1] \text{ and } k = 0, 1, \ldots, n \\ 0 \qquad\qquad\qquad\qquad \text{otherwise} \end{cases} \qquad (7.131)$$

Thus the posterior density of P is beta with parameters $r = k + 1$ and $s = n - k + 1$.   ◆

Example 7.4.5. Expectation and Variance of Beta Variables.

The expected value is very easy to compute, because the relevant integral produces another beta function. Thus, if X is beta with parameters r and s, then

$$E(X) = \frac{1}{B(r,s)} \int_0^1 x \cdot x^{r-1}(1-x)^{s-1}\,dx = \frac{B(r+1,s)}{B(r,s)}$$
$$= \frac{\Gamma(r+1)\Gamma(s)}{\Gamma(r+s+1)} \cdot \frac{\Gamma(r+s)}{\Gamma(r)\Gamma(s)} = \frac{r}{r+s}. \qquad (7.132)$$

Similarly,

$$Var(X) = \frac{rs}{(r+s)^2(r+s+1)}. \qquad (7.133)$$

We leave the proof of this formula as Exercise 7.4.18.   ◆

Exercises

Exercise 7.4.1.

Find the probability of obtaining, in i.i.d., parameter p Bernoulli trials, r successes before s failures.

Exercise 7.4.2.

Let N_r be the number of i.i.d., parameter p Bernoulli trials needed to produce either r successes or r failures, whichever occurs first. Find the p.f. of N_r.

Exercise 7.4.3.

Let Y_r be the number of failures in i.i.d., parameter p Bernoulli trials before the rth success. Find the p.f. of Y_r.

Exercise 7.4.4.

1. A die is rolled until six shows up for the second time. What is the probability that no more than eight rolls are needed?
2. How many rolls are needed to make the probability of getting the second 6 on or before the last roll exceed $1/2$?

Exercise 7.4.5.

For any positive integers r and s, let X_r be the number of i.i.d. Bernoulli trials with parameter p up to and including the rth success and X_{r+s} their number up to and including the $(r + s)$th success. Find the joint p.f. of X_r and X_{r+s}.

Exercise 7.4.6.

Let S_m denote the number of successes in the first m of $m + n$ i.i.d. parameter p Bernoulli trials for any positive integers m and n. Find the p.f. of S_m under the condition that the rth success, for any positive $r \leq m + n$, occurs on the $(m + n)$th trial.

Exercise 7.4.7.

Show that, for $\alpha \geq 1$ the mode of the gamma density (that is the x-value where $f(x)$ takes on its maximum) is $\frac{\alpha - 1}{\lambda}$.

Exercise 7.4.8.

Sketch the gamma density for the following (α, λ) pairs: $(1, 1)$, $(1, 2)$, $(2, 1)$, $(1, 1/2)$, $(1/2, 1)$, $(1/2, 2)$, $(4, 4)$.

Exercise 7.4.9.

For a gamma random variable T with arbitrary α and λ, prove:

1. $E\left(T^k\right) = \frac{\alpha(\alpha+1)\cdots(\alpha+k-1)}{\lambda^k}$ for any positive integer k.
2. $Var\left(T\right) = \frac{\alpha}{\lambda^2}$.
3. The m.g.f. of T is $\psi\left(t\right) = \left(\frac{\lambda}{\lambda-t}\right)^\alpha$ for $t < \lambda$.

Exercise 7.4.10.

Prove Corollary 7.4.3.

Exercise 7.4.11.

Prove Equation 7.120.

Exercise 7.4.12.

Choose a point at random in the plane, with its coordinates X and Y independent standard normal random variables. What is the probability that the point is inside the unit circle?

Exercise 7.4.13.

Let X and Y be i.i.d $N\left(0, \sigma^2\right)$ normal random variables. Find the density of $U = X^2 + Y^2$.

Exercise 7.4.14.

Let $X_1, X_2, \ldots, X_n$, be i.i.d $N\left(0, \sigma^2\right)$ normal random variables. Find the density of $V = \sum_{i=1}^{n} X_i^2$.

Exercise 7.4.15.

Let $X_1, X_2, \ldots, X_n$, be i.i.d uniform random variables on $[0, 1]$. Show that $Y = \max\left(X_1, X_2, \ldots, X_n\right)$ and $Z = \min\left(X_1, X_2, \ldots, X_n\right)$ are beta and find their parameters r and s.

Exercise 7.4.16.

Show that the mode of the distribution given by Equation 7.131 (that is, the p-value where $f_{P|X}\left(p|k\right)$ takes on its maximum) is the relative frequency $\frac{k}{n}$. (You need to treat the cases $k = 0$ and $k = n$ separately from the others!)

Exercise 7.4.17.

Sketch the beta density for the following (r, s) pairs: $(1, 2)$, $(2, 1)$, $(2, 2)$, $(1, 3)$, $(1/2, 1)$, $(11, 21)$.

Exercise 7.4.18.

For a beta random variable X with arbitrary r and s, prove

1. $E\left(X^k\right) = \frac{r(r+1)\cdots(r+k-1)}{(r+s)(r+s+1)\cdots(r+s+k-1)}$ for any positive integer k.
2. $Var\left(X\right) = \frac{rs}{(r+s)^2(r+s+1)}$.

Exercise 7.4.19.

Modify Example 7.4.4 by taking the prior distribution to be beta with arbitrary, known values r and s. Find the posterior distribution of the event A if in $n \geq 1$ independent performances of the same experiment we obtain $k \leq n$ successes.

7.5 Multivariate Normal Random Variables

In many applications, we have to deal simultaneously with two or more normal random variables whose joint distribution is a direct generalization of the normal distribution. For example, the height and weight of a randomly selected person is such a pair and so are the test scores of a student on two exams in a math course and the heights of a randomly selected father-son pair. Also, in statistical samples from a normally distributed population, the joint observations follow a multivariate normal distribution.

We take a somewhat indirect, but mathematically convenient, approach to defining bivariate random variables. We start with two independent standard normal random variables and transform them linearly.

Definition 7.5.1. *Bivariate Normal Random Variables.* *Let Z_1 and Z_2 be independent standard normal random variables and $a_{11}, a_{12}, a_{21}, a_{22}, b_1$ and b_2 any constants satisfying $a_{11}^2 + a_{12}^2 \neq 0$, $a_{21}^2 + a_{22}^2 \neq 0$ and $a_{11}a_{22} - a_{12}a_{21} \neq 0$. Then*

$$X_1 = a_{11}Z_1 + a_{12}Z_2 + b_1 \text{ and}$$
$$X_2 = a_{21}Z_1 + a_{22}Z_2 + b_2 \tag{7.134}$$

are said to form a bivariate normal pair.

By Theorems 7.2.4 and 7.2.6, the marginals X_1 and X_2 are (univariate) normal with means $\mu_1 = b_1$ and $\mu_2 = b_2$ and variances $\sigma_1^2 = a_{11}^2 + a_{12}^2$ and $\sigma_2^2 = a_{21}^2 + a_{22}^2$, respectively. Furthermore, $\sigma_{1,2} = Cov(X_1, X_2) = a_{11}a_{21} + a_{12}a_{22}$, by the definition of Z_1 and Z_2 as independent standard normal random variables, and the correlation coefficient of X_1 and X_2 is $\rho = (a_{11}a_{21} + a_{12}a_{22})/\sigma_1\sigma_2$. Note that $\rho \neq \pm1$ by Corollary 6.4.1 and the requirement that $a_{11}a_{22} - a_{12}a_{21} \neq 0$.

In Theorem 6.4.2, we saw that for independent random variables X and Y whose expectations exist, $Cov(X, Y) = 0$. One of the most important properties of bivariate normal random variables is that the converse of this fact holds for them:

Theorem 7.5.1. *For Bivariate Normal Random Variables, Zero Covariance Implies Independence.* *If X_1 and X_2 are bivariate normal, then $Cov(X_1, X_2) = 0$ implies their independence.*

Proof. We are going to use the bivariate moment generating function

$$\psi(s, t) = E\left(e^{sX_1 + tX_2}\right) \tag{7.135}$$

to prove this theorem.

By Theorem 7.2.6, $Y = sX_1 + tX_2$ is normal, because it is a linear combination of the original, independent, random variables Z_1 and Z_2. Clearly, it has mean $\mu_Y = s\mu_1 + t\mu_2$ and variance $\sigma_Y^2 = s^2\sigma_1^2 + t^2\sigma_2^2 + 2st\sigma_{1,2}$. (Here we wrote $\sigma_{1,2}$ for $Cov(X_1, X_2)$.) Thus, by the definition of the m.g.f. of Y as $\psi_Y(t) = E\left(e^{tY}\right)$ and by Equation 7.47,

$$\psi(s,t) = \psi_Y(1) = e^{s\mu_1 + t\mu_2 + \left(s^2\sigma_1^2 + t^2\sigma_2^2 + 2st\sigma_{1,2}\right)/2}. \tag{7.136}$$

Hence, if $\sigma_{1,2} = 0$, then $\psi(s,t)$ factors as

$$\psi(s,t) = e^{s\mu_1 + s^2\sigma_1^2/2}e^{t\mu_2 + t^2\sigma_2^2/2}, \tag{7.137}$$

which is the product of the moment generating functions of X_1 and X_2.

Now, if X_1 and X_2 are independent, then, clearly,

$$\psi(s,t) = E\left(e^{sX_1 + tX_2}\right) = E\left(e^{sX_1}e^{tX_2}\right) = E\left(e^{sX_1}\right)E\left(e^{tX_2}\right)$$
$$= e^{s\mu_1 + s^2\sigma_1^2/2}e^{t\mu_2 + t^2\sigma_2^2/2}, \tag{7.138}$$

the same function as the one we obtained above from the assumption $\sigma_{1,2} = 0$. By the uniqueness of moment generating functions, which holds in the two-dimensional case as well, X_1 and X_2 must therefore be independent if $\sigma_{1,2} = 0$.   ∎

Note that this theorem does not say that if X_1 and X_2 are only separately, rather than jointly, normal, then $Cov(X_1, X_2) = 0$ implies their independence. That statement is not true in general, as Exercise 7.5.4 shows.

Define two new standard normal random variables as

$$Y_1 = \frac{a_{11}}{\sigma_1}Z_1 + \frac{a_{12}}{\sigma_1}Z_2 \text{ and}$$
$$Y_2 = \frac{1}{\sqrt{1-\rho^2}}\left[\left(\frac{a_{21}}{\sigma_2} - \frac{a_{11}\rho}{\sigma_1}\right)Z_1 + \left(\frac{a_{22}}{\sigma_2} - \frac{a_{12}\rho}{\sigma_1}\right)Z_2\right] \tag{7.139}$$

One can check by some straightforward calculations (7.5.1) that Y_1 and Y_2 are indeed standard normal and $Cov(Y_1, Y_2) = 0$. Thus, by Theorem 7.5.1, Y_1 and Y_2 are independent. Furthermore, we can write X_1 and X_2 in terms of Y_1 and Y_2 as

$$X_1 = \sigma_1 Y_1 + \mu_1, \text{ and}$$
$$X_2 = \sigma_2\left(\rho Y_1 + \sqrt{1-\rho^2}Y_2\right) + \mu_2 \tag{7.140}$$

From here we can easily obtain the conditional expectation, variance and density of X_2 given $X_1 = x_1$, because, if $X_1 = x_1$, then $Y_1 = (x_1 - \mu_1)/\sigma_1$ and

$$X_2 = \sigma_2\left(\rho\frac{x_1 - \mu_1}{\sigma_1} + \sqrt{1-\rho^2}Y_2\right) + \mu_2. \tag{7.141}$$

Thus, since Y_2 is independent of Y_1, it is unaffected by the condition $X_1 = x_1$, and so, under this condition, X_2 is normal with mean

$$E\left(X_2 | X_1 = x_1\right) = \mu_2 + \rho\sigma_2 \frac{x_1 - \mu_1}{\sigma_1} \qquad (7.142)$$

and variance

$$Var\left(X_2 | X_1 = x_1\right) = \left(1 - \rho^2\right)\sigma_2^2. \qquad (7.143)$$

Hence

$$f_{X_2 | X_1}\left(x_2, x_1\right) = \frac{1}{\sqrt{2\pi\left(1 - \rho^2\right)}\sigma_2} \exp \frac{-\left(x_2 - \mu_2 - \rho\sigma_2 \frac{x_1 - \mu_1}{\sigma_1}\right)^2}{2\left(1 - \rho^2\right)\sigma_2^2}. \qquad (7.144)$$

Observe that $x_2 = E\left(X_2 | X_1 = x_1\right)$ is a linear function of x_1. The graph of this function in the $x_1 x_2$-plane is called the *regression line or least squares line of X_2 on X_1*, and its equation can also be written in the form

$$\frac{x_2 - \mu_2}{\sigma_2} = \rho \frac{x_1 - \mu_1}{\sigma_1}, \qquad (7.145)$$

or as

$$z_2 = \rho z_1, \qquad (7.146)$$

where we wrote z_1 and z_2 for the standardizations of x_1 and x_2. Thus, the regression line goes through "the point of averages" (μ_1, μ_2) and has, in standard units, slope ρ.

The name "regression" was coined by Francis Galton in the 1880s. He studied the dependence of children's height on the height of their parents, and noticed that the children of tall parents tended to be shorter and the children of short parents taller than their parents, in spite of the fact that the distribution of heights was the same in both generations. He named this effect "*regression to the mean*" or the "*regression effect.*" He multiplied all female heights by 1.08 and considered X_1 to be the mid-parent height and X_2 their grown-up child's height and noticed that (X_1, X_2) is a bivariate normal pair. Thus, if we take $\mu_1 = \mu_2 = \mu$ and $\sigma_1 = \sigma_2 = \sigma$ in Equation 7.145, then it becomes

$$x_2 - \mu = \rho\left(x_1 - \mu\right). \qquad (7.147)$$

Since ρ is between 0 and 1, this equation shows that

$$|x_2 - \mu| < |x_1 - \mu|, \qquad (7.148)$$

unless $x_1 = x_2 = \mu$, exactly what he observed.

The explanation of the effect is that the height of a person is partly inherited and partly due to random environmental causes during development. Both parts contribute to the regression effect: The height of an individual is controlled by a large number of inherited *recessive* genes, but not all of these are expressed in a person. When both parents are tall, only a fraction of the expressed increased-height genes coincide, and in their children only those will be expressed in which they inherit the increased-height form from both parents Thus, in the children, fewer of such genes will increase their height, than in their parents. On the other hand, the environmental effect is purely random and is centered at the population mean, and so part of the tallness of a parent is just due to luck, which will not be inherited, and for the child, luck is random around the mean. In fact there is a regression effect in the reverse direction as well: very tall children have generally shorter parents and very short children taller parents.

Note that, by Equation 7.145, the regression effect occurs not just when $\sigma_1 = \sigma_2$, but also whenever $\sigma_1 > \rho\sigma_2$ even if $\mu_1 \neq \mu_2$, but it does not occur if $\sigma_1 \leq \rho\sigma_2$. However, in standard units, it does occur for all nondegenerate values of the parameters.

Regression to the mean occurs in many situations. For example, a student may score very well on an exam but may not score so well on an equally difficult second exam, because part of the reason for the first high score may be luck and the second exam may not be as lucky as the first one. Similarly, a stock may perform very well one year but not so well the next year. This may be explained again by attributing much of the first high performance to good luck and the second worse performance to bad luck.

Mistaking the regression effect for something causal is called the *regression fallacy*. For instance, the student who scored poorly on an exam and scored better on a second one may not have really improved but may have just had better luck. Sometimes the regression fallacy can lead to bad policies or absurd conclusions. For example, a famous psychologist, Daniel Kahneman, observed a situation in which an instructor concluded that praise is counterproductive and berating is effective. This happened at an Israeli flight school, where cadets were praised when they did well and were screamed at when they executed some maneuver badly. Most of the time, the praised cadets did worse the next time and the berated ones did better. But the change was just the regression effect, and attributing it to the praise or berating is the regression fallacy.

Returning to the mathematics, note that the conditional variance $Var(X_2|X_1 = x_1)$ is the same for every value of x_1. Statisticians call this property of the bivariate normal distribution *homoscedasticity* (Greek for "same scatter"). Other bivariate distributions generally do not have this property; such distributions, for which $Var(X_2|X_1 = x_1)$ is a nonconstant function of x_1, are called *heteroscedastic*.

Multiplying the conditional density in Equation 7.144 by the marginal density

$$f_{X_1}(x_1) = \frac{1}{\sqrt{2\pi}\sigma_1} \exp \frac{-(x_1 - \mu_1)^2}{2\sigma_1^2}, \qquad (7.149)$$

we obtain the joint density of X_1 and X_2. Thus, we have proved the following theorem:

Theorem 7.5.2. *Bivariate Normal Density.* *If X_1 and X_2 form a bivariate normal pair with variances σ_1, σ_2 and correlation coefficient $\rho \neq \pm 1$, then their joint density is given by*

$$f(x_1, x_2) = \frac{1}{2\pi\sqrt{(1-\rho^2)}\sigma_1\sigma_2}. \qquad (7.150)$$

$$\exp\left\{\frac{-1}{2(1-\rho^2)}\left[\left(\frac{x_1 - \mu_1}{\sigma_1}\right)^2 - 2\rho\left(\frac{x_1 - \mu_1}{\sigma_1}\right)\left(\frac{x_2 - \mu_2}{\sigma_2}\right) + \left(\frac{x_2 - \mu_2}{\sigma_2}\right)^2\right]\right\}.$$

Clearly, if X_1 and X_2 have a joint density like this one, then we can write them as in Equations 7.140 in terms of standard normal Y_1 and Y_2, which shows that X_1 and X_2 are a bivariate normal pair. In fact, many books define bivariate normal pairs as random variables that have a joint density of this form.

Notice the symmetry of $f(x_1, x_2)$ with respect to interchanging the subscripts 1 and 2. Consequently, the conditional expectation, variance, and density of X_1 given $X_2 = x_2$ can be obtained from the foregoing conditional expressions simply by interchanging the subscripts 1 and 2.

Example 7.5.1. Level Curves and Regression Line.

Assume that the parameters in the example of the heights of parents and children are $\mu_1 = \mu_2 = \mu = 69''$, $\sigma_1 = \sigma_2 = \sigma = 2.6''$ and $\rho = 2/3$.. Then the level curves of the bivariate density are ellipses given by the equation

$$(x_1 - 69)^2 - \frac{4}{3}(x_1 - 69)(x_2 - 69) + (x_2 - 69)^2 = c, \qquad (7.151)$$

for various values of c. In Figure 7.5 we have plotted three of these curves together with the corresponding regression line

$$x_2 - 69 = \frac{2}{3}(x_1 - 69). \qquad (7.152)$$

We have also drawn a vertical line at $x_1 = 71$ to show that its segment inside an ellipse is bisected by the regression line. This observation can be explained by the fact that the conditional density in Equation 7.144 for any x_1 is normal and so it is centered at its mean, which is a point on the regression line. (See also Exercise 7.5.12.)

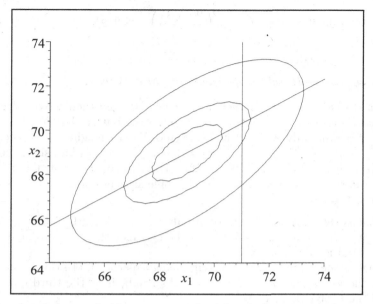

Fig. 7.5. Level curves of a bivariate normal p.d.f. and the corresponding regression line.

Example 7.5.2. Two Exams.

The scores on two successive exams taken by the students of the same large class usually approximate a bivariate normal distribution. Assume that X_1 and X_2, the scores of a randomly selected student on two exams, are bivariate normal with $\mu_1 = \mu_2 = 70$, $\sigma_1 = \sigma_2 = 12$ and $\rho = 0.70$. Suppose a student scored 90 on the first exam, what is his expected score on the second exam and what is the probability that he will score 90 or more on the second exam?

The conditional expected score on the second exam is given by Equation 7.142:

$$E\left(X_2|X_1 = 90\right) = 70 + 0.70 \cdot 12 \cdot \frac{90 - 70}{12} = 84. \tag{7.153}$$

Notice that the high score of 90 on the first exam gives a mean prediction of only 84 on the second exam. This result is an instance of the regression effect, which, as mentioned above, is universal for bivariate normal variables: given an "extreme" value of one of the variables (as measured in standard units), the expected value of the other variable will be less extreme.

The conditional variance of X_2 is given by Equation 7.143:

$$Var\left(X_2|X_1 = 90\right) = \left(1 - 0.70^2\right)12^2 = 73.44. \tag{7.154}$$

Thus, under the condition $X_1 = 90$, X_2 is normal with $\mu = 84$ and $\sigma = \sqrt{73.44} \approx 8.57$. Therefore

$$P\left(X_2 \geq 90 | X_1 = 90\right) = 1 - \Phi\left(\frac{90 - 84}{8.57}\right) \approx 0.242.$$ (7.155)

♦

Example 7.5.3. Heights of Husbands and Wives.

In a statistical study, the heights of husbands and their wives were measured and found to have a bivariate normal distribution. Let X_1 denote the height of a randomly selected husband and X_2 the height of his wife, with $\mu_1 = 68$", $\mu_2 = 64$", $\sigma_1 = 4$", $\sigma_2 = 3.6$", $\rho = 0.25$. (The slight positive correlation can be attributed to the fact that, to some extent, taller people tend to marry taller ones, and shorter people shorter ones.)

Given these data,

a) what is the expected height of a man whose wife is 61" tall,
b) what is the probability of the wife being taller than her husband, if the husband is of average height,
c) what is the probability of the wife being taller than the third quartile of all the wives' heights, if her husband's height is at the third quartile of all the husbands' heights?

a) The conditional expected height of a man whose wife is 61" tall is given by Equation 7.142, with the subscripts switched:

$$E\left(X_1 | X_2 = 61\right) = 68 + 0.25 \cdot 4 \cdot \frac{61 - 64}{3.6} \approx 67.17.$$ (7.156)

b) The conditional variance of X_2, if the husband is of average height, is given by Equation 7.143:

$$Var\left(X_2 | X_1 = 68\right) = \left(1 - 0.25^2\right) 3.6^2 \approx 12.15$$ (7.157)

Thus, under the condition $X_1 = 68$, X_2 is normal with $\mu = 64$ and $\sigma = \sqrt{12.15} \approx 3.49$. Therefore

$$P\left(X_2 > 68 | X_1 = 68\right) = 1 - \Phi\left(\frac{68 - 64}{3.49}\right) \approx 0.126.$$ (7.158)

c) The z-value for the third quartile is, from $P(Z \leq z) = .75$, $z_{.75} \approx 0.6745$. Thus, the third quartile of all the wives' heights is $x_{2,.75} = \mu_2 + \sigma_2 z_{.75} \approx 64 + 3.6 \cdot 0.6745 \approx 66.428$" and the third quartile of all the husbands' heights is $x_{1,.75} = \mu_1 + \sigma_1 z_{.75} \approx 68 + 4 \cdot 0.6745 \approx 70.698$". Hence, under the condition $X_1 = 70.698$, X_2 is normal with $\mu \approx 64 + 0.25 \cdot 3.6 \cdot \frac{70.698 - 68}{4} \approx 64.607$ and $\sigma = \sqrt{\left(1 - 0.25^2\right) 3.6^2} \approx 3.4857$. Therefore

$$P\left(X_2 > 66.428 | X_1 = 70.698\right) \approx 1 - \Phi\left(\frac{66.428 - 64.607}{3.4857}\right)$$
$$\approx 1 - \Phi\left(0.5224\right) \approx 0.321.$$ (7.159)

Note that if we write X_1 and X_2 in standard units as Z_1 and Z_2, then the condition $X_1 = 70.698$ corresponds to $Z_1 = 0.6745$ and under this condition, by Equation 7.145, Z_2 is normal with $\mu = \rho z_1 \approx 0.25 \cdot 0.6745 \approx 0.1686$ and $\sigma = \sqrt{(1 - 0.25^2)} \approx 0.96825$. Thus,

$$
\begin{aligned}
P(X_2 > x_{2,.75} | X_1 = x_{1,.75}) &= P(Z_2 > z_{.75} | Z_1 = z_{.75}) \\
&\approx 1 - \Phi\left(\frac{0.6745 - 0.1686}{0.96825}\right) \\
&\approx 1 - \Phi(0.5224) \approx 0.321, \qquad (7.160)
\end{aligned}
$$

the same as before. As this calculation in standard units shows, the result does not depend on $\mu_1, \mu_2, \sigma_1, \sigma_2$, but only on ρ. ♦

Example 7.5.4. Density with a Homogeneous Quadratic Exponent.

Let X_1 and X_2 have a joint density of the form

$$
f(x_1, x_2) = C \exp \frac{-1}{2}(x_1^2 - 2x_1 x_2 + 4x_2^2), \qquad (7.161)
$$

where C is an appropriate constant. Show that (X_1, X_2) is a bivariate normal pair and find its parameters and C.

This problem could be solved by integration, but it is much easier to just compare the exponents in Equations 7.150 and 7.161, which is what we shall do.

First, clearly, $\mu_1 = \mu_2 = 0$, and the equality of the exponents requires that we solve

$$
\frac{1}{(1 - \rho^2)}\left(\frac{x_1^2}{\sigma_1^2} - 2\rho\frac{x_1 x_2}{\sigma_1 \sigma_2} + \frac{x_2^2}{\sigma_2^2}\right) = ax_1^2 - 2bx_1 x_2 + cx_2^2 \quad \text{for all } x_1, x_2, \quad (7.162)
$$

for the unknowns σ_1, σ_2 and ρ, with $a = b = 1$ and $c = 4$. Hence

$$
a = \frac{1}{(1 - \rho^2)\sigma_1^2}, \quad b = \frac{\rho}{(1 - \rho^2)\sigma_1 \sigma_2}, \quad c = \frac{1}{(1 - \rho^2)\sigma_2^2} \qquad (7.163)
$$

and so

$$
\rho^2 = \frac{b^2}{ac} = \frac{1}{4}, \quad 1 - \rho^2 = \frac{ac - b^2}{ac} = \frac{3}{4} \qquad (7.164)
$$

and

$$
\sigma_1^2 = \frac{1}{a(1 - \rho^2)} = \frac{c}{ac - b^2} = \frac{4}{3}, \quad \sigma_2^2 = \frac{1}{c(1 - \rho^2)} = \frac{a}{ac - b^2} = \frac{1}{3}. \qquad (7.165)
$$

Since also $\sigma_1 > 0$, $\sigma_2 > 0$ and $\text{sign}(\rho) = \text{sign}(b)$, we obtain $\sigma_1 = \frac{2}{\sqrt{3}}$, $\sigma_2 = \frac{1}{\sqrt{3}}$, $\rho = \frac{1}{2}$ and $C = \frac{1}{2\pi\sqrt{(1-\rho^2)}\sigma_1 \sigma_2} = \frac{\sqrt{3}}{2\pi}$.

Thus, we have found the values of the parameters that make the density given by Equation 7.161 correspond to a bivariate normal density, thereby also showing that (X_1, X_2) is a bivariate normal pair.   ♦

The method of the foregoing example yields the following generalization:

Theorem 7.5.3. *Bivariate Normal Density with General Quadratic* **Exponent.** *A pair of random variables (X_1, X_2) is bivariate normal if and only if its density is of the form*

$$f(x_1, x_2)$$
$$= C \exp \left(\frac{-1}{2} \left[a \left(x_1 - \mu_1 \right)^2 - 2b \left(x_1 - \mu_1 \right) \left(x_2 - \mu_2 \right) + c \left(x_2 - \mu_2 \right)^2 \right] \right),$$

$$(7.166)$$

for any constants a, b, c, satisfying[10] $a > 0$ and $ac - b^2 > 0$, and $C = \frac{\sqrt{ac - b^2}}{2\pi}$.

Example 7.5.5. *Density with an Inhomogeneous Quadratic Exponent.*

Let X_1 and X_2 have a joint density of the form

$$f(x_1, x_2) = A \exp \left(\frac{-1}{2} (x_1^2 - 2x_1 x_2 + 4x_2^2 - 4x_1 + 10x_2) \right), \qquad (7.167)$$

where A is an appropriate constant. Show that (X_1, X_2) is a bivariate normal pair and find its parameters and A.

First, we want to put $f(x_1, x_2)$ in the form of Equation 7.166. If we expand the terms in Equation 7.166 and compare the result with Equation 7.167. Since variances and covariances do not depend on the values of μ_1 and μ_2, we can set $\mu_1 = \mu_2 = 0$ in Equation 7.166. Thus we find that σ_1, σ_2 and ρ depend only on the quadratic terms, and therefore, together with $a = b = 1$ and $c = 4$, are the same as in Example 7.5.4.

To find μ_1 and μ_2, we may compare the first degree terms of the exponents in Equations 7.166 and 7.167. So, we must have

$$(-2a\mu_1 + 2b\mu_2) x_1 = -4x_1 \quad \text{and} \quad (-2c\mu_2 + 2b\mu_1) x_2 = 10x_2, \qquad (7.168)$$

that is,

$$-\mu_1 + \mu_2 = -2 \quad \text{and} \quad \mu_1 - 4\mu_2 = 5. \qquad (7.169)$$

Thus, $\mu_1 = 1$ and $\mu_2 = -1$ and

$$f(x_1, x_2) = C \exp \left(\frac{-1}{2} \left[(x_1 - 1)^2 - 2 (x_1 - 1) (x_2 + 1) + 4 (x_2 + 1)^2 \right] \right),$$

$$(7.170)$$

---

[10] A quadratic form (that is, a polynomial with quadratic terms only) whose coefficients satisfy these conditions is called *positive definite*, because its values are then positive for any choice of x_1 and x_2. (See any linear algebra text.)

where $C = \frac{\sqrt{3}}{2\pi}$ as in Example 7.5.4, and A is C times the exponential of the constant term in the quadratic expression above, that is, $A = \frac{\sqrt{3}}{2\pi} \exp\left(\frac{-1}{2} \left[(-1)^2 - 2(-1)(1) + 4(1)^2 \right] \right) = \frac{\sqrt{3}}{2\pi} e^{-7/2}.$

Thus, by putting $f(x_1, x_2)$ in the form of Equation 7.166, we have shown that (X_1, X_2) is a bivariate normal pair and found all of its parameters.   ◆

We have another straightforward consequence of Definition 7.5.1:

Theorem 7.5.4. *Linear Combinations of Bivariate Normals are Bivariate Normal.* *If X_1 and X_2 form a bivariate normal pair and, for any constants $c_{11}, c_{12}, c_{21}, c_{22}, d_1$ and d_2 satisfying $c_{11}^2 + c_{12}^2 \neq 0, c_{21}^2 + c_{22}^2 \neq 0$ and $c_{11}c_{22} - c_{12}c_{21} \neq 0$,*

$$T_1 = c_{11}X_1 + c_{12}X_2 + d_1 \text{ and}$$
$$T_2 = c_{21}X_1 + c_{22}X_2 + d_2, \tag{7.171}$$

then T_1 and T_2 are a bivariate normal pair, too.

Proof. Substituting X_1 and X_2 from Equations 7.134 into the definition of T_1 and T_2, we get the latter as linear functions of Z_1 and Z_2, which shows that they are a bivariate normal pair, too.   ∎

Corollary 7.5.1. *Existence of Independent Linear Combinations.* *If X_1 and X_2 form a bivariate normal pair, then there exist constants $c_{11}, c_{12}, c_{21}, c_{22}$, satisfying $c_{11}^2 + c_{12}^2 \neq 0, c_{21}^2 + c_{22}^2 \neq 0$ and $c_{11}c_{22} - c_{12}c_{21} \neq 0$, such that*

$$T_1 = c_{11}X_1 + c_{12}X_2 \text{ and}$$
$$T_2 = c_{21}X_1 + c_{22}X_2 \tag{7.172}$$

are independent normal random variables.

Proof. Suppose X_1 and X_2 are given in terms of their parameters $\sigma_1 > 0$, $\sigma_2 > 0, \rho, \mu_1, \mu_2$. If $\rho = 0$, then X_1 and X_2 are themselves independent, and so assume that $\rho \neq 0$. Clearly,

$$\begin{aligned} Cov(T_1, T_2) &= c_{11}c_{21}Var(X_1) \\ &\quad + (c_{11}c_{22} + c_{12}c_{21}) Cov(X_1, X_2) + c_{12}c_{22}Var(X_2) \\ &= c_{11}c_{21}\sigma_1^2 + (c_{11}c_{22} + c_{12}c_{21}) \sigma_1\sigma_2\rho + c_{12}c_{22}\sigma_2^2. \end{aligned} \tag{7.173}$$

If we set, for instance, $c_{21} = 0, c_{11} = c_{22} = 1$ and $c_{12} = -\rho\sigma_1/\sigma_2$, then $Cov(T_1, T_2) = 0$ and the inequalities required of the c_{ij}'s are also satisfied, and so T_1 and T_2 are independent. There exist infinitely many other solutions as well. One, a rotation, given in Exercise 7.5.5, is especially interesting.   ∎

Definition 7.5.1 can easily be generalized to more than two variables:

Definition 7.5.2. *Multivariate Normal Random Variables. For any integers* $m, n \geq 2$, *let* $Z_1, Z_2, \ldots, Z_n$ *be independent standard normal random variables and* a_{ij} *and* b_i, *for all* $i = 1, 2, \ldots, m$, $j = 1, 2, \ldots, n$, *any constants satisfying* $\sum_{j=1}^{n} a_{ij}^2 \neq 0$ *for all* i. *Then the random variables*

$$X_i = \sum_{j=1}^{n} a_{ij} Z_j + b_i \quad \text{for} \quad i = 1, 2, \ldots, m \qquad (7.174)$$

are said to form a multivariate normal m-tuple.

Note that the joint distribution of the X_i may be less than n-dimensional. (It is n-dimensional if and only if the matrix (a_{ij}) has rank n.)

Next, we state some theorems about multivariate normal random variables without proof.

Theorem 7.5.5. *Two Linearly Independent Linear Combinations of Independent Normals Are Bivariate Normal. Any two of the* X_i *defined above, say* X_i *and* X_k, *form a bivariate normal pair, provided that neither* $X_i - b_i$ *nor* $X_k - b_k$ *is a scalar multiple of the other.*

Theorem 7.5.6. *For Multivariate Normal Random Variables Zero Covariances Imply Independence. If* $X_1, X_2, \ldots, X_m$ *form a multivariate normal m-tuple and* $Cov(X_i, X_k) = 0$ *for all* i, k, *then* $X_1, X_2, \ldots, X_m$ *are totally independent.*

Theorem 7.5.7. *Density Function of Multivariate Normal Random Variables.* $X_1, X_2, \ldots, X_m$ *form a multivariate normal m-tuple if and only if their joint density is of the form*

$$f(x_1, x_2, \ldots, x_m) = C \exp\left[-\frac{1}{2} \sum_{i=1}^{m} \sum_{k=1}^{m} c_{ik}(x_i - \mu_i)(x_k - \mu_k)\right], \quad (7.175)$$

where the μ_i *and* μ_k *are any constants and the* c_{ik} *are such that the quadratic form* $\sum_{i=1}^{m} \sum_{k=1}^{m} c_{ik}(x_i - \mu_i)(x_k - \mu_k)$ *is positive semidefinite[11] and* C *is a normalizing constant.*

The last theorem could be sharpened by giving an explicit formula for C and relating the c_{ik} to the covariances (the μ_i are the expected values), but even to state these relations would require concepts from linear algebra and the proof would require multivariable calculus. We leave such matters to more advanced books.

---

[11] A quadratic form is positive semidefinite if its value is ≥ 0 for all arguments. For instance, in two dimensions $(x + y)^2$ is positive semidefinite.

Exercises

Exercise 7.5.1.

Show that Y_1 and Y_2 defined by Equations 7.139 are standard normal and $Cov(Y_1, Y_2) = 0$.

Exercise 7.5.2.

Let (X_1, X_2) be a bivariate normal pair with parameters $\mu_1 = 2$, $\mu_2 = -1, \sigma_1 = 3, \sigma_2 = 2$ and $\rho = 0.8$. Find

1. $f(x_1, x_2)$,
2. $E(X_2|X_1 = x_1)$ and $E(X_1|X_2 = x_2)$,
3. $Var(X_2|X_1 = x_1)$ and $Var(X_1|X_2 = x_2)$,
4. $f_{X_2|X_1}(x_2|x_1)$ and $f_{X_1|X_2}(x_1|x_2)$,
5. $f_{X_1}(x_1)$ and $f_{X_2}(x_2)$.

Exercise 7.5.3.

Let X_1 and X_2 have a joint density of the form

$$f(x_1, x_2) = A \exp \frac{-1}{2}(x_1^2 + x_1 x_2 + 2x_2^2 - 2x_1 + 6x_2), \tag{7.176}$$

where A is an appropriate constant. Show that (X_1, X_2) is a bivariate normal pair and find its parameters and A.

Exercise 7.5.4.

Show that if (X, Y) has the joint density

$$f(x, y) = \frac{1}{2\pi} \left[\left(\sqrt{2}e^{-x^2/2} - e^{-x^2} \right) e^{-y^2} + \left(\sqrt{2}e^{-y^2/2} - e^{-y^2} \right) e^{-x^2} \right], \tag{7.177}$$

which is not bivariate normal and then X and Y have standard normal marginal densities and their covariance is zero, but they are not independent.

Exercise 7.5.5.

Let (X_1, X_2) be a bivariate normal pair with $Cov(X_1, X_2) \neq 0$. Show that the rotation

$$T_1 = X_1 \cos\theta - X_2 \sin\theta$$
$$T_2 = X_1 \sin\theta + X_2 \cos\theta \tag{7.178}$$

by the angle θ results in independent normal T_1 and T_2 if and only if

$$\cot 2\theta = \frac{Var(X_2) - Var(X_1)}{2Cov(X_1, X_2)}. \tag{7.179}$$

Exercise 7.5.6.

What is the probability that the average score of a randomly selected student in the two exams of Example 7.5.2 will be over 80?

Exercise 7.5.7.

What is the 90th percentile score in the second exam of Example 7.5.2 for those students who scored 80 on the first exam?

Exercise 7.5.8.

The heights and weights of a large number of men were found to have a bivariate normal distribution with $\rho = 0.7$. If a randomly selected man's height from this population is at the third quartile, then what is the percentile rank of the expected value of his weight under this condition?

Exercise 7.5.9.

Prove that a pair (X_1, X_2) of random variables is bivariate normal if and only if $Y = aX_1 + bX_2$ is normal for every choice of constants a and b, not both zero.

Exercise 7.5.10.

What is the probability of the wife being taller than her husband for a randomly selected couple from the population described in Example 7.5.3 (without any restriction on the husband's height)?

Exercise 7.5.11.

Let (X_1, X_2) be a bivariate normal pair with parameters $\mu_1 = 2$, $\mu_2 = -1, \sigma_1 = 3, \sigma_2 = 2$ and $\rho = 0.8$. Find the parameters and the joint density of $U_1 = X_1 + 2X_2$ and $U_2 = X_1 - 2X_2 + 1$.

Exercise 7.5.12.

Prove algebraically that the center of any vertical chord of an ellipse given by Equation 7.151 lies on the regression line given by Equation 7.152.

8. The Elements of Mathematical Statistics

8.1 Estimation

In probability theory, we always assumed that we knew some probabilities, and we computed other probabilities or related quantities from those. On the other hand, in mathematical statistics, we use observed data to compute probabilities or related quantities or to make decisions or predictions.

The problems of mathematical statistics are classified as *parametric or nonparametric*, depending on how much we know or assume about the distribution of the data. In parametric problems, we assume that the distribution belongs to a given family, for instance, that the data are observations of values of a normal random variable, and we want to determine a parameter or parameters, such as μ or σ. In nonparametric problems, we make no assumption about the distribution and want to determine either single quantities like $E(X)$ or the whole distribution, that is, $F(x)$ or $f(x)$, or to use the data for decisions or predictions.

We begin with some essential terminology. First, we restate and somewhat expand Definition 7.2.3:

Definition 8.1.1. *Random Sample, Statistic, Sample Mean, and Sample Variance.* *n independent and identically distributed (abbreviated: i.i.d.) random variables $X_1, \ldots, X_n$ are said to form a random sample of size n from their common distribution. Any function $g(X_1, \ldots, X_n)$ of the sample variables is called a statistic. The particular statistics $\overline{X}_n = \frac{1}{n} \sum X_i$ and $\widehat{\Sigma}^2 = \frac{1}{n} \sum_{i=1}^{n} \left(X_i - \overline{X}_n \right)^2$ are called the sample mean and sample variance, respectively. The probability distribution of a statistic is sometimes called a sampling distribution.*

Suppose that the common p.f. or p.d.f. of the X_i is $f(x)$ or, if we want to indicate the dependence on a parameter, $f(x; \theta)$. We shall denote the joint p.f. or p.d.f. of $(X_1, \ldots, X_n)$ by $f_n(\mathbf{x})$ or $f_n(\mathbf{x}; \theta)$, where we use the vector abbreviations $\mathbf{x} = (x_1, \ldots, x_n)$ for the possible values of $\mathbf{X} = (X_1, \ldots, X_n)$. We shall use the general notation θ for vector-valued parameters, too, for example, for (μ, σ) in case of normal X_i.

© Springer International Publishing Switzerland 2016
G. Schay, *Introduction to Probability with Statistical Applications*,
DOI 10.1007/978-3-319-30620-9_8

Definition 8.1.2. *Estimator and Estimate. Given a random sample* **X** *whose distribution depends on an unknown parameter* θ, *a statistic* $g(\mathbf{X})$ *is called an estimator of* θ *if, for any observed value* **x** *of* **X**, $g(\mathbf{x})$ *is considered to be an estimate of* θ. *The estimator* $g(\mathbf{X})$ *is a random variable, and to emphasize its connection to* θ, *we sometimes denote it by* $\widehat{\Theta}$. *The observed value* $g(\mathbf{x})$ *is a number (or a vector), which we also denote by* $\widehat{\theta}$.

The most commonly used method for obtaining estimators and estimates is the following.

Definition 8.1.3. *Method of Maximum Likelihood. Consider a random sample* **X** *whose distribution depends on an unknown parameter* θ. *For any fixed* **x**, *the function* $f_n(\mathbf{x}; \theta)$ *regarded as a function* $L(\theta)$ *of* θ, *is called the likelihood function of* θ. *A value* $\widehat{\theta}$ *of* θ *that maximizes* $L(\theta)$ *is called a maximum likelihood estimate of* θ. *In many important applications,* $\widehat{\theta}$ *exists, is unique, and is a function of* **x**. *For* $\widehat{\theta} = g(\mathbf{x})$, *we call the random variable* $\widehat{\Theta} = g(\mathbf{X})$ *the maximum likelihood estimator of* θ. *We abbreviate both maximum likelihood estimate and maximum likelihood estimator as MLE.*

The reasoning behind this method is that we feel that the most likely value, among the various possible values of θ, should be one that makes the probability (or probability density) of the observed **x** as high as possible. For example, consider a sample of just one observation x from a normal distribution with unknown mean μ (the general parameter θ is now μ) and known σ. If we observe $x = 2$, then among the p.d.f. curves shown in Figure 8.1, the right-most one, with $\mu = 2$, is the most likely to have generated this x, and so we choose $\widehat{\mu} = 2$, because that choice gives the highest probability to X being near the observed $x = 2$.

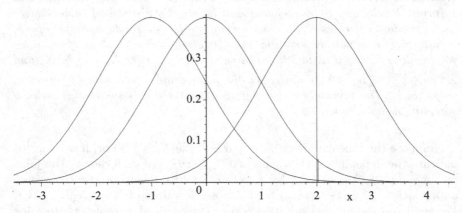

Fig. 8.1. Three possible p.d.f.'s from which the observation $x = 2$ was generated.

Next, we present several examples of the use of the method.

Example 8.1.1. Estimating the Probability of an Event.

Consider any event A in any probability space and let p denote its unknown probability. Let X be a Bernoulli r.v. with parameter $\theta = p$ so that $X = 1$ if A occurs and 0 otherwise. (Such an X is called the *indicator function* of A.) To estimate p, we perform the underlying experiment n times and observe the corresponding i.i.d. Bernoulli random variables $X_1, \ldots, X_n$. The p.f. of X_i can be written as

$$f(x_i; p) = p^{x_i} (1 - p)^{1 - x_i} \text{ for } x_i = 0, 1. \tag{8.1}$$

Hence the likelihood function of p is

$$L(p) = f_n(\mathbf{x}; p) = \prod_{i=1}^{n} p^{x_i} (1 - p)^{1 - x_i} = p^{\sum x_i} (1 - p)^{n - \sum x_i}. \tag{8.2}$$

To find the maximum of $L(p)$, we may differentiate $\ln(L(p))$:

$$\ln(L(p)) = \sum x_i \ln p + \left(n - \sum x_i\right) \ln(1 - p), \tag{8.3}$$

and

$$\frac{d}{dp} \ln(L(p)) = \sum x_i \frac{1}{p} - \left(n - \sum x_i\right) \frac{1}{1 - p}. \tag{8.4}$$

Setting this expression to 0, dividing by n, and writing $\bar{x}_n = \frac{1}{n} \sum x_i$, we get

$$\bar{x}_n \frac{1}{p} - (1 - \bar{x}_n) \frac{1}{1 - p} = 0. \tag{8.5}$$

This equation gives the critical value $p = \bar{x}_n$. The second derivative would show that L has a maximum there, as required. Thus, our maximum likelihood estimate of p is $\hat{p} = \bar{x}_n$, and the corresponding maximum likelihood estimator is $\hat{P} = \bar{X}_n$.   ♦

The estimator above is used to estimate the mean (if it exists) of arbitrary random variables as well and has two noteworthy properties:

1. It is *unbiased*, that is, its expected value is the true (though usually unknown) value of the parameter: $E(\bar{X}_n) = \mu$, by Equation 6.88. (In the Bernoulli case $\mu = p$.)
2. It is *consistent*, meaning that it converges to μ in probability as $n \to \infty$, that is, $\lim_{n \to \infty} P(|\bar{X}_n - \mu| < \varepsilon) = 1$, by the law of large numbers, Theorem 6.2.7.

Example 8.1.2. Estimating the Mean of a Normal Distribution with Known Variance.

For a random sample from an $N(\mu, \sigma^2)$ distribution with known σ and unknown μ, the likelihood function of μ is

$$L(\mu) = f_n(\mathbf{x}; \mu) = \prod_{i=1}^{n} \frac{1}{\sqrt{2\pi}\sigma} e^{-(x_i - \mu)^2/2\sigma^2} = \left(\frac{1}{\sqrt{2\pi}\sigma}\right)^n e^{-\sum(x_i - \mu)^2/2\sigma^2}.$$

$$(8.6)$$

Clearly, this function takes on its maximum when

$$g(\mu) = \frac{1}{2\sigma^2} \sum_{i=1}^{n} (x_i - \mu)^2 \tag{8.7}$$

is minimum. Differentiating and setting $g'(\mu)$ to zero, we get

$$g'(\mu) = -\frac{1}{2\sigma^2} \sum_{i=1}^{n} 2(x_i - \mu) = 0, \tag{8.8}$$

from which

$$\sum_{i=1}^{n} x_i - n\mu = 0. \tag{8.9}$$

Hence $\mu = \frac{1}{n} \sum x_i = \overline{x}_n$ is the critical value of μ. Since $g''(\mu) = \frac{n}{\sigma^2} > 0$, the function g has a minimum and the function L a maximum at $\widehat{\mu} = \overline{x}_n$.

Thus, again, the maximum likelihood estimate is $\overline{x}_n$, and the maximum likelihood estimator is $\widehat{M} = \overline{X}_n$, with the same two properties that were mentioned at the end of the preceding example. ♦

Example 8.1.3. Estimating the Mean and Variance of a Normal Distribution.

For a random sample from an $N(\mu, \sigma^2)$ distribution with unknown μ and σ, the likelihood function is a function of two variables, or, in other words, the parameter θ may be regarded as the two-dimensional vector (μ, σ^2) or as (μ, σ). Thus

$$L(\mu, \sigma) = f_n(\mathbf{x}; \mu, \sigma) = \prod_{i=1}^{n} \frac{1}{\sqrt{2\pi}\sigma} e^{-(x_i - \mu)^2/2\sigma^2} = \left(\frac{1}{\sqrt{2\pi}\sigma}\right)^n e^{-\sum(x_i - \mu)^2/2\sigma^2}.$$

$$(8.10)$$

Now, we need to set the two partial derivatives of L equal to zero and solve the resulting two equations simultaneously. The solution of $\frac{\partial L}{\partial \mu} = 0$ turns out to be independent of σ and exactly the same as in the previous example. So, we get $\widehat{\mu} = \overline{x}_n$ again.

To solve $\frac{\partial L}{\partial \sigma} = 0$, we use logarithmic differentiation:

$$\ln L\left(\mu, \sigma\right) = -n \ln \sqrt{2\pi} - n \ln \sigma - \frac{1}{2\sigma^2} \sum_{i=1}^{n} \left(x_i - \mu\right)^2 \tag{8.11}$$

and

$$\frac{\partial}{\partial \sigma} \ln L\left(\mu, \sigma\right) = -\frac{n}{\sigma} + \frac{1}{\sigma^3} \sum_{i=1}^{n} \left(x_i - \mu\right)^2 = 0, \tag{8.12}$$

which yields

$$\sigma = \left(\frac{1}{n} \sum_{i=1}^{n} \left(x_i - \mu\right)^2\right)^{1/2}. \tag{8.13}$$

Using the second derivative test for functions of two variables, we could show that L has a maximum at these values of μ and σ.

Hence the MLE of the standard deviation is

$$\widehat{\sigma} = \left(\frac{1}{n} \sum_{i=1}^{n} \left(x_i - \overline{x}_n\right)^2\right)^{1/2}. \tag{8.14}$$

We leave it as an exercise to show that the MLE $\widehat{\sigma^2}$ of the variance equals $\widehat{\sigma}^2$.

Next, we are going to show that the corresponding estimator, the sample variance

$$\widehat{\Sigma}^2 = \frac{1}{n} \sum_{i=1}^{n} \left(X_i - \overline{X}_n\right)^2 \tag{8.15}$$

is biased.

Let us first reformulate the sum in the above expression:

$$\sum_{i=1}^{n} \left(X_i - \overline{X}_n\right)^2 = \sum_{i=1}^{n} X_i^2 - 2\overline{X}_n \sum_{i=1}^{n} X_i + n\overline{X}_n^2. \tag{8.16}$$

Substituting $\sum_{i=1}^{n} X_i = n\overline{X}_n$ in the middle term on the right, we get

$$\sum_{i=1}^{n} \left(X_i - \overline{X}_n\right)^2 = \sum_{i=1}^{n} X_i^2 - n\overline{X}_n^2. \tag{8.17}$$

Hence

$$E\left(\frac{1}{n} \sum_{i=1}^{n} \left(X_i - \overline{X}_n\right)^2\right) = \frac{1}{n} \sum_{i=1}^{n} E\left(X_i^2\right) - E\left(\overline{X}_n^2\right). \tag{8.18}$$

Using Equations 6.75 and 6.88, we obtain

$$E\left(\frac{1}{n}\sum_{i=1}^{n}(X_i - \overline{X}_n)^2\right) = \frac{1}{n}\sum_{i=1}^{n}\left[Var\left(X_i^2\right) + \mu^2\right] - \left[Var\left(\overline{X}_n^2\right) + \mu^2\right]$$

$$= \frac{1}{n}\sum_{i=1}^{n}\left[\sigma^2 + \mu^2\right] - \frac{\sigma^2}{n} - \mu^2 = \frac{n-1}{n}\sigma^2.$$

$$(8.19)$$

As the above formula shows, we can define an unbiased estimator of the variance by

$$\widehat{V} = \frac{n}{n-1}\widehat{\Sigma}^2 = \frac{1}{n-1}\sum_{i=1}^{n}(X_i - \overline{X}_n)^2, \tag{8.20}$$

and this is the estimator used by most statisticians, together with the corresponding estimate $\widehat{v}$ instead of $\widehat{\sigma}^2$. In fact, many books call this $\widehat{V}$ the sample variance, and most statistical calculators have keys for both $\widehat{v}$ and $\widehat{\sigma}^2$. For large n, the two estimates differ very little, and the choice is really arbitrary anyway. In principle, however, $\widehat{\sigma}^2$ seems more natural, and although $\widehat{V}$ is an unbiased estimator of the variance, $\sqrt{\widehat{V}}$ is not an unbiased estimator of the standard deviation.[1]   ◆

Let us note that the sample variance (in either form), just as the sample mean, is used to estimate the variance (if it exists) of arbitrary random variables and not just of normal ones. The proof above shows that $\widehat{V}$ is unbiased for those too, and both $\widehat{\Sigma}^2$ and $\widehat{V}$ are consistent by the central limit theorem.

Example 8.1.4. Estimating the Upper Bound of a Uniform Distribution.

Let X be uniform on the interval $[0, \theta]$, with the value of the parameter θ unknown. Then the p.f. of X is given by

$$f(x; \theta) = \begin{cases} \frac{1}{\theta} & \text{if } 0 \leq x \leq \theta \\ 0 & \text{otherwise,} \end{cases} \tag{8.21}$$

and so the likelihood function is given by

$$L(\theta) = \begin{cases} \frac{1}{\theta^n} & \text{if } 0 \leq x_i \leq \theta \text{ for } i = 1, \ldots n \\ 0 & \text{otherwise.} \end{cases} \tag{8.22}$$

Since $\frac{1}{\theta^n}$ is a decreasing function of θ, its maximum occurs at the smallest value of θ that the inequalities $x_i \leq \theta$ allow. (Recall from calculus that

---

[1] The requirement that an estimator be unbiased can lead to absurd results. See M. Hardy, An illuminating counterexample, *Am. Math. Monthly* **110** (2003) 234–238.

the maximum of a continuous function on a closed interval may occur at an endpoint of the interval, rather than at a critical point.) Thus the MLE estimate of θ must be the largest observed value x_i, that is,

$$\widehat{\theta} = \max\{x_1, \ldots, x_n\}. \tag{8.23}$$

Note, however, the curious fact that if X were defined to be uniform on the open interval $(0, \theta)$, rather than on the closed interval, then the MLE would not exist, because then we would have to maximize $L(\theta)$ subject to the conditions $0 < x_i < \theta$ and so $\max\{x_1, \ldots, x_n\}$ would not be a possible value for θ. ◆

For all its many successes and popularity, the method of maximum likelihood does not always work. In some cases, the maximum does not exist or is not unique.

Often another method, the *method of moments* is used to find estimators. This method consists of expressing a parameter as a function of the moments of the r.v. and using the same function of the sample moments as an estimator of the parameter.

Example 8.1.5. Estimating the Parameter of an Exponential Distribution.

Consider an exponential r.v. X with parameter λ. Then, by 6.1.5, $\lambda = 1/E(X)$. Hence, according to the method of moments, we estimate λ by $\widehat{\Lambda} = 1/\overline{X}_n$. On the other hand, by Equation 6.80, $\lambda = 1/SD(X)$ as well, and so we could also estimate λ by $1/\widehat{\Sigma}$. ◆

Example 8.1.6. Estimating the Parameter of a Poisson Distribution.

Consider a Poisson r.v. X with parameter λ. Then, by Theorem 7.1.2 $\lambda = E(X) = Var(X)$. Thus, the method of moments suggests the estimator $\widehat{\Lambda} = \overline{X}_n$ or $\widehat{\Lambda} = \widehat{\Sigma}^2$. ◆

Another popular method for obtaining estimators is based on Bayes' theorem, but we shall not discuss it here.

On the other hand, even the best estimator can be off the mark. For instance, if we toss a fair coin, say, ten times, we may easily get six heads, and so by Example 8.1.1, we would estimate p as 0.6. Consequently, we want to know how much confidence we can have in such an estimate. This question is usually answered by constructing intervals around the estimate so that these intervals cover the true value of the parameter with a given high probability. In other words, we construct interval estimates instead of point estimates.

Example 8.1.7. Interval Estimates of the Mean of a Normal Distribution with Known Variance.

Let X_i be i.i.d. $N(\mu, \sigma^2)$ random variables for $i = 1, \ldots, n$. Then, by Corollary 7.2.3, the sample mean $\overline{X}_n$ is $N(\mu, \sigma^2/n)$, and so

$$P\left(\left|\frac{\overline{X}_n - \mu}{\sigma/\sqrt{n}}\right| < c\right) = 2\Phi(c) - 1 \tag{8.24}$$

for any $c > 0$, or, equivalently,

$$P\left(\overline{X}_n - c\frac{\sigma}{\sqrt{n}} < \mu < \overline{X}_n + c\frac{\sigma}{\sqrt{n}}\right) = 2\Phi(c) - 1. \tag{8.25}$$

If we assume that σ is known and μ is unknown, then Equation 8.25 can be interpreted as saying that the random interval $\left(\overline{X}_n - c\frac{\sigma}{\sqrt{n}}, \overline{X}_n + c\frac{\sigma}{\sqrt{n}}\right)$ contains the unknown, but fixed, parameter μ with probability $2\Phi(c) - 1$. We must emphasize that this statement is different from our usual probability statements in which we were concerned with a random variable falling in a fixed interval. Here the μ is fixed, and the endpoints of the interval are random variables, because $\overline{X}_n$ is a statistic computed from a random sample.

Now, if we observe a value $\overline{x}_n$ of $\overline{X}_n$, then the fixed, and no longer random, interval

$$\left(\overline{x}_n - c\frac{\sigma}{\sqrt{n}}, \overline{x}_n + c\frac{\sigma}{\sqrt{n}}\right) \tag{8.26}$$

is called a *confidence interval* for μ with *confidence coefficient or level* $\gamma = 2\Phi(c) - 1$, or a 100γ *percent confidence interval*. We cannot say that μ falls in this interval with *probability* γ, because neither μ nor the interval is random, and that is why we use the word "confidence" rather than "probability." The corresponding probability statement, Equation 8.25, implies that if we observe many such confidence intervals from different samples, that is, with different observed values for $\overline{x}_n$, then approximately 100γ percent of them will contain μ. Whether a single such interval will contain μ or not, we usually cannot say. What we can always say is that, by its definition, our interval is a member of a large set of similar potential intervals, 100γ percent of which do contain μ.

It was natural for us to start our discussion of confidence intervals with an arbitrary value for c, but, in applications, it is more common to start with given confidence coefficients γ. Then c can be computed as

$$c = \Phi^{-1}\left(\frac{\gamma + 1}{2}\right). \tag{8.27}$$

Thus, for instance, if we want a 95% confidence interval for μ, then $\gamma = .95$ yields $c = \Phi^{-1}\left(\frac{1.95}{2}\right) = \Phi^{-1}(.975) \approx 1.96$, that is, c will be the 97.5th percentile of the standard normal distribution, which is approximately 1.96. Hence, for an observed sample mean $\overline{x}_n$, the interval $\left(\overline{x}_n - 1.96\frac{\sigma}{\sqrt{n}}, \overline{x}_n + 1.96\frac{\sigma}{\sqrt{n}}\right)$ is a 95% confidence interval for μ.   ♦

We generalize the concepts introduced in the above example as follows.

Definition 8.1.4. *Confidence Intervals*. *Consider a random sample X whose distribution depends on an unknown parameter θ and two statistics $A = g_1(\mathbf{X})$ and $B = g_2(\mathbf{X})$ with $A < B$. If a and b are any observed values of A and B and $P(A < \theta < B) = \gamma$, then (a, b) is called a 100γ percent confidence interval for θ, and γ the confidence coefficient or confidence level of the interval (a, b).[2] If $A = -\infty$ or $B = \infty$, then $(-\infty, b)$ and (a, ∞) are called one-sided confidence intervals.*

The construction in Example 8.1.7 of confidence intervals for the mean of a normal distribution can be used for the mean of other distributions or with unknown σ, in case of *large samples*, when the CLT is applicable. In case σ is unknown, we just use $\widehat{\sigma}$ from Equation 8.14 in Equation 8.25.

Example 8.1.8. Confidence Interval for the Probability of an Event.

As in Example 8.1.1, consider any event A in any probability space and let p denote its unknown probability. Let X be a Bernoulli r.v. with parameter p so that $X = 1$ if A occurs and 0 otherwise. To estimate p, we perform the underlying experiment n times and observe the corresponding i.i.d. Bernoulli random variables $X_1, \ldots, X_n$. As in Example 8.1.1, let $\widehat{P} = \overline{X}_n$ and $\widehat{p} = \overline{x}_n$. We use the sample variance (Equation 8.15) as the estimator of the variance of X. In the present case, $\sum_{i=1}^n X_i^2 = \sum_{i=1}^n X_i$ because each X_i is 0 or 1, and so

$$\widehat{\Sigma}^2 = \frac{1}{n} \sum_{i=1}^n \left(X_i - \overline{X}_n\right)^2 = \frac{1}{n} \sum_{i=1}^n X_i^2 - \overline{X}_n^2$$

$$= \frac{1}{n} \sum_{i=1}^n X_i - \overline{X}_n^2 = \overline{X}_n - \overline{X}_n^2 = \widehat{P}\left(1 - \widehat{P}\right). \tag{8.28}$$

Having observed the values $x_1, \ldots, x_n$, we use the corresponding estimate

$$\widehat{\sigma}^2 = \widehat{p}(1 - \widehat{p}) \tag{8.29}$$

of σ^2. Notice that this estimate is the same as that which we would get by replacing p in Equation 6.89 by $\widehat{p}$.

Now, if n is large, then, by the de Moivre-Laplace theorem, the distribution of $\widehat{P} = \overline{X}_n$ is approximately normal, and so, for any $c > 0$, the interval

$$\left(\widehat{p} - c\sqrt{\frac{\widehat{p}\,(1 - \widehat{p})}{n}},\ \widehat{p} + c\sqrt{\frac{\widehat{p}\,(1 - \widehat{p})}{n}}\right) \tag{8.30}$$

from Example 8.1.7 is an approximate confidence interval for p with confidence level $\gamma = 2\Phi(c) - 1$, provided both endpoints lie between 0 and 1.

---

[2] Some people use a slightly different terminology. They call the random interval (A, B) a confidence interval and (a, b) its observed value.

If one of the endpoints lies outside $[0, 1]$, then, since p is a probability, we use a one-sided confidence interval. For instance, if $\widehat{p} + c\sqrt{\frac{\widehat{p}(1-\widehat{p})}{n}} > 1$, then the interval

$$\left(\widehat{p} - c\sqrt{\frac{\widehat{p}(1-\widehat{p})}{n}},\ 1 \right) \tag{8.31}$$

is an approximate confidence interval for p with confidence level $\gamma = 1 - \Phi(-c) = \Phi(c)$, because

$$\gamma = \mathrm{P}\left(\widehat{P} > \widehat{p} - c\sqrt{\frac{\widehat{p}(1-\widehat{p})}{n}} \right) = \mathrm{P}\left(\frac{\widehat{P} - \widehat{p}}{\sqrt{\frac{\widehat{p}(1-\widehat{p})}{n}}} > -c \right) \approx \mathrm{P}(Z > -c). \tag{8.32}$$

$\blacklozenge$

As can be seen from the general definition, a confidence interval does not have to be symmetric about the estimate. In the examples above, however, the symmetric confidence interval was the shortest one. On the other hand, in some applications, we are interested in one-sided confidence intervals, as in the example below.

Example 8.1.9. Voter Poll.

Suppose a politician obtains a poll that shows that 52% of 400 likely voters, randomly selected from a much larger population, would vote for him. What confidence can he have that he would win the election, assuming that there are no changes in voter sentiment until election day?

We need the same setup as in Example 8.1.8. We know the sample size n and the proportion $\widehat{p}$ of favorable voters in the sample and want to find the confidence level of the winning interval[3] $(.50, \infty)$ for the proportion p of favorable voters in the voting population.

The normal approximation gives

$$\mathrm{P}\left(\frac{\overline{X}_n - p}{\sigma/\sqrt{n}} < c \right) = \Phi(c), \tag{8.33}$$

for any $c > 0$, or, equivalently,

$$\mathrm{P}\left(\overline{X}_n - c\frac{\sigma}{\sqrt{n}} < p \right) = \Phi(c). \tag{8.34}$$

Thus, with $\widehat{p} = \overline{x}_n$ and $\widehat{\sigma}^2 = \widehat{p}(1 - \widehat{p})$, the interval $\left(\widehat{p} - c\sqrt{\frac{\widehat{p}(1-\widehat{p})}{n}},\ \infty \right)$ is a $\gamma = \Phi(c)$ level confidence interval for p.

---

[3] Of course, a probability cannot be greater than 1, and so the upper limit of the interval should be 1 rather than ∞, but the normal approximation gives only a minuscule probability to the $(1, \infty)$ interval, and we may therefore ignore this issue.

From the given data, $\widehat{p} = .52$ and $n = 400$ and so

$$\left(\widehat{p} - c\sqrt{\frac{\widehat{p}(1-\widehat{p})}{n}}, \ \infty\right) = (0.52 - 0.02498c, \ \infty). \tag{8.35}$$

The politician wants to know the confidence level of the $(0.50, \infty)$ interval. So, we need to solve $0.52 - 0.02498c = 0.50$, and we get $c = 0.80$ and $\gamma = \Phi(0.80) = 0.788$. Thus, by this poll, he can have approximately 78.8% confidence in winning the election.   ◆

Exercises

Exercise 8.1.1.

In Equation 8.158, replace σ^2 by v and differentiate with respect to v to show that the MLE $\widehat{\sigma^2} = \widehat{v}$ of the variance equals $\widehat{\sigma}^2$.

Exercise 8.1.2.

Find the MLE for the parameter λ of an exponential r.v.

Exercise 8.1.3.

Show that $\widehat{\sigma}^2$, as given by Equation 8.14 for a normal r.v. X and n distinct values $x_1, \ldots, x_n$, equals the variance of a discrete r.v. X^* with n distinct, equally likely possible values $x_1, \ldots, x_n$.

Exercise 8.1.4.

Find the MLE $\widehat{\lambda}$ for the parameter λ of a Poisson r.v. (Note that this MLE does not exist if all observed values equal 0.)

Exercise 8.1.5.

Let X be a continuous r.v. whose p.d.f., for $\lambda > 0$, is given by

$$f(x; \lambda) = \begin{cases} \lambda x^{\lambda-1} & \text{if } 0 < x < 1 \\ 0 & \text{otherwise.} \end{cases} \tag{8.36}$$

a)   Find the MLE for the parameter λ.
b)   Find an estimator for λ by the method of moments. (Hint: First compute $E(X)$.)

Exercise 8.1.6.

Let X be uniform on the interval $[\theta_1, \theta_2]$. Find the MLEs of θ_1 and θ_2. (Hint: The extrema occur at the endpoints of an interval.)

Exercise 8.1.7.

Let X be uniform on the interval $(0, \theta)$. Show that $\Theta = \frac{n+1}{n} \max(X_1, \ldots, X_n)$ is an unbiased estimator of θ.

Exercise 8.1.8.

A random sample of 50 cigarettes of a certain brand of cigarettes is tested for nicotine content. The measurements result in a sample mean $\widehat{\mu} = 20\text{mg}$ and sample SD $\widehat{\sigma} = 4\text{mg}$. Find 90, 95, and 99% confidence intervals for the unknown mean nicotine content μ of this brand, using the normal approximation.

Exercise 8.1.9.

A random sample of 500 likely voters in a city is polled, and 285 are found to be Democrats. Find 90, 95, and 99% approximate confidence intervals for the percentage of Democrats in the city.

Exercise 8.1.10.

In a certain city, the mathematics SAT scores of a random sample of 100 students are found to have mean $\widehat{\mu}_1 = 520$ in 2002, and of another random sample of 100 students, $\widehat{\mu}_2 = 533$ in 2003, with the same SD $\widehat{\sigma}_1 = \widehat{\sigma}_2 = 60$ in both years. The question is whether there is a real increase in the average score *for the whole city*, or is the increase due only to chance fluctuation in the samples. Find the confidence level of the one-sided confidence interval $(0, \infty)$ for the difference $\mu = \mu_2 - \mu_1$. Use the normal approximation and $\widehat{\sigma}^2 = \widehat{\sigma}_1^2 + \widehat{\sigma}_2^2$. What conclusion can you draw from the result?

8.2 Testing Hypotheses

In many applications, we do not need to estimate the value of a parameter θ, we just need to decide which of two nonoverlapping sets, Ω_0 or Ω_A, it is likely to lie in. The assumption that it falls in Ω_0 is called the *null hypothesis* H_0 and that it falls in Ω_A, the *alternative hypothesis* H_A.

Often, we want to test some treatment of the population under study, and then the null hypothesis corresponds to the assumption that the treatment has no effect on the value of the parameter, while the alternative hypothesis to the assumption that the treatment has an effect. In other cases, we may compare two groups and then the null hypothesis corresponds to the assumption that there is no difference between certain parameters for the two groups, while the alternative hypothesis corresponds to the assumption that there is a difference. Based on a test statistic Y from sample data, we wish to accept one of these hypotheses for the population(s) and reject the other. In this section, we consider only one-point sets for H_0, of the form $\Omega_0 = \{\theta_0\}$. Such a hypothesis is called *simple*. Any hypothesis that corresponds to more than one θ value is called *composite*. H_A is mostly considered to be composite, with Ω_A of the form $\{\theta|\theta < \theta_0\}$, $\{\theta|\theta > \theta_0\}$ or $\{\theta|\theta \neq \theta_0\}$. The distribution of a test statistic Y under the assumption H_0 is called its *null distribution*.

Example 8.2.1. Cold Remedy.

Suppose a drug company wants to test the effectiveness of a proposed new drug for reducing the duration of the common cold. The drug is given to $n = 100$ randomly selected patients at the onset of their symptoms. Suppose that the length of the illness in untreated patients has mean $\mu_0 = 7$ days and SD $\sigma = 1.5$ days. Let $\overline{X}$ denote the average length of the cold in a sample of 100 treated patients, and, say, we observe $\overline{X} = 5.2$ in the actual sample. This example fits in the general scheme by the identifications $\theta = \mu$ and $Y = \overline{X}$.

The question is: Is this reduction just chance variation due to randomness in the sample, or is it real, that is, due to the drug? In other words, we want to decide whether the result $\overline{X} = 5.2$ is more likely to indicate that the sample comes from a *population* with mean $\mu = 7$ or one with reduced mean $\mu < 7$. (One may think of this population as the millions of possible users of this drug. Would they see a reduced duration on average?) More precisely, we assume that $\overline{X}$ is normally distributed (by the CLT, even if the individual durations are not) with SD $\sigma/\sqrt{n}$, but with an unknown mean μ. Then we want to decide, based on the observed value of the estimator $\overline{X}$ of μ, which of the hypotheses $\mu = \mu_0$ or $\mu < \mu_0$ to accept. The first of these conditions is H_0 and the second one is H_A. (To be continued.)   ◆

Example 8.2.2. Weight Reduction.

We want to test the effectiveness of a new drug for weight reduction and administer it, say, to a random sample of 36 adult women for a month. Let $\overline{X}$ denote the average weight loss (as a positive value) of these women from the beginning to the end of the month and $\widehat{\Sigma}$ the sample SD of the weight losses. Suppose we observe $\overline{X} = 1.5$ lbs. and $\widehat{\Sigma} = 4$ lbs. By the CLT, we assume that the average weight loss of these women is normally distributed, and we estimate σ by the observed value of $\widehat{\sigma} = 4$ lbs. The mean weight loss μ (of a hypothetical population, from which the sample is drawn) is unknown, and we want to decide whether the observed $\overline{X}$ value supports the null hypothesis $\mu = 0$, that is, whether the observed average weight reduction is just chance variation due to randomness in the sample or it supports the alternative hypothesis $\mu > 0$, that the reduction is real, that is, caused by the drug. (To be continued.)   ♦

Now, how do we decide which hypothesis to accept and which to reject? We use a test statistic Y, like $\overline{X}$ in the examples above, which is an estimator of the unknown parameter θ, and designate a set C such that we reject H_0 and accept H_A when the observed value of Y falls in C, and accept H_0 and reject H_A otherwise. We allow no third choice.[4] This procedure is called a *(statistical) test*, and the set C is called the *rejection region* (we reject H_0) *or the critical region of the test*.[5]

The set C is usually taken to be an interval of the type $[c, \infty)$ or $(-\infty, c]$ or the union of two such intervals. Which of these types of sets to use as C, is determined by the alternative hypothesis. If H_A is of the form $\mu < \mu_0$, as in Example 8.2.1, then H_A is supported by small values of $\overline{X}$, and so we take C to be of the form $(-\infty, c]$. On the other hand, as in Example 8.2.2, if H_A is of the form $\mu > \mu_0$, then H_A is supported by large values of $\overline{X}$, and so we take C to be of the form $[c, \infty)$. Finally, if H_A is of the form $\mu \neq \mu_0$, then we take $C = [\mu_0 + c, \infty) \cup (-\infty, \mu_0 - c]$. (In this section, we assume that H_0 is of the form $\mu = \mu_0$.)

To complete the description of a test, we still need to determine the value of the constant c in the definition of the critical region. We determine c from the probability of making the wrong decision.

There are two types of wrong decisions that we can make: rejecting H_0, when it is actually true, which is called an *error of type 1*, and accepting H_0 when H_A is true, which is called an *error of type 2*. In the examples above, a type 1 error would mean that we accept an ineffective drug, while a type 2 error would mean that we reject a good drug. The usual procedure is to prescribe a small value α for the probability of an error of type 1 and devise a test, that is, a rejection region C, such that the probability of Y falling in C is α if H_0 is true. The probability α of a type 1 error is called the *level of significance* of our test. Thus

$$\alpha = P\left(\overline{X} \in C | H_0\right) = P(\textit{type 1 error}). \tag{8.37}$$

α is traditionally set to be 5% or 1%, and then, knowing the distribution of $\overline{X}$ when H_0 is true, we use Equation 8.37 to determine the set C.

---

[4] However, some statisticians prefer to say "do not reject H_0" rather than "accept H_0" when the test gives just weak evidence in favor of H_0, and they imply or say that the test is inconclusive or that further testing is needed.

[5] In some books, the rejection region is defined to be the set in the n-dimensional space of sample data that corresponds to C.

For $\alpha = 5\%$, we call the observed value of Y, if it does fall in C, *statistically significant*, and if α is set at 1% and Y is observed to fall in C, then we call the result (that is, the observed value of Y, supporting H_A) *highly significant*.

In most cases, statistical tests are based on consideration of type 1 errors alone, as described above. The reason for this is, that we usually want to prove that a new procedure or drug is effective and we publish or use it only if the statistical test rejects H_0. But in that case we can commit only a type 1 error, i.e., we reject H_0 wrongly. In fact, most medical or psychological journals will accept only statistically significant results.

Nevertheless, in some situations, we want or have to accept H_0 and then type 2 errors may arise. We shall discuss them in the next section.

We are now ready to set up the tests for our earlier examples.

Example 8.2.3. Cold Remedy, Continued.

Our test statistic is $\overline{X}$, which we take to be a normal r.v., because n is sufficiently large for the CLT to apply. Because of this use of the CLT, a test of this kind is called a *large-sample Z-test*.

We are interested in the probability of a type 1 error, that is, of wrongly rejecting $H_0 : \mu = 7$, that the drug is worthless, when it actually *is* worthless. So, we assume that H_0 is true and take the parameters of the distribution of $\overline{X}$ to be μ_0 and $\sigma/\sqrt{n}$, which in this case are 7 and $1.5/\sqrt{100} = 0.15$, respectively. Since H_A is of the form $\mu < \mu_0$, we take the rejection region to be of the form $(-\infty, c]$, that is, we reject H_0 if $\overline{X} \leq c$. We determine c from the requirement

$$P\left(\overline{X} \leq c | H_0\right) = \alpha. \tag{8.38}$$

Setting $\alpha = 1\%$, we have

$$P\left(\overline{X} \leq c | H_0\right) = P\left(\frac{\overline{X} - 7}{0.15} \leq \frac{c - 7}{0.15}\right) \approx \Phi\left(\frac{c - 7}{0.15}\right) = .01. \tag{8.39}$$

Hence

$$\frac{c - 7}{0.15} = \Phi^{-1}(.01) \approx -2.33 \tag{8.40}$$

and

$$c \approx 7 - 2.33 \cdot 0.15 \approx 6.65. \tag{8.41}$$

Thus, the observed value $\overline{X} = 5.2$ is $\leq c$, and this result is highly significant. In other words, the null hypothesis, that the result is due to chance, is rejected, and the drug is declared effective. (Whether the reduction of the mean length of the illness from 7 to 5.2 days is important or not, is a different matter, on which statistical theory has nothing to say. We must not mistake statistical significance for importance. The terminology is misleading: a highly significant result may be quite unimportant; its statistical significance just means that the effect is very likely real and not just due to chance.) ♦

Example 8.2.4. Weight Reduction, Continued.

Our test statistic is again $\overline{X}$, which we assume to be normal with mean $\mu_0 = 0$ and SD $\widehat{\sigma}/\sqrt{n} = 4/6$. Since H_A is of the form $\mu > \mu_0$, we take the rejection region to be of the form $[c, \infty)$, that is, we reject H_0 if $\overline{X} \geq c$. We determine c from the requirement

$$\mathrm{P}\left(\overline{X} \geq c|H_0\right) = \alpha. \tag{8.42}$$

Setting $\alpha = 1\%$, we have

$$\mathrm{P}\left(\overline{X} \geq c|H_0\right) \approx \mathrm{P}\left(\frac{\overline{X} - 0}{0.667} \geq \frac{c - 0}{0.667}\right) \approx 1 - \Phi\left(\frac{c}{0.667}\right) = .01. \tag{8.43}$$

Hence

$$\frac{c}{0.667} = \Phi^{-1}(.99) \approx 2.33 \tag{8.44}$$

and

$$c \approx 2.33 \cdot 0.667 \approx 1.55. \tag{8.45}$$

Thus, the observed value $\overline{X} = 1.5$ is $< c$, and this result is not highly significant. At this 1% level, we accept H_0.

On the other hand, setting $\alpha = 5\%$, we determine c from

$$\mathrm{P}\left(\overline{X} \geq c|H_0\right) \approx \mathrm{P}\left(\frac{\overline{X} - 0}{0.667} \geq \frac{c - 0}{0.667}\right) \approx 1 - \Phi\left(\frac{c}{0.667}\right) = .05, \tag{8.46}$$

and we get

$$\frac{c}{0.667} = \Phi^{-1}(.95) \approx 1.645 \tag{8.47}$$

and

$$c \approx 1.645 \cdot 0.667 \approx 1.10. \tag{8.48}$$

Thus, the observed value $\overline{X} = 1.5$ is $\geq c$, and so this result is significant, though, as we have seen above, not highly significant.

In other words, the null hypothesis, that the result is due to chance, is rejected at the $\alpha = 5\%$ level but accepted at the $\alpha = 1\%$ level. The drug may be declared probably effective, but perhaps more testing, that is, a larger sample, is required.

Presenting the result of a test only as the rejection or acceptance of the null hypothesis at a certain level of significance, does not make full use of the information available from the observed value of the test statistic. For instance, in this example, the observed value $\overline{X} = 1.5$ was very close to the $c = 1.55$ value required for a highly significant result, but this information is lost if we merely report what happens at the 1% and 5% levels. In order to convey the maximum amount of information available from an observation, we usually report the lowest significance level at which the observation would lead to a rejection of H_0. Thus, we report $P\left(\overline{X} \geq c | H_0\right)$ for the observed value c of $\overline{X}$. This probability is called the *observed significance level or P-value of the result*. In the last example it is

$$P\left(\overline{X} \geq 1.55 | H_0\right) \approx P\left(\frac{\overline{X} - 0}{0.667} \geq \frac{1.55 - 0}{0.667}\right) \approx 1 - \Phi\left(\frac{1.55}{0.667}\right) \approx 0.0102.$$
(8.49)

◆

In general, we make the following definition:

Definition 8.2.1. P-Value. *The observed significance level or P-value of a result involving a test statistic Y is defined as $P(Y \in C | H_0)$ with the critical region C being determined by the observed value c of Y.*

In Example 8.2.1, for instance, with $c = 5.2$, the P-value is

$$P\left(\overline{X} \leq 5.2 | H_0\right) = P\left(\frac{\overline{X} - 7}{0.15} \leq \frac{5.2 - 7}{0.15}\right) \approx \Phi(-12) \approx 0,$$
(8.50)

which calls for the rejection of H_0 with virtual certainty, in contrast to the relatively anemic 1% obtained above in the weight reduction example.

We summarize the Z-test in the following definition:

Definition 8.2.2. Z-Test. *We use this test for the unknown mean μ of a population if we have a) a random sample of any size from a normal distribution with known σ or b) a large random sample from any distribution so that $\overline{X}$ is nearly normal by the CLT. The null hypothesis is $H_0 : \mu = \mu_0$, where μ_0 is the μ-value we want to test against one of the alternative hypotheses $H_A : \mu > \mu_0, \mu < \mu_0$, or $\mu \neq \mu_0$. The test statistic is*

$$Z = \frac{\overline{X} - \mu_0}{\sigma / \sqrt{n}}$$
(8.51)

in case (a), and

$$Z = \frac{\overline{X} - \mu_0}{\hat{\sigma} / \sqrt{n}}$$
(8.52)

in case (b), where $\hat{\sigma}$ is given by Equation 8.14. Let z denote the observed value of Z, that is, the value computed from the actual sample, that is $z = \frac{\overline{x} - \mu_0}{\sigma / \sqrt{n}}$ or $z = \frac{\overline{x} - \mu_0}{\hat{\sigma} / \sqrt{n}}$, where $\overline{x}$ is the observed value of the random variable $\overline{X}$.

Then, for $H_A : \mu < \mu_0$ the P-value is $\Phi(z)$, for $H_A : \mu > \mu_0$ the P-value is $1 - \Phi(z)$, and for $H_A : \mu \neq \mu_0$ the P-value is $2(1 - \Phi(|z|))$.

We reject H_0 if the P-value is small and accept it otherwise.

Example 8.2.5. Testing Fairness of a Coin; a Two-Tailed Test.

Suppose we want to test whether a certain coin is fair or not. We toss it $n = 100$ times and want to use the relative frequency $\overline{X}$ of heads obtained as our test statistic. We want to find the rejection region that results in a level of significance $\alpha = .05$.

$\overline{X}$ is binomial, but by the CLT, we can approximate its distribution with a normal distribution with parameters $\mu = p = \mathrm{P}(H)$ and $\sigma = \sqrt{pq/100}$. The hypotheses we want to test are $H_0 : p = .5$ and $H_A : p \neq .5$. Thus, for H_0, $\sigma = \sqrt{\frac{1}{2} \cdot \frac{1}{2} \cdot \frac{1}{100}} = .05$. The rejection region should be of the form $C = (-\infty, .5 - c] \cup [.5 + c, \infty) = \overline{(.5 - c, .5 + c)}$. The requirement $\alpha = .05$ translates into finding c such that

$$\mathrm{P}(C|H_0) = \mathrm{P}(|\overline{X} - .5| > c) = \mathrm{P}\left(\frac{|\overline{X} - .5|}{.05} > \frac{c}{.05}\right)$$

$$\approx \mathrm{P}\left(|Z| > \frac{c}{.05}\right) = 2\left(1 - \Phi\left(\frac{c}{.05}\right)\right) = .05, \quad (8.53)$$

or

$$\Phi\left(\frac{c}{.05}\right) = 0.975. \quad (8.54)$$

Hence $c/.05 \approx 1.96$ and $c \approx .098$. So, we accept H_0, that is, declare the coin fair, if $\overline{X}$ falls in the interval $(0.402, 0.598)$, and reject H_0 otherwise. ◆

In many statistical tests, we have to use distributions other than the normal, as in the next example.

Example 8.2.6. Sex Bias in a Jury.

Suppose the 12 members of a jury were selected randomly from a large pool of potential jurors consisting of an equal number of men and women, and the jury ends up with three women and nine men. We wish to test the hypothesis H_0 that the probability p of selecting a woman is $p_0 = 1/2$, versus the alternative H_A that $p < 1/2$. Note that H_0 means that the jury is randomly selected from the general population, about half of which consists of women, and H_A means that the selection is done from a subpopulation from which some women are excluded.

The test statistic we use is the number X of women in the jury. This X is binomial, and under the assumption H_0, it has parameters $n = 12$ and $p_0 = 1/2$. The rejection region is of the form $\{x \leq c\}$, and to obtain the P-value for the actual jury, we must use $c = 3$. Thus, the P-value is

$$\mathrm{P}(X \leq 3|H_0) = \sum_{k=0}^{3} \binom{12}{k} \left(\frac{1}{2}\right)^{12} \approx .073. \quad (8.55)$$

So, although the probability is low, it is possible that there was no sex bias in this jury selection, and we accept the null hypothesis. To be more certain, one way or the other, we would have to examine more juries selected by the same process.   ◆

In cases where the sample is small, the distribution is unknown, and the evidence seems to point very strongly against the null hypothesis; we may use Chebyshev's inequality to estimate the P-value, as in the next example.

Example 8.2.7. Age of First Marriage in Ancient Rome.

Lelis, Percy, and Verstraete[6] studied the ages of Roman historical figures at the time of their first marriage. They did this to refute earlier improbably high age estimates that were based on funerary inscriptions. Others had found that for women, the epitaphs were written by their fathers up to an average age of 19 and after that by their husbands and jumped to the conclusion that women first married at an average age of 19. (A similar estimate of 26 was obtained for men.)

From the historical record, the ages at first marriage of 26 women were 11, 12, 12, 12, 12, 13, 13, 13, 13, 13, 14, 14, 14, 14, 14, 14, 15, 15, 15, 15, 15, 15, 16, 16, 17, 17.

The mean of these numbers is 14.0, and the standard deviation is 1.57.

A random sample of size 26 is just barely large enough to assume that the average is normally distributed with standard deviation $\frac{1.57}{\sqrt{26}} \approx 0.31$; nevertheless, we first assume this but then obtain another estimate without this assumption as well.

This sample, however, is a sample of convenience. We may assume though that it is close to a random sample, at least from the population of upper class women. We also assume that marriage customs remained steady during the centuries covered. (For this reason, we omitted three women for whom records were available from the Christian era.)

We take the null hypothesis to be that the average is 19 and the alternative that it is less. With the above assumptions, we can compute the P-value, that is, the probability that the mean in the sample turns out to be 14 or less if the population mean is 19, as

$$P\left(\overline{X} \leq 14\right) = P\left(\tfrac{\overline{X}-19}{0.31} \leq \tfrac{14-19}{0.31}\right) \approx \Phi\left(\tfrac{14-19}{0.31}\right) \approx \Phi\left(-16\right) \approx 0.$$

Thus, the null hypothesis must be rejected with practical certainty, unless the assumptions can be shown to be invalid.

The ridiculously low number we obtained depends heavily on the validity of the normal approximation, which is questionable. We can avoid it and compute an estimate for the P-value by using Chebyshev's inequality (see Theorem 6.2.6) instead, which is valid for any distribution. Using the latter,

---

[6] A. A. Lelis, W. A. Percy and B. C. Verstraete, *The Age of Marriage in Ancient Rome* (The Edwin Mellen Press, 2003)

we have $P(|\overline{X}_n - \mu| > \varepsilon) = P(|\overline{X}_n - 19| > 5) \leq \frac{\sigma^2}{n\varepsilon^2} \approx \frac{1.57^2}{26 \cdot 5^2} \approx 3.8 \times 10^{-3}$.
This estimate, though very crude (in the sense that the true P-value is probably much lower), is much more reliable than the one above, and it is still sufficiently small to enable us to conclude that the null hypothesis, of an average age 19 at first marriage, is untenable.

So, how can one explain the evidence of the tombstones? Apparently, people were commemorated by their fathers if possible, whether they were married or not at the time of their deaths, and only after the death of the father (who often died fairly young) did this duty fall to the spouse.   ♦

In many applications, we analyze the difference of paired observations. For instance, the difference in the blood pressure of people before and after administering a drug can be used to test the effectiveness of the drug. Similarly, differences in twins (both people and animals) are often used to investigate effects of drugs, when one twin is treated and the other is not. The genetic similarity of the twins ensures that the observed effect is primarily due to the drug and not to other factors.

Example 8.2.8. Smoking and Bone Density.

The effect of smoking on bone density was investigated by studying pairs of twin women.[7] A reduction in bone density is an indicator of osteoporosis, a serious disease, mainly of elderly women, which frequently results in bone fractures. Among other results, the bone density of the lumbar spine of 41 twin pairs was measured, the twins of each pair differing by 5 or more pack-years of smoking. (Pack-years of smoking was defined as the lifetime tobacco use, calculated by the number of years smoked times the average number of cigarettes smoked per day, divided by 20.) The following mean bone densities were obtained (SE means the SD of the mean):

	Lighter smoker (g/cm^2)	Heavier smoker (g/cm^2)	Difference
Mean ± SE	0.795 ± 0.020	0.759 ± 0.021	0.036 ± 0.014

The null hypothesis was that the mean bone densities μ_2 and μ_1 of the two *populations*, the heavier and the lighter smokers, are equal, that is, $\mu = \mu_1 - \mu_2 = 0$ and the alternative that $\mu > 0$. The test statistic is the mean difference in the sample, which is large enough for the normal approximation to apply. Thus, the observed z-value is $z = \frac{0.036}{0.014} \approx 2.57$, and so the P-value is $P(Z_2 - Z_1 > 2.57) \approx 1 - \Phi(2.57) \approx 0.005$, a highly significant result. Apparently, smoking does cause osteoporosis.   ♦

We shall return to hypothesis testing with different parameters and distributions in later sections.

---

[7] J. L. Hopper and E. Seeman, The Bone Density of Female Twins Discordant for Tobacco Use. NEJM, Feb. 14, 1994.

Exercises

In all questions below where a decision is required, formulate a null and an alternative hypothesis for a population parameter, set up a test statistic and a rejection region, compute the P-value, and draw a conclusion whether to accept or reject the null hypothesis. Use the normal approximation where possible.

Exercise 8.2.1.

At a certain school, there are many sections of calculus classes. On the common final exam, the average grade is 66 and the SD is 24. In a section of 32 students (who were randomly assigned to this section), the average turns out to be only 53. Is this explainable by chance or does this class likely come from a population with a lower mean, due to some real effect, like bad teaching, illness, or drug use?

Exercise 8.2.2.

On a large farm, the cows weigh on the average 520 kilograms. A special diet is tried for 50 randomly selected cows to increase their weight, which is then observed to have an average of 528 kilograms and an SD of 25 kilograms. Is the diet effective?

Exercise 8.2.3.

Assume that a special diet is tried for 50 randomly selected cows and their weight is observed to increase an average of 10 kilograms with an SD of 20 kilograms. Is the diet effective?

Exercise 8.2.4.

In a certain large town, 10% of the population is black and 90% is white. A jury pool of 50, supposedly randomly selected people, turns out to be all white. Is there evidence of racial discrimination here?

Exercise 8.2.5.

As in Example 8.2.7, Lelis et al. also considered 83 Roman men in order to refute two earlier estimates of 28 and 24 years for the average age at first marriage. Although this was a sample of convenience, obtained from historical records of famous men, we may assume that it is a random sample. They found a sample mean of 21.17 and sample standard deviation 5.47. Does this sample refute the null hypotheses of $\mu = 28$ and $\mu = 24$?

Exercise 8.2.6.

Suppose a customer wants to buy a large lot of computer chips. The manufacturer claims that 99% of the chips are defect-free. The customer tests a random sample of $n = 50$ of them and finds one defective chip in the sample. Use the binomial distribution for the number X of nondefective chips in the sample to test $H_0 : p_0 = 0.99$ versus the alternative $H_A : p < 0.99$, where p is the probability that a chip is nondefective. Is the claim acceptable?

Exercise 8.2.7.

Let $X_1, \ldots, X_n$ be a random sample with $n = 100$, from a population with $\sigma = 2$ and unknown mean μ. Test $H_0 : \mu = \mu_0$ versus $H_A : \mu \neq \mu_0$ by rejecting H_0 if $|\overline{X} - \mu_0| > c$ for some c. Find c such that the P-value is 0.05.

8.3 The Power Function of a Test

In the preceding section, we discussed type 1 errors, that is, errors committed when H_0 is true but is erroneously rejected. Here we are going to consider errors of type 2, that is, errors committed when H_0 is erroneously accepted although H_A is true. Whenever we accept H_0, we should consider the possibility of a type 2 error.

Since H_A is usually composite, that is, corresponds to more than just a single value of the parameter, we cannot compute the probability of a type 2 error without specifying which value of θ in Ω_A this probability $\beta(\theta)$ is computed for. Thus, with Y denoting the test statistic and C its critical region (where we reject H_0),

$$\beta(\theta) = P(\text{type 2 error } | \theta \in \Omega_A) = P\left(Y \in \overline{C} | \theta \in \Omega_A\right). \tag{8.56}$$

$\beta(\theta)$ is sometimes called the *size* of a type 2 error for the given value θ.

Definition 8.3.1. Power Function. *The power function of a test is the function given by*

$$\pi(\theta) = P(Y \in C | \theta) \text{ for } \theta \in \Omega_0 \cup \Omega_A \tag{8.57}$$

and the function given by $1 - \pi(\theta)$ *the operating characteristic function of the test.*

The reason for the name "power function" is that for $\theta \in \Omega_A$ the value of the function measures how likely it is that we reject H_0 when it should indeed be rejected, that is, how powerful the test is for such θ.

The name "operating characteristic function" comes from applying such tests to acceptance sampling, that is, to deciding whether to accept a lot of certain manufactured items, by counting the number x of nondefectives in the sample and accepting the lot if x is greater than some prescribed value and rejecting it otherwise. Accepting the lot corresponds to accepting H_0 that the manufacturing process operates well enough, and $1 - \pi(\theta) = P\left(Y \in \overline{C}|\theta\right)$ is its probability. For $\theta \in \Omega_A$, $1 - \pi(\theta) = \beta(\theta)$.

Clearly, if H_0 is simple, that is, $\Omega_0 = \{\theta_0\}$, then

$$\pi(\theta) = \begin{cases} \alpha & \text{if } \theta \in \Omega_0 \\ 1 - \beta(\theta) & \text{if } \theta \in \Omega_A \end{cases}. \tag{8.58}$$

Example 8.3.1. Cold Remedy, Continued.

Let us determine the power function for the test discussed in Examples 8.2.1 and 8.2.3. In these examples $n = 100$, $\theta_0 = \mu_0 = 7$, and $Y = \overline{X}$, which is approximately normally distributed with parameters $\theta = \mu$, and $SD = .15$. For $\alpha = .01$, we obtained the rejection region $C = \{\overline{x} : \overline{x} \leq 6.65\}$. Thus,

$$\pi(\mu) = P\left(\overline{X} \in C|\mu\right) = P\left(\overline{X} \leq 6.65|\mu\right)$$

$$= P\left(\frac{\overline{X} - \mu}{0.15} \leq \frac{6.65 - \mu}{0.15}\right) \approx \Phi\left(\frac{6.65 - \mu}{0.15}\right) \quad \text{for } \mu \leq 7, \tag{8.59}$$

and the graph of this power function is given in Figure 8.2.

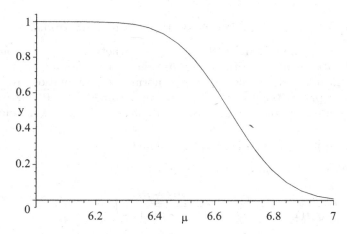

Fig. 8.2. Graph of $y = \pi(\mu)$.

Let us examine a few values of $\pi(\mu)$ as shown in the graph.

For $\mu = 7$, $\pi(7) \approx .01 = \alpha = P(\text{type 1 error}) = $ probability of accepting a worthless drug as effective.

At $\mu = 6.65$, the boundary of the rejection region, $\pi(6.65) = .5$, which is reasonable, because a slightly higher μ would lead to an incorrect acceptance of H_0 and a slightly lower μ to a correct rejection of H_0. Thus, at $\mu = 6.65$ we are just as likely to make a correct decision as an incorrect one.

For μ-values between 6.65 and 7, the probability of (an incorrect) rejection of H_0 decreases as it should, because μ is getting closer to $\mu_0 = 7$.

For $\mu \leq 6.2$, $\pi(\mu)$ is almost 1. Apparently, 6.2 is sufficiently far from $\mu_0 = 7$, so that the test (correctly) rejects H_0 with virtual certainty.

For μ-values between 6.2 and 6.65, the probability of (a correct) rejection of H_0 decreases from 1 to .5, because the closer the true value of μ is to 6.65, the less likely it becomes that the test will reject H_0.

Note that we could extend $\pi(\mu)$ to μ-values greater than 7, but doing so would make sense only if we changed H_0 from $\mu = 7$ to $\mu \geq 7$. In this changed H_0, we would have $\pi(\mu) < .01 = \alpha$ for $\mu > 7$.   ◆

Notice that the rejection region C and the power function $\pi(\theta)$ do not depend on the exact form of H_0 and H_A. For example, the same C and $\pi(\theta)$ that we had in the example above could describe a test for deciding between $H_0 : 6.9 \leq \mu \leq 7$ and $H_A : \mu \leq 6.9$ as well.

In general, whether H_0 is composite or not, we define the *size of the test* to be

$$\alpha = \sup_{\theta \in \Omega_0} \pi(\theta) = \mathrm{lub}\, \mathrm{P}\,(\text{type 1 error}). \tag{8.60}$$

In case of a simple H_0, that is, for $H_0 : \theta = \theta_0$, this definition reduces to $\alpha = \pi(\theta_0)$.

Example 8.3.2. Testing Fairness of a Coin, Continued.

Here we continue Example 8.2.5. We test whether a certain coin is fair or not. We toss it $n = 100$ times and use the relative frequency $\overline{X}$ of heads obtained as our test statistic with the normal approximation, to test the value of the parameter $\theta = p = \mathrm{P}(H)$. We obtained $C = \overline{(0.402, 0.598)}$ as the rejection region for $\alpha = .05$. Now we want to find the power function for this test.

By definition $\pi(p) = \mathrm{P}\left(\overline{X} \in C | p\right)$, and so

$$\pi(p) = \mathrm{P}\left(\overline{X} \leq 0.402 | p\right) + \mathrm{P}\left(\overline{X} \geq 0.598 | p\right)$$

$$= \mathrm{P}\left(\frac{\overline{X} - p}{\sqrt{p(1-p)/100}} \leq \frac{0.402 - p}{\sqrt{p(1-p)/100}}\right)$$

$$+ \mathrm{P}\left(\frac{\overline{X} - p}{\sqrt{p(1-p)/100}} \geq \frac{0.598 - p}{\sqrt{p(1-p)/100}}\right)$$

$$\approx \Phi\left(\frac{0.402 - p}{\sqrt{p(1-p)/100}}\right) + 1 - \Phi\left(\frac{0.598 - p}{\sqrt{p(1-p)/100}}\right). \tag{8.61}$$

As can be seen in Figure 8.3, $\pi(p)$ has its minimum $\alpha = .05$ at $p = \frac{1}{2}$ and equals 0.5 at the boundary points $p = 0.402$ and $p = 0.598$ of the rejection region. For $p \leq 0.3$ and $p \geq 0.7$, a correct rejection of H_0 occurs with probability practically 1, that is, the probability $\beta(p) = 1 - \pi(p)$ of a type 2 error is near zero there.

♦

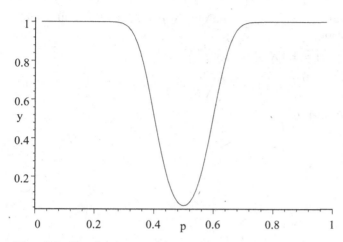

Fig. 8.3. Graph of $y = \pi(p)$.

When we design a test, we want to make the probabilities of both types of errors small, that is, we want a power function that is small on Ω_0 and large on Ω_A. Generally, we have a choice of the values of two variables: the sample size n and the boundary value c of the rejection region, and so, if we do not fix n in advance as in the preceding examples, then we can prescribe the size $\beta(\theta)$ of the type 2 error at some point in the rejection region in addition to prescribing α. This procedure is illustrated in the next example.

Example 8.3.3. Cold Remedy, Again.

As in Example 8.3.1, we assume $\theta_0 = \mu_0 = 7$, and an approximately normal $T = \overline{X}$ with mean $\theta = \mu$ but $SD = 1.5/\sqrt{n}$. With the rejection region of the form $C = (-\infty, c)$, we want to determine c and n such that $\alpha = .01$ and $\beta(6) = .01$ as well. These conditions amount to

$$P\left(\overline{X} \leq c \mid \mu = 7\right) = P\left(\frac{\overline{X} - 7}{1.5/\sqrt{n}} \leq \frac{c - 7}{1.5/\sqrt{n}} \,\middle|\, \mu = 7\right) \approx \Phi\left(\frac{c - 7}{1.5/\sqrt{n}}\right) = .01$$
(8.62)

and

$$P\left(\overline{X} \geq c \mid \mu = 6\right) = P\left(\frac{\overline{X} - 6}{1.5/\sqrt{n}} \geq \frac{c - 6}{1.5/\sqrt{n}} \,\middle|\, \mu = 6\right) \approx 1 - \Phi\left(\frac{c - 6}{1.5/\sqrt{n}}\right) = .01.$$
(8.63)

Hence

$$\frac{c - 7}{1.5/\sqrt{n}} = \Phi^{-1}(.01) = -2.3263 \tag{8.64}$$

and

$$\frac{c - 6}{1.5/\sqrt{n}} = \Phi^{-1}(.99) = 2.3263. \tag{8.65}$$

These two equations are solved (approximately) by $c = 6.5$ and $n = 49$, which yield the power function

$$\pi(\mu) = P\left(\overline{X} \in C | \mu\right) = P\left(\overline{X} \le 6.5 | \mu\right)$$

$$= P\left(\frac{\overline{X} - \mu}{1.5/7} \le \frac{6.5 - \mu}{1.5/7}\right) \approx \Phi\left(\frac{6.5 - \mu}{1.5/7}\right) \text{ for } \mu \le 7, \tag{8.66}$$

whose graph is shown in Figure 8.4.

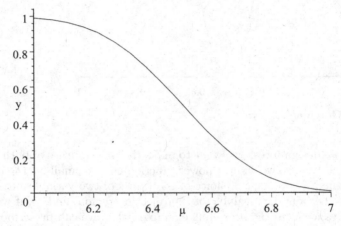

Fig. 8.4. Graph of $y = \pi(\mu)$.

This graph is much flatter than Figure 8.2, because here we are satisfied with less accuracy than in Example 8.3.1. Here we required $\beta(6) = .01$, but in Example 8.3.1, we had $\beta(6) \approx 10^{-5}$. On the other hand, in the present case, we can get away with a smaller sample, which is often a useful advantage. ◆

Exercises

Exercise 8.3.1.

a)   In Example 8.3.1, what is the meaning of a type 2 error?

b)   What is the probability that we accept the drug as effective if $\mu = 6.5$?

Exercise 8.3.2.

a)   In Example 8.3.2, what is the meaning of a type 2 error?
b)   What is the probability that we accept the coin as fair if $p = 0.55$?

Exercise 8.3.3.

As in Exercise 8.2.1, consider a large school where there are many sections of calculus classes, and on the common final exam, the average grade is 66, the SD is 24, and a certain section has 32 students. We want to test whether the given section comes from the same population or one with a lower average but with the same SD, that is, test $H_0 : \mu = 66$ against $H_A : \mu < 66$. Find the rejection region that results in a level of significance $\alpha = .05$ and find and plot the power function for this test.

Exercise 8.3.4.

As in Exercise 8.2.2, consider a special diet for n cows randomly selected from a population of cows weighing on average 500 kilograms with an SD of 25 kilograms. Find the critical region and the sample size n for a test, in terms of the average weight $\overline{X}$ of the cows in the sample, to measure the effectiveness of the diet, by deciding between $H_0 : \mu = 500$ against $H_A : \mu > 500$, with level of significance $\alpha = .05$ and $\beta(515) = .05$. Find and plot the power function for this test, using the normal approximation.

Exercise 8.3.5.

Suppose a customer wants to buy a large lot of computer memory chips and tests a random sample of $n = 12$ of them. He rejects the lot if there is more than one defective chip in the sample and accepts it otherwise. Use the binomial distribution to find and plot the operating characteristic function of this test as a function of the probability p of a chip being nondefective.

8.4 Sampling from Normally Distributed Populations

As mentioned before, in real life, many populations have a normal or close to normal distribution. Consequently, statistical methods devised for such populations are very important in applications.

In Corollary 7.2.3, we saw that the sample mean of a normal population is normally distributed, and in Example 8.1.7, we gave confidence intervals based on $\overline{X}$ for an unknown μ when σ was known.

Here we shall discuss sampling when both μ and σ are unknown. In this case, we use the MLE estimators $\overline{X}$ and $\widehat{\Sigma}^2$ for μ and σ^2 (see Example 8.1.3) and first want to prove that, surprisingly, they are independent of each other, in spite of the fact that they are both functions of the same r.v.'s X_i.

Before proving this theorem, we present two lemmas.

Lemma 8.4.1. *For a random sample from an $N(\mu, \sigma^2)$ distribution, $X_i - \overline{X}$ and $\overline{X}$ are uncorrelated.*

Proof. From Corollary 7.2.3, we know that $E\left(\overline{X}\right) = \mu$, and so $E\left(X_i - \overline{X}\right) = \mu - \mu = 0$. Let us change over to the new variables $Y_i = X_i - \mu$. Then $\overline{Y} = \frac{1}{n}\sum_{j=1}^{n} Y_j = \overline{X} - \mu$ and

$$Cov\left(X_i - \overline{X}, \overline{X}\right) = E\left(\left(X_i - \overline{X}\right)\left(\overline{X} - \mu\right)\right) = E\left(\left(Y_i - \overline{Y}\right)\overline{Y}\right)$$
$$= E\left(Y_i\overline{Y}\right) - E\left(\overline{Y}^2\right). \tag{8.67}$$

Now, $E\left(Y_iY_j\right) = E\left(Y_i\right)E\left(Y_j\right) = 0$, if $i \neq j$, and $E\left(Y_i^2\right) = \sigma^2$. Thus,

$$E\left(Y_i\overline{Y}\right) = \frac{1}{n}\sum_{j=1}^{n} E\left(Y_iY_j\right) = \frac{1}{n}E\left(Y_i^2\right) = \frac{\sigma^2}{n}. \tag{8.68}$$

Also, from Corollary 7.2.3

$$E\left(\overline{Y}^2\right) = Var\left(\overline{X}\right) = \frac{\sigma^2}{n}, \tag{8.69}$$

and therefore

$$Cov\left(X_i - \overline{X}, \overline{X}\right) = 0. \tag{8.70}$$

∎

Lemma 8.4.2. *If, for any integer $n > 1$, $(X_1, X_2, \ldots, X_n)$ form a multivariate normal n-tuple and $Cov\left(X_i, X_n\right) = 0$ for all $i \neq n$, then X_n is independent of $(X_1, X_2, \ldots, X_{n-1})$.*

Proof. The proof will be similar to that of Theorem 7.5.1. We are going to use the multivariate moment generating function

$$\psi_{1,2,\ldots,n}\left(s_1, s_2, \ldots, s_n\right) = E\left(\exp\left(\sum_{i=1}^{n} s_iX_i\right)\right). \tag{8.71}$$

By Theorem 7.2.6, $Y = \sum_{i=1}^{n} s_iX_i$ is normal, because it is a linear combination of the original independent random variables Z_i. Clearly, it has mean

$$\mu_Y = \sum_{i=1}^{n} s_i\mu_i \tag{8.72}$$

and variance

$$\sigma_Y^2 = E\left(\left(\sum_{i=1}^{n} s_i\left(X_i - \mu_i\right)\right)^2\right) = E\left(\left(\sum_{i=1}^{n} s_i\left(X_i - \mu_i\right)\right)\left(\sum_{j=1}^{n} s_j\left(X_j - \mu_j\right)\right)\right)$$
$$= E\left(\sum_{i=1}^{n}\sum_{j=1}^{n} s_is_j\left(X_i - \mu_i\right)\left(X_j - \mu_j\right)\right) = \sum_{i=1}^{n}\sum_{j=1}^{n} s_is_j\sigma_{ij}. \tag{8.73}$$

Here $\sigma_{ij} = Cov\,(X_i, X_j)$ if $i \neq j$, and $\sigma_{ii} = Var\,(X_i)$. Thus, by the definition of the m.g.f. of Y as $\psi_Y(t) = E\left(e^{tY}\right)$ and by Equation 7.47,

$$\psi_{1,2,\ldots,n}\,(s_1, s_2, \ldots, s_n) = \psi_Y(1) = \exp\left(\sum_{i=1}^{n} s_i\mu_i + \frac{1}{2}\sum_{i=1}^{n}\sum_{j=1}^{n} s_i s_j \sigma_{ij}\right).$$

(8.74)

Now, separating the terms with a subscript n from the others, we get

$$\psi_{1,2,\ldots,n}\,(s_1, s_2, \ldots, s_n) = \exp\left(\sum_{i=1}^{n-1} s_i\mu_i + \frac{1}{2}\sum_{i=1}^{n-1}\sum_{j=1}^{n-1} s_i s_j \sigma_{ij} + s_n\mu_n + \frac{1}{2}s_n^2\sigma_{nn}\right),$$

(8.75)

because we assumed $\sigma_{in} = \sigma_{ni} = 0$. Hence $\psi_{1,2,\ldots,n}\,(s_1, s_2, \ldots, s_n)$ factors as

$$\exp\left(\sum_{i=1}^{n-1} s_i\mu_i + \frac{1}{2}\sum_{i=1}^{n-1}\sum_{j=1}^{n-1} s_i s_j \sigma_{ij}\right)\exp\left(s_n\mu_n + \frac{1}{2}s_n^2\sigma_{nn}\right)$$

$$= \psi_{1,2,\ldots,n-1}\,(s_1, s_2, \ldots, s_{n-1})\,\psi_n\,(s_n),$$

(8.76)

which is the product of the moment generating functions of $(X_1, X_2, \ldots, X_{n-1})$ and X_n.

Now, if $(X_1, X_2, \ldots, X_{n-1})$ and X_n are independent, then their joint m.g.f. factors into precisely the same product. So, by the uniqueness of moment generating functions, which holds in the n-dimensional case as well, $(X_1, X_2, \ldots, X_{n-1})$ and X_n must be independent if $\sigma_{in} = 0$ for all i.   ∎

We are now ready to prove the promised theorem.

Theorem 8.4.1. *Independence of the Sample Mean and Variance.*
For a random sample from an $N(\mu, \sigma^2)$ distribution, the sample mean $\overline{X} = \frac{1}{n}\sum_{i=1}^{n} X_i$ and the sample variance $\widehat{\Sigma}^2 = \frac{1}{n}\sum_{i=1}^{n}\left(X_i - \overline{X}\right)^2$ are independent.

Proof. $\overline{X}$ and each $X_i - \overline{X}$ can be written as linear combinations of the standardizations of the i.i.d. normal X_i variables and have therefore a multivariate normal distribution. By Lemma 8.4.1, $Cov\left(X_i - \overline{X}, \overline{X}\right) = 0$ for all i and, by Lemma 8.4.2 applied to the $n+1$ variables $X_i - \overline{X}$ and $\overline{X}$, we obtain that $\left(X_1 - \overline{X}, X_2 - \overline{X}, \ldots, X_n - \overline{X}\right)$ and $\overline{X}$ are independent. Hence, by an obvious extension of Theorem 5.5.7 to $n+1$ variables, $\overline{X}$ is independent of $\widehat{\Sigma}^2 = \frac{1}{n}\sum_{i=1}^{n}\left(X_i - \overline{X}\right)^2$.   ∎

Next, we turn to obtaining the distribution of $\widehat{\Sigma}^2$.

First, note that the sum $\sum_{i=1}^{n} (X_i - \mu)^2 / \sigma^2$ is a chi-square random variable with n degrees of freedom. (See Definition 7.4.3.) Interestingly, the use of $\overline{X}$, in place of μ, in the definition of $\widehat{\Sigma}^2$ just reduces the number of degrees of freedom by 1 and leaves the distribution chi-square:

Theorem 8.4.2. *Distribution of the Sample Variance.* *For a random sample from an* $N(\mu, \sigma^2)$ *distribution, the scaled sample variance* $\frac{n\widehat{\Sigma}^2}{\sigma^2} = \frac{1}{\sigma^2}\sum_{i=1}^{n} (X_i - \overline{X})^2$ *is a chi-square random variable with* $n - 1$ *degrees of freedom.*

Proof. We can write

$$\widehat{\Sigma}^2 = \frac{1}{n}\sum_{i=1}^{n} (X_i - \overline{X})^2 = \frac{1}{n}\sum_{i=1}^{n} \left[(X_i - \mu) - (\overline{X} - \mu)\right]^2$$

$$= \frac{1}{n}\sum_{i=1}^{n} (X_i - \mu)^2 - \frac{2}{n}(\overline{X} - \mu)\sum_{i=1}^{n} (X_i - \mu) + (\overline{X} - \mu)^2, \qquad (8.77)$$

and simplifying on the right, we get

$$\widehat{\Sigma}^2 = \frac{1}{n}\sum_{i=1}^{n} (X_i - \mu)^2 - (\overline{X} - \mu)^2. \qquad (8.78)$$

Multiplying both sides by n/σ^2 and rearranging result in

$$\frac{n\widehat{\Sigma}^2}{\sigma^2} + \left(\frac{\overline{X} - \mu}{\sigma/\sqrt{n}}\right)^2 = \sum_{i=1}^{n} \left(\frac{X_i - \mu}{\sigma}\right)^2. \qquad (8.79)$$

The terms under the summation sign are the squares of independent standard normal random variables, and so their sum is chi-square with n degrees of freedom. The two terms on the left are independent, and the second term is chi-square with 1 degree of freedom. If we denote the m.g.f. of $\frac{n\widehat{\Sigma}^2}{\sigma^2}$ by $\psi(t)$, then, by Theorem 6.3.2 and Example 7.4.3,

$$\psi(t)(1 - 2t)^{-1/2} = (1 - 2t)^{-n/2} \quad \text{for } t < \frac{1}{2}. \qquad (8.80)$$

Hence

$$\psi(t) = (1 - 2t)^{-(n-1)/2} \quad \text{for } t < \frac{1}{2}, \qquad (8.81)$$

which is the m.g.f. of a chi-square random variable with $n - 1$ degrees of freedom. ∎

Example 8.4.1. Confidence Interval for the SD of Weights of Packages.

For manufacturers of various packages, it is important to know the variability of the weight around the nominal value. For example, assume that the weight of a 1 lb. package of sugar is normally distributed with unknown σ. (It doesn't matter whether we know μ or not.) We take a random sample of $n = 20$ such packages and observe the value $\hat{\sigma} = 1.2$ oz. for the sample SD $\hat{\Sigma}$. Find 90% confidence limits for σ.

By Theorem 8.4.2, $\frac{20\hat{\Sigma}^2}{\sigma^2}$ has a chi-square distribution with 19 degrees of freedom. To find 90% confidence limits for σ, we may obtain, from a table or by computer, the 5th and the 95th percentiles of the chi-square distribution with 19 degrees of freedom, that is, look up the numbers $\chi^2_{.05}$ and $\chi^2_{.95}$ such that

$$P\left(\chi^2_{19} \leq \chi^2_{.05}\right) = 0.05 \tag{8.82}$$

and

$$P\left(\chi^2_{19} \leq \chi^2_{.95}\right) = 0.95. \tag{8.83}$$

We find $\chi^2_{.05} \approx 10.12$ and $\chi^2_{.95} \approx 30.14$. Therefore,

$$P\left(10.12 < \frac{20\hat{\Sigma}^2}{\sigma^2} \leq 30.14\right) \approx 0.90. \tag{8.84}$$

For $\hat{\Sigma} = 1.2$ the double inequality becomes

$$10.12 < \frac{20 \cdot 1.2^2}{\sigma^2} \leq 30.14, \tag{8.85}$$

which can be solved for σ to give, approximately,

$$0.98 \leq \sigma < 1.69. \tag{8.86}$$

♦

As we have seen, the distribution of the statistic $\frac{n\hat{\Sigma}^2}{\sigma^2}$, used to estimate the variance, does not depend on μ. On the other hand, the distribution of the estimator $\overline{X}$ for μ depends on both μ and σ, and so, it is not suitable for constructing confidence intervals or tests for μ if σ is not known.

William S. Gosset, writing under the pseudonym Student (because his employer, the Guinness brewing company did not want the competition to learn that such methods were useful in the brewery business) in 1908 introduced the statistic

$$T = \frac{\overline{X} - \mu}{\hat{\Sigma}/\sqrt{n-1}}, \tag{8.87}$$

(named *Student's T with n − 1 degrees of freedom*) which is analogous to the $Z = \frac{(\overline{X} - \mu)}{\sigma/\sqrt{n}}$ statistic but does not depend on σ. It is widely used for constructing confidence intervals or tests for μ from small samples (approximately, $n \leq 30$) from a normal or nearly normal population with unknown σ. For larger samples, the central limit theorem applies and we can use Z with $\hat{\sigma}$ in place of σ, as in Examples 8.1.7 and 8.2.3. In fact, the density of T approaches the density of Z as $n \to \infty$.

Next, we are going to derive the density of T in several steps.

Theorem 8.4.3. *Density of a Ratio.* *If X and Y are independent continuous random variables with density functions f_X and f_Y, respectively, and $f_Y(y) = 0$ for $y \leq 0$, then the density of $U = X/Y$ is given by*

$$f_U(u) = \int_0^\infty y f_Y(y) f_X(yu)\, dy. \tag{8.88}$$

Proof. We have

$$F_U(u) = \mathrm{P}\left(\frac{X}{Y} \leq u\right) = \iint_{x/y \leq u} f_X(x) f_Y(y)\, dx\, dy$$

$$= \int_0^\infty f_Y(y) \left(\int_{-\infty}^{yu} f_X(x)\, dx\right) dy. \tag{8.89}$$

Hence, by differentiating under the first integral sign on the right and using the chain rule and the first part of the fundamental theorem of calculus, we obtain

$$f_U(u) = F'_U(u) = \int_0^\infty f_Y(y) \left(\frac{d}{d(yu)} \int_{-\infty}^{yu} f_X(x)\, dx\right) \frac{\partial(yu)}{\partial u}\, dy$$

$$= \int_0^\infty f_Y(y)\, [f_X(yu)\, y]\, dy. \tag{8.90}$$

∎

Theorem 8.4.4. *Density of $\sqrt{n}Z/\chi_n$ for Independent Z and χ_n.* *If Z is standard normal and χ_n is chi with n degrees of freedom and they are independent of each other, then*

$$U = \frac{\sqrt{n}Z}{\chi_n} \tag{8.91}$$

has density

$$f_U(u) = \frac{\Gamma\left(\frac{n+1}{2}\right)}{\sqrt{n\pi}\,\Gamma\left(\frac{n}{2}\right)} \left(1 + \frac{u^2}{n}\right)^{-(n+1)/2} \qquad for \quad -\infty < u < \infty. \tag{8.92}$$

Proof. Apply Theorem 8.4.3 to $U = \frac{\sqrt{n}Z}{X_n}$. The density of $X = \sqrt{n}Z$ is

$$f_X(x) = \frac{1}{\sqrt{2\pi n}} e^{-x^2/2n} \quad \text{for} \quad -\infty < x < \infty \tag{8.93}$$

and, by Corollary 7.4.3, the density of $Y = \chi_n$ is

$$f_Y(x) = \frac{2}{2^{n/2}\Gamma(n/2)} x^{n-1} e^{-\frac{x^2}{2}} \quad \text{for} \quad 0 < x < \infty. \tag{8.94}$$

Thus,

$$f_U(u) = \int_0^\infty y \frac{2}{2^{n/2}\Gamma(n/2)} y^{n-1} e^{-\frac{y^2}{2}} \frac{1}{\sqrt{2\pi n}} e^{-\frac{(yu)^2}{2n}} dy$$

$$= \frac{2}{2^{n/2}\Gamma(n/2)\sqrt{2\pi n}} \int_0^\infty y^n e^{-\frac{y^2}{2}\left(1+\frac{u^2}{n}\right)} dy. \tag{8.95}$$

Upon substituting $t = \frac{y^2}{2}\left(1 + \frac{u^2}{n}\right)$ in the last integral, we get

$$f_U(u) = \frac{\left(1 + \frac{u^2}{n}\right)^{-(n+1)/2}}{\sqrt{n\pi}\Gamma\left(\frac{n}{2}\right)} \int_0^\infty t^{\frac{n-1}{2}} e^{-t} dt. \tag{8.96}$$

Here, by the definition of the Γ-function (257), the integral equals $\Gamma\left(\frac{n+1}{2}\right)$, yielding the desired result. ∎

Theorem 8.4.5. *Distribution of* T. *For a random sample of size n from an $N(\mu, \sigma^2)$ distribution, Student's statistic*

$$T = \frac{\overline{X} - \mu}{\widehat{\Sigma}/\sqrt{n-1}} \tag{8.97}$$

with $n-1$ degrees of freedom, has the same distribution as the random variable $U = \sqrt{n-1}Z/\chi_{n-1}$ for independent Z and χ_{n-1}, and its density is

$$f_T(t) = \frac{\Gamma\left(\frac{n}{2}\right)}{\sqrt{(n-1)\pi}\Gamma\left(\frac{n-1}{2}\right)} \left(1 + \frac{t^2}{n-1}\right)^{-n/2} \quad \text{for} \quad -\infty < t < \infty. \tag{8.98}$$

Proof. By Corollary 7.2.3, $Z = (\overline{X} - \mu)\sqrt{n}/\sigma$ is standard normal, and by Theorem 8.4.2, $\chi_{n-1} = \widehat{\Sigma}\sqrt{n}/\sigma$ is chi with $n - 1$ degrees of freedom. Also, they are independent, by Theorem 8.4.1. Thus, Theorem 8.4.4 applied to these variables, with $n - 1$ in place of n, yields the statement of the theorem. ∎

The density given by Equation 8.98 is called *Student's t-density with $n-1$ degrees of freedom.* The values of the corresponding distribution function are usually obtained from tables or by computer from statistical software.

Note that $E(T)$ does not exist for $n = 1$ degree of freedom (it is the Cauchy distribution) and, by symmetry, $E(T) = 0$ for $n > 1$ degrees of freedom.

Also note that this density does not depend on μ and σ, and so it is suitable for constructing confidence intervals or tests for μ if σ is not known.

As mentioned at the end of Example 8.1.3, most statisticians use

$$\widehat{V} = \frac{n}{n-1}\widehat{\Sigma}^2 = \frac{1}{n-1}\sum_{i=1}^{n}(X_i - \overline{X}_n)^2 \tag{8.99}$$

as an estimator of the unknown variance of a normal population, instead of $\widehat{\Sigma}^2$. Using the corresponding estimator

$$\Sigma^+ = \sqrt{\frac{n}{n-1}}\,\widehat{\Sigma} = \left(\frac{1}{n-1}\sum_{i=1}^{n}(X_i - \overline{X}_n)^2\right)^{1/2} \tag{8.100}$$

for the standard deviation, we can write Student's T as

$$T = \frac{\overline{X} - \mu}{\Sigma^+/\sqrt{n}}. \tag{8.101}$$

This way of writing T brings it into closer analogy to the statistic

$$Z = \frac{\overline{X} - \mu}{\sigma/\sqrt{n}}, \tag{8.102}$$

used for estimating μ when σ is known.

Example 8.4.2. Confidence Interval for the Mean Weight of Packages.

As in Example 8.4.1, assume that the weight of a 1 lb. package of sugar is normally distributed with unknown σ and consider a random sample of $n = 20$ such packages and observe the values $\overline{x} = 16.1$ oz. and $\widehat{\sigma} = 1.2$ oz. for the sample mean $\overline{X}$ and SD $\widehat{\Sigma}$. Find 90% confidence limits for μ.

By Theorem 8.4.5, $T = (\overline{X} - \mu)\sqrt{n-1}/\widehat{\Sigma}$ has the t-distribution with 19 degrees of freedom in this case. Thus, we need to determine two numbers t_1 and t_2 such that $P(t_1 < T < t_2) = .90$ for this distribution. It is customary to choose $t_1 = -t_2 = -t$. Then, by the left-right symmetry of the t-distributions, we want to find t such that $P(T < -t) = .05$. From a t-table we obtain $t \approx 1.7291$, and so

$$P\left(-1.7291 < \frac{\overline{X} - \mu}{\widehat{\Sigma}/\sqrt{19}} < 1.7291\right) = .90, \tag{8.103}$$

or, equivalently,

$$P\left(\overline{X} - 1.7291\frac{\widehat{\Sigma}}{\sqrt{19}} < \mu < \overline{X} + 1.7291\frac{\widehat{\Sigma}}{\sqrt{19}}\right) = .90. \tag{8.104}$$

Substituting the observed values $\overline{x} = 16.1$ and $\widehat{\sigma} = 1.2$ for $\overline{X}$ and $\widehat{\Sigma}$, we get

$$15.624 < \mu < 16.576 \tag{8.105}$$

as a 90% confidence interval for μ.   ◆

Example 8.4.3. Small Sample Test for Weight Reduction.

As in Example 8.2.2, we want to test the effectiveness of a new drug for weight reduction and administer it, this time, to a random sample of just 10 adult women for a month. We assume that the weight loss (as a positive value), from the beginning to the end of the month, of each of these women is i.i.d. normal. Let $\overline{X}$ denote the average weight loss and $\widehat{\Sigma}$ the sample SD of the weight losses. Suppose we observe $\overline{x} = 1.5$ lbs. and $\widehat{\sigma} = 4$ lbs. We estimate σ by the observed value of $\widehat{\sigma} = 4$. The mean weight loss μ is unknown, and we want to find the extent to which the observed $\overline{X}$ value supports the null hypothesis $\mu = 0$, that is, to find the P-value of the observed average weight reduction.

Since the sample is small, σ is unknown, and the population is normal, we may use the T-statistic with mean $\mu_0 = 0$ and with 9 degrees of freedom for our test. Since H_A is of the form $\mu > \mu_0$, we take the rejection region to be of the form $[1.5, \infty)$, that is, we reject H_0 if $\overline{X} \geq 1.5$ or, equivalently, if $T \geq t$, where

$$t = \frac{\overline{x} - \mu_0}{\widehat{\sigma}/\sqrt{n-1}} = \frac{1.5 - 0}{4/\sqrt{9}} = 1.125. \tag{8.106}$$

From a t-table,

$$P\left(T \geq 1.125\right) = .145. \tag{8.107}$$

This P-value is fairly high, which means that the probability of an erroneous rejection of the null hypothesis would be high, or, in other words, our observed result can well be explained by the null-hypothesis: the weight reduction is not statistically significant.   ◆

The test of the preceding example is called the *t-test* or *Student's t-test* and is used for hypotheses involving the mean μ of a normal population when the sample is small ($n \leq 30$) and the SD is unknown and is estimated from the sample.

Exercises

Exercise 8.4.1.

The lifetimes of five light bulbs of a certain type are measured and are found to be 850, 920, 945, 1008, and 1022 hours, respectively. Assuming that the lifetimes are normally distributed, find 95% confidence intervals for μ and σ.

Exercise 8.4.2.

In high-precision measurements, repeated results usually vary due to uncontrollable and unknown factors. Scientists generally adopt the Gauss model for such measurements, according to which the measured data are like samples from a normally distributed population. Suppose a grain of salt is measured three times and is found to weigh 254, 276, and 229 micrograms, respectively. Assuming the Gauss model, with μ being the true weight and σ unknown, find a 95% confidence interval for μ, centered at $\bar{x}$.

Exercise 8.4.3.

At the service counter of a department store, a sign says that the average service time is 2.5 minutes. To test this claim, five customers were observed, and their service times turned out to be 140, 166, 177, 132, and 189 seconds, respectively. Assuming a normal distribution for the service times, test $H_0 : \mu = 150$ sec. against $H_A : \mu > 150$ sec. Find the P-value and draw a conclusion whether the store's claim is acceptable or not.

Exercise 8.4.4.

A new car model is claimed to run at 40 miles/gallon on the highway. Five such cars were tested, and the following fuel efficiencies were found: 42, 36, 39, 41, and 37 miles/gallon. Assuming a normal distribution for the fuel efficiencies, test $H_0 : \mu = 40$ against $H_A : \mu < 40$. Find the P-value and draw a conclusion whether the claim is acceptable or not.

Exercise 8.4.5.

Prove that the density $f_T(t)$ of T, given by Equation 8.98, tends to the standard normal density $\varphi(t)$ as $n \to \infty$.

Exercise 8.4.6.

In Example 8.2.8, we cited a study of twins (Footnote 7), in which the following mean bone densities of the lumbar spine of 20 twin pairs were also measured for twins of each pair differing by 20 or more pack-years of smoking (rather than just 5 pack-years, as discussed earlier):

	Lighter smoker (g/cm^2)	Heavier smoker (g/cm^2)	Difference
Mean $\pm$ SE	0.794 ± 0.032	0.726 ± 0.032	0.068 ± 0.020

Assuming normally distributed data, do a t-test for the effect of smoking on bone density.

Exercise 8.4.7.

Prove that for T with $n > 2$ degrees of freedom $Var\,(T) = \frac{n}{n-2}$. Hint: use the fact that the $U = \frac{\sqrt{n}Z}{\chi_n}$ in Theorem 8.4.4 has a t-distribution with n degrees of freedom and Z and χ_n are independent.

8.5 Chi-Square Tests

In various applications, where a statistical experiment may result in several, not just two, possible outcomes, chi-square distributions provide tests for proving or disproving an underlying theoretical prediction of the observations.

Example 8.5.1. Pea Color.

In 1865, an Austrian monk, Gregor Mendel, published a revolutionary scientific article in which he proposed a theory for the inheritance of certain characteristics of pea plants, on the basis of what we now call genes and he called entities. This was truly remarkable, because he arrived at his theory by cross-breeding experiments, without ever being able to see genes under a microscope. Among other things, he found that when he crossed purebred yellow-seeded with purebred green-seeded plants, then all the hybrid seeds turned out yellow, but when he crossed these hybrids with each other, then about 75% of the seeds turned out yellow and 25% green.

He explained this observation as follows: there are two variants (alleles) of a gene that determine seed color: say, g and y. Each seed contains two of these variants, and the seeds containing gy, yg and yy are yellow and those containing gg are green. (We call y dominant and g recessive.) Every ordinary cell of a plant contains the same pair as the seed from which it grew, but the sex cells (sperm and egg) get only one of these genes, by splitting the pair of ordinary cells. The purebred parents have gene pairs gg and yy, and so their sex cells have g and y, respectively. (Purebred plants with only yellow seeds can be produced by crossing yellows with each other over several generations, until no greens are produced.) Thus, the first-generation hybrids all get a g from one parent and a y from the other, resulting in type gy or yg. (Actually, these two types are the same; we just need to distinguish them from each other for the purpose of computing probabilities, as we did for two coins.) The seeds of these hybrids all look yellow.

In the next generation, when crossing first-generation hybrids with each other, each parent may contribute a g or a y to each sex cell with equal probability, and when those mate at random, we get all four possible pairs with equal probability. Since three pairs gy, yg, and yy look yellow and only one, gg, looks green, $p = \mathrm{P}(\text{yellow}) = 3/4 = p_0$ and $q = \mathrm{P}(\text{green}) = 1/4 = q_0$.

Suppose that to test the theory, we grow $n = 1000$ second-generation hybrid seeds and obtain $n_1 = 775$ yellow and $n_2 = 225$ green seeds. We take $H_0 : p = p_0$ and $q = q_0$, and $H_A : p \neq p_0$ and $q \neq q_0$. Karl Pearson in 1900 suggested using the following statistic for such problems:

$$K^2 = \frac{(N_1 - np_0)^2}{np_0} + \frac{(N_2 - nq_0)^2}{nq_0}, \tag{8.108}$$

where N_1 and N_2 are the random variables whose observed values are n_1 and n_2.

The reasons for choosing this form are that $(N_1 - np_0)^2$ and $(N_2 - nq_0)^2$ measure the magnitude of the deviations of the actual from the expected values, and we should consider only their sizes *relative* to the expected values. A large value of $(N_1 - np_0)^2$ indicates a relatively bigger discrepancy from the expectation when np_0 is small, than when it is large. The fractions take care of this consideration. (The fact that the numerators are squared but the denominators are not, may seem strange, but it makes the mathematics come out right.)

This statistic is especially useful when there are more than two possible outcomes. In the present case, we could just use the Z-test for p (assuming large n), since q is determined by p. (See Exercise 8.5.1.) We may, however, use K^2, as well. Substituting $q_0 = 1 - p_0$ and $N_2 = n - N_1$ in Equation 8.108, we obtain

$$K^2 = (N_1 - np_0)^2 \left(\frac{1}{np_0} + \frac{1}{nq_0} \right) \tag{8.109}$$

or, equivalently,

$$K^2 = \frac{(N_1 - np_0)^2}{np_0 q_0}. \tag{8.110}$$

By the CLT, the distribution of $(N_1 - np_0)/\sqrt{np_0 q_0}$ tends to the standard normal as $n \to \infty$, and so the distribution of K^2 tends to the chi-square distribution with one degree of freedom. Thus, using the chi-square table and the given values of n, n_1, p_0, q_0, we get the large-sample approximation of the P-value of the test as

$$P\left(K^2 \geq \frac{(n_1 - np_0)^2}{np_0 q_0}\right) = P\left(K^2 \geq \frac{(775 - 750)^2}{1000 \cdot \frac{3}{4} \cdot \frac{1}{4}}\right)$$

$$\approx P\left(\chi_1^2 \geq 3.33\right) \approx .068. \tag{8.111}$$

Hence, the null hypothesis can be accepted.   ♦

Observe that, because of the relation $N_1 + N_2 = n$ together with $p_0 + q_0 = 1$, the sum of two dependent square terms in Equation 8.108 reduces to just one such term in Equation 8.110. In other words, only one of N_1 or N_2 is free to vary. Similarly, if there are $k > 2$ possible outcomes, the relation expressing the fact that the sample size is n, and correspondingly the sum of the probabilities is 1, produces $k - 1$ *independent* square terms in the generalization of Equation 8.108. In fact, if we also use the data to estimate r parameters of the given distribution $p_{01}, p_{02}, \ldots, p_{0k}$ (examples of this will follow), then the number of independent terms turns out to be $k - 1 - r$, which is also the number of degrees of freedom for the limiting chi-square random variable. Thus, the number of degrees of freedom equals the number of independent random variables N_i that are free to vary. Hence the name "degrees of freedom."

We summarize all this as follows:

**Definition 8.5.1. *Chi-Square Test for a Finite Distribution.* ** *Suppose we consider an experiment with $k \geq 2$ possible outcomes, with unknown probabilities $p_1, p_2, \ldots, p_k$, and we want to decide between two hypotheses $H_0 : p_i = p_{0i}$ for all $i = 1, 2, \ldots, k$ and $H_A : p_i \neq p_{0i}$ for some $i = 1, 2, \ldots, k$, where $p_{01}, p_{02}, \ldots, p_{0k}$ are given.*

We consider n independent repetitions of the experiment with the random variables N_i denoting the number of times the ith outcome occurs, for $i = 1, 2, \ldots, k$, where $\sum_{i=1}^{k} N_i = n$. We use the test statistic

$$K^2 = \sum_{i=1}^{k} \frac{(N_i - np_{0i})^2}{np_{0i}}. \tag{8.112}$$

It can be proved that the distribution of K^2 tends to the chi-square distribution with $k - 1$ degrees of freedom. Furthermore, if we also use the data to estimate r parameters of the given distribution $p_{01}, p_{02}, \ldots, p_{0k}$, then the distribution of K^2 tends to the chi-square distribution with $k - 1 - r$ degrees of freedom.. Thus, *we obtain the P-value of the test approximately,*[8] *for large n, by using the chi-square table for* $P = P(\chi^2 \geq \hat{\chi}^2)$, *where*

$$\hat{\chi}^2 = \sum_{i=1}^{k} \frac{(n_i - np_{0i})^2}{np_{0i}} \tag{8.113}$$

is the observed value of K^2. If P is less than 0.05, we say that H_A is significant and if it is less than 0.01, highly significant and accept H_A, and otherwise we accept H_0. In particular, a small value of $\hat{\chi}^2$ that leads to a large P-value is strong evidence in favor of H_0 (provided that the data are really from a random sample). ∎

It may be helpful to remember Formula 8.113 in words as

$$\hat{\chi}^2 = \sum_{all\ categories} \frac{(\text{observed frequency} - \text{expected frequency})^2}{\text{expected frequency}}, \tag{8.114}$$

where the expected frequencies are based on H_0.

Example 8.5.2. Are Murders Poisson Distributed?

In a certain state, the following table shows the number n_i of weeks in three years with i murders. Can we model these numbers with a Poisson distribution?

i	0	1	2	3	4	5	6	7	8
n_i	4	12	23	34	33	23	16	8	3

First, we have to determine the parameter of the Poisson distribution with the best fit to these data. Since for a Poisson random variable $\lambda = E(X)$, we should use $\bar{x}$, that is, the average number of murders per week, as the estimate of λ. Thus we choose

$$\lambda = \frac{1}{156} \sum_{i=1}^{7} i n_i$$

$$= \frac{1}{156}(0 \cdot 4 + 1 \cdot 12 + 2 \cdot 23 + 3 \cdot 34 + 4 \cdot 33 + 5 \cdot 23 + 6 \cdot 16 + 7 \cdot 8 + 8 \cdot 3)$$

$$= 3.7372 \tag{8.115}$$

and so,

$$p_{0i} = \frac{3.7372^i e^{-3.7372}}{i!} \quad \text{for } i = 0, 1, 2, \ldots. \tag{8.116}$$

---

[8] A rough rule of thumb is that n should be large enough so that $np_{0i} \geq 10$ for each i, although some authors go as low as 5 instead of 10 and, for large k, even allow a few np_{0i} to be close to 1.

Some of the numbers np_{0i} are less than 10, and so, to safely use the chi-square approximation, we lump those together into two categories and tabulate the expected frequencies as follows:

i	0, 1	2	3	4	5	6	7, 8, ...
np_{0i}	17.603	25.95	32.327	30.203	22.575	14.061	13.279

Since we now have $k = 7$ categories and $r = 1$ estimated parameter, we use chi-square with 5 degrees of freedom. Thus,

$$\hat{\chi}^2 \approx \frac{(16 - 17.603)^2}{17.603} + \frac{(23 - 25.95)^2}{25.95} + \frac{(34 - 32.327)^2}{32.327} + \frac{(33 - 30.203)^2}{30.203}$$
$$+ \frac{(23 - 22.575)^2}{22.575} + \frac{(16 - 14.061)^2}{14.061} + \frac{(11 - 13.279)^2}{13.279}$$
$$\approx 1.4935 \tag{8.117}$$

and from a table, the corresponding P-value is $P(\chi_5^2 \geq \hat{\chi}^2) \approx 0.91$. Thus, we have very strong evidence for accepting H_0, that is, that the data came from a Poisson distribution with $\lambda = 3.7372$, except that there may be some distortion within the lumped categories.   ♦

In the example above, we tested whether the data represented a random sample from a Poisson distributed population. In general, a test for deciding whether the distribution of a population is a specified one is called a *test for goodness of fit*. For discrete distributions and large n, the chi-square test can be used for this purpose as above. For continuous distributions, we reduce the problem to a discrete one by partitioning the domain into a finite number of intervals and approximating the continuous distribution by the discrete distribution given by the probabilities of the intervals. Although this approximation may mask some features of the original distribution, it is still widely used in many applications as in the following example.

Example 8.5.3. Grades.

The grades assigned by a certain professor in several calculus classes were distributed according to the following table:

Points	(85,100]	(70,85]	(55,70]	(40,55]	[0,40]
Grade	A	B	C	D	F
Frequ.	45	56	157	83	52

Do these grades represent a random sample from an underlying normal distribution?

To answer this question, first we have to estimate μ and σ of the best fitting normal distribution, and then we may use a chi-square test as follows. First, $n = 393$, and we estimate μ and σ, by using the midpoints of the class-intervals, as

$$\bar{x} = \frac{1}{393}\,(92.5 \cdot 45 + 77.5 \cdot 56 + 62.5 \cdot 157 + 47.5 \cdot 83 + 20 \cdot 52) = 59.28,$$

$$(8.118)$$

and

$$\hat{\sigma} = \frac{1}{\sqrt{393}}[(92.5 - 59.28)^2 \cdot 45 + (77.5 - 59.28)^2 \cdot 56 + (62.5 - 59.28)^2 \cdot 157$$

$$+ (47.5 - 59.28)^2 \cdot 83 + (20 - 59.28)^2 \cdot 52]^{1/2}$$

$$= 20.28.$$

$$(8.119)$$

Using the normal distribution with these parameters, we get the probabilities p_{0i} for the class-intervals as

$$P\,((85, 100]) = \Phi\left(\frac{100 - 59.28}{20.28}\right) - \Phi\left(\frac{85 - 59.28}{20.28}\right) \approx 0.10, \qquad (8.120)$$

$$P\,((70, 85]) = \Phi\left(\frac{85 - 59.28}{20.28}\right) - \Phi\left(\frac{70 - 59.28}{20.28}\right) \approx 0.20, \qquad (8.121)$$

$$P\,((55, 70]) = \Phi\left(\frac{70 - 59.28}{20.28}\right) - \Phi\left(\frac{55 - 59.28}{20.28}\right) \approx 0.29, \qquad (8.122)$$

$$P\,((40, 55]) = \Phi\left(\frac{55 - 59.28}{20.28}\right) - \Phi\left(\frac{40 - 59.28}{20.28}\right) \approx 0.25, \qquad (8.123)$$

$$P\,((40, 55]) = \Phi\left(\frac{40 - 59.28}{20.28}\right) - \Phi\left(\frac{0 - 59.28}{20.28}\right) \approx 0.16, \qquad (8.124)$$

and the expected numbers np_{0i} as

Points	(85,100]	(70,85]	(55,70]	(40,55]	[0,40]
np_{0i}	39.3	78.6	114	98.3	63

Thus,

$$\hat{\chi}^2 \approx \frac{(45 - 39.3)^2}{39.3} + \frac{(56 - 78.6)^2}{78.6} + \frac{(157 - 114)^2}{114} + \frac{(83 - 98.3)^2}{98.3} + \frac{(52 - 63)^2}{63} \approx 27.8.$$

$$(8.125)$$

and the number of degrees of freedom is $5 - 1 - 2 = 2$, because we had five categories and estimated two parameters from the data. Hence, a chi-square probability computation gives the P-value $P(\chi_2^2 \geq \hat{\chi}^2) \approx 10^{-6}$. Thus, we reject the null hypothesis, that the distribution is normal, with a very high degree of confidence.   ♦

The chi-square test can also be used for testing independence of two distributions. We illustrate how by an example, first.

Example 8.5.4. Age and Party of Voters.

We take a random sample of 500 voters in a certain town and want to determine whether the age and party affiliation categories, as discussed in Examples 4.3.3 and 4.3.6, are independent of each other in the population. Thus, we take H_0 to be the hypothesis that each age category is independent of each party category and H_A that they are not independent. We shall give a quantitative formulation of these hypotheses below.

Suppose the sample yields the following observed frequency table for this two-way classification, also called a *contingency table*:

Age\ Party	Republican	Democrat	Independent	Any affiliation
Under 30	41	52	60	153
30 to 50	55	64	60	179
Over 50	48	53	67	168
Any age	144	169	187	500

First, we convert this table to a table of relative frequencies, by dividing each entry by 500:

Age\ Party	Republican	Democrat	Independent	Any affiliation
Under 30	.082	.104	.120	.306
30 to 50	.110	.128	.120	.358
Over 50	.096	.106	.134	.336
Any age	.288	.338	.374	1.000

These numbers represent the probabilities that, given the sample, a randomly chosen one of the 500 persons would fall in the appropriate category. Now, under the assumption of independence, the joint probabilities would be the products of the marginal probabilities. For instance, we would have P(Under 30 ∩ Republican) = .306 · .288 = .088128. We show these products in the next table:

Age\ Party	Republican	Democrat	Independent	Any affiliation
Under 30	.088128	.103428	.114444	.306
30 to 50	.103104	.121004	.133892	.358
Over 50	.096768	.113568	.125664	.336
Any age	.288	.338	.374	1.000

Thus, the promised quantitative expression of H_0 is the assumption that the joint probabilities in the population (not in the sample) of the cross classification are the nine joint probabilities in the table above.

Hence, the expected frequencies under H_0 are 500 times these probabilities, as given below:

Age\ Party	Republican	Democrat	Independent	Any affiliation
Under 30	44.064	51.714	57.222	153
30 to 50	51.552	60.502	66.946	179
Over 50	48.384	56.784	62.832	168
Any age	144	169	187	500

Consequently,

$$\hat{\chi}^2 \approx \frac{(41-44.064)^2}{44.064} + \frac{(52-51.714)^2}{51.714} + \frac{(60-57.222)^2}{57.222}$$
$$+ \frac{(55-51.552)^2}{51.552} + \frac{(64-60.502)^2}{60.502} + \frac{(60-66.946)^2}{66.946}$$
$$+ \frac{(48-48.384)^2}{48.384} + \frac{(53-56.784)^2}{56.784} + \frac{(67-62.832)^2}{62.832}$$
$$\approx 2.035. \tag{8.126}$$

The number of degrees of freedom is $9 - 4 - 1 = 4$, because the number of terms is $k = 9$, and the marginal probabilities may be regarded as parameters estimated from the data and $r = 4$ of them determine all six (any two of the row-sums determine the third one, and the same is true for the column-sums). Hence, a chi-square probability computation gives the P-value $P(\chi_4^2 \geq \hat{\chi}^2) \approx .73$, which suggests the acceptance of the independence hypothesis.[9]   ♦

The method of the example above can be generalized to arbitrary two-way classifications:

Theorem 8.5.1. *Chi-Square Test for Independence from Contingency Tables.* *Suppose we want to test the independence of two kinds of categories in a population, with a categories of the first kind and b of the second. We take a random sample of size n and construct a size $a \times b$ contingency table from the observed $k = ab$ joint frequencies n_{ij}. We convert this table to a table of relative frequencies $r_{ij} = n_{ij}/n$ and compute the row and column sums $r_i = \sum_{j=1}^{b} r_{ij}$ and $s_j = \sum_{i=1}^{a} r_{ij}$. We define the k joint probabilities $p_{0,ij} = r_i s_j$ and take H_0 to be the hypothesis that the unknown joint probabilities p_{ij} in the population satisfy $p_{ij} = p_{0,ij}$ for all $i = 1, 2, \ldots, a$ and $j = 1, 2, \ldots, b$, that is, that the relative frequencies r_{ij} come from the probability distribution $p_{0,ij}$, which represents independent categories with the appropriate marginals r_i and s_j.*

We define

$$K^2 = \sum_{i=1}^{a} \sum_{i=1}^{b} \frac{(N_{ij} - np_{0,ij})^2}{np_{0,ij}}, \tag{8.127}$$

---

[9] The result could be explained by non-independent distributions as well, but a computation of type 2 errors would be hopeless because of the various ways non-independence can occur.

where the N_{ij} is the random variable whose observed values are the n_{ij}. The r_i and s_j are parameters estimated from the data but, since $\sum r_i = 1$ and $\sum s_j = 1$, we need to estimate only $r = a + b - 2$ parameters. Thus, the distribution of K^2 tends to the chi-square distribution with $k - 1 - r = (a - 1)(b - 1)$ degrees of freedom. We obtain the P-value of the test approximately, for large n, by using a chi-square table for $P = P(\chi^2 \geq \hat{\chi}^2)$, where

$$\hat{\chi}^2 = \sum_{i=1}^{a} \sum_{i=1}^{b} \frac{(n_{ij} - np_{0,ij})^2}{np_{0,ij}} \tag{8.128}$$

is the observed value of K^2.

There exists still another use of chi-square, which is very similar to the one above: testing contingency tables for *homogeneity*. In such problems, we have several subpopulations and a sample of prescribed size from each, and we want to test whether the probability distribution over a set of categories is the same in each subpopulation. If it is, then we call the population *homogeneous* over the subpopulations with respect to the distribution over the given categories. For example, we could modify Example 8.5.4 by taking the three age groups as the subpopulations, deciding how many we wish to sample from each group and testing whether the distribution of party affiliation is the same in each age group, that is, whether the population is homogeneous over age with respect to party affiliation. We do such a modification of Example 8.5.4 next:

Example 8.5.5. Testing Homogeneity of Party Distribution Over Age Groups.

Suppose we decide to sample 150 voters under 30, 200 voters between 30 and 50, and 250 voters over 50 and want to test whether the distribution of party affiliation is the same in each age group of the population. We observe the following sample data:

Age\ Party	Republican	Democrat	Independent	Any affiliation
Under 30	41	55	54	150
30 to 50	52	66	82	200
Over 50	61	83	106	250
Any age	154	204	242	600

Under H_0, which is the assumption of homogeneity, the most likely probability distribution of party affiliation can be obtained by dividing each column sum by 600, and then the expected frequencies can be computed by multiplying these fractions by each row sum. Thus, for instance $P(\text{Republican}) = \frac{154}{600}$ and $E(n(\text{Under } 30 \cap \text{Republican})) = \frac{154}{600} \cdot 150 = 38.5$. As this calculation shows, under the present H_0, the expected frequencies are computed exactly as in the test for independence, and we get them as

Age\ Party	Republican	Democrat	Independent	Any affiliation
Under 30	38.50	51.00	60.50	150
30 to 50	51.33	68.00	80.67	200
Over 50	64.17	85.00	100.83	250
Any age	154	204	242	600

Thus,

$$\hat{\chi}^2 \approx \frac{(41 - 38.50)^2}{38.50} + \frac{(55 - 51)^2}{51} + \frac{(54 - 60.5)^2}{60.5}$$
$$+ \frac{(52 - 51.33)^2}{51.33} + \frac{(66 - 68)^2}{68} + \frac{(82 - 80.67)^2}{80.67}$$
$$+ \frac{(61 - 64.17)^2}{64.17} + \frac{(83 - 85)^2}{85} + \frac{(106 - 100.83)^2}{100.83}$$
$$\approx 1.73 \tag{8.129}$$

The number of independent data is $k = 6$, because in each row, the three frequencies must add up to the given row sum and so only two are free to vary. We estimated $r = 2$ independent column sums as parameters. Thus, the number of degrees of freedom is $k - r = 4$,.the same as in Example 8.5.4. (We do not need to subtract 1, because the fact that the sum of the joint frequencies is 600 has already been used in eliminating a column sum.) Hence, a chi-square probability computation gives the P-value $P(\chi_4^2 \geq \hat{\chi}^2) \approx .78$. This is strong evidence for accepting the hypothesis of homogeneity.   ◆

We can generalize Example 8.5.5:

Theorem 8.5.2. *Chi-Square Test for Homogeneity.* *If a population is made up of several subpopulations and we want to test whether the distribution over certain categories is the same in each subpopulation, then we take a sample of prescribed size from each subpopulation and construct a contingency table from the observed frequencies for each category in each subpopulation. We compute chi-square and the number of degrees of freedom exactly as in the test for independence and draw conclusions in the same way.*

Exercises

Exercise 8.5.1.

Use the Z-test for p in Example 8.5.1 (assuming large n) instead of the chi-square test, and show that it leads to the sáme P-value.

Exercise 8.5.2.

The grades assigned by a certain professor in several calculus classes to $n = 420$ students were distributed according to the following table:

Points	(85,100]	(70,85]	(55,70]	(40,55]	[0,40]
Grade	A	B	C	D	F
Frequ.	40	96	138	99	47

Over the years, the calculus grades in the department have been normally distributed with $\mu = 63$ and $\sigma = 18$. Use a chi-square test to determine whether the professor's grades may be considered to be a random sample from the same population.

Exercise 8.5.3.

Explain why in the chi-square test for homogeneity, just as in the chi-square test for independence, the number of degrees of freedom is $(a - 1)(b - 1)$, where now a is the number of subpopulations and b the number of categories.

Exercise 8.5.4.

Assume the same data as in Example 8.5.5 and set up a chi-square test to test the homogeneity of age distribution over party affiliation, that is, test whether this sample indicates the same age distribution in each party. What general conclusion can you draw from this example?

Exercise 8.5.5.

My calculator produced the following list of twenty random numbers:
.366, .428, .852, .602, .852, .598, .766, .627, .432, .939,
.618, .217, .002, .060, .391, .004, .099, .288, .630, .499.
Does this sample support the hypothesis that the calculator generates random numbers from the uniform distribution (apart from rounding) over the interval $[0, 1]$?

Exercise 8.5.6.

In an office, the numbers of incoming phone calls in thirty ten-minute periods were observed to be $3, 2, 1, 0, 0, 1, 4, 0, 0, 1, 1, 1, 2, 3, 2, 0, 0, 1, 2, 1, 1, 1, 2, 2,$ $3, 3, 2, 1, 4, 1$.
Does this sample support the hypothesis that the number of calls follows a Poisson distribution?

Exercise 8.5.7.

The following table shows the numbers of students distributed according to grade and sex in some of my recent elementary statistics classes:

Sex\ Grade	A	B	C	D	F	P
M	5	6	4	4	7	5
F	9	11	9	11	9	8

Does this sample support the hypothesis that in such classes grades are independent of sex?

8.6 Two-Sample Tests

In many situations, we want to compare statistics gathered from two samples.

For example, in testing medications, patients are assigned at random to two groups whenever possible: the treatment group, in which patients get the new drug to be tested, and the control group, in which patients get no treatment. To avoid bias, the assignment is usually double blind, that is, neither the patients, nor the physicians know who is in which group. To ensure this blindness, the patients in the control group are given something like a sugar pill (called a placebo), that has no effect, and both the patient and the administering physician are kept in the dark by an administrator who keeps a secret record of who got a real pill and who got a fake one.

Other two-sample situations involve comparisons between analogous results, like exam scores, incomes, prices, various health statistics, etc. in different years, or between different groups, like men and women or Republicans and Democrats, and so on.

Comparing the means of two independent normal or two arbitrary, large samples is very easy:

Definition 8.6.1. *Two-Sample Z-Test.* *In this test, we compare the unknown means μ_1 and μ_2 of two populations, using two independent samples either a) of arbitrary sizes n_1 and n_2 from two normal distributions with known σ_1 and σ_2 or b) of large sizes n_1 and n_2 from any distributions so that $\overline{X}_1$ and $\overline{X}_2$ are nearly normal by the CLT. The null hypothesis is $H_0 : \mu_1 = \mu_2$, or equivalently, $\mu = \mu_1 - \mu_2 = 0$ and we want to test against one of the alternative hypotheses $H_A : \mu > 0, \mu < 0,$ or $\mu \neq 0$. Thus, in these two cases, $\overline{X}_1 - \overline{X}_2$ is normal or may be taken as normal. Hence, writing $\hat{\sigma}_1^2$ and $\hat{\sigma}_2^2$ for the observed sample variances, we use the test statistics, standard normal under H_0,*

$$Z = \frac{\overline{X}_1 - \overline{X}_2}{\sigma_{\overline{X}}} \tag{8.130}$$

in case (a), and

$$Z = \frac{\overline{X}_1 - \overline{X}_2}{\widehat{\sigma}_{\overline{X}}} \tag{8.131}$$

in case (b), where

$$\sigma_{\overline{X}} = \sqrt{\frac{\sigma_1^2}{n_1} + \frac{\sigma_2^2}{n_2}} \tag{8.132}$$

and

$$\widehat{\sigma}_{\overline{X}} = \sqrt{\frac{\widehat{\sigma}_1^2}{n_1} + \frac{\widehat{\sigma}_2^2}{n_2}}. \tag{8.133}$$

From this point on, we proceed exactly as in the one-sample Z-test.

We can verify that the distribution of the Z above is standard normal in case (a) and nearly so in case (b):

Under H_0, in case (a), $\overline{X}_1$ and $\overline{X}_2$ are independent sample means of samples of sizes n_1 and n_2 from two normal populations with common mean μ and standard deviations σ_1 and σ_2. Thus, the means of $\overline{X}_1$ and $\overline{X}_2$ are both the same μ and their variances are $\frac{\sigma^2}{n_1}$ and $\frac{\sigma^2}{n_2}$. Hence, $\overline{X}_1 - \overline{X}_2$ is normal with mean 0 and standard deviation $\sigma_{\overline{X}} = \sqrt{\frac{\sigma_1^2}{n_1} + \frac{\sigma_2^2}{n_2}}$, and so $Z = \frac{\overline{X}_1 - \overline{X}_2}{\sigma_{\overline{X}}}$ is standard normal. In case (b), we just need to replace σ_1 and σ_2 by their estimates from the samples.

Example 8.6.1. Exam Scores of Men and Women.

On a calculus test at a certain large school, a random sample of 25 women had a mean score of 64 and SD of 14 and a random sample of 25 men had a mean score of 60 and SD of 12. Can we conclude that the women at this school do better in calculus?

We use a large-sample Z-test for the difference $\mu = \mu_1 - \mu_2$ of the two mean scores, with μ_1 denoting the women's mean score in the population and μ_2 that of the men. We take $H_0 : \mu = 0$ and $H_A : \mu > 0$. The test statistic is $\overline{X}_1 - \overline{X}_2$, with $\overline{X}_1$ denoting the women's mean score in the sample and $\overline{X}_2$ that of the men. The rejection region is $\{\overline{x}_1 - \overline{x}_2 \geq 64 - 60\}$. We may assume that, under H_0, $\overline{X}_1 - \overline{X}_2$ is approximately normal with SD $\sqrt{\frac{14^2 + 12^2}{25}} \approx 3.7$. Thus, we obtain the P-value as $P\left(\overline{X} - \overline{Y} \geq 4 | H_0\right) = P\left(\frac{\overline{X} - \overline{Y}}{3.7} \geq \frac{4}{3.7}\right) \approx 1 - \Phi\left(\frac{4}{3.7}\right) \approx 0.14$. Consequently, we accept the null hypothesis that the men and women have the same average score in the *population*; the discrepancy in the samples is probably just due to chance caused by the random selection process.   ♦

Example 8.6.2. Osteoarthritis Treatment.

D.O. Clegg et al.[10] have studied the effects of the popular supplements glucosamine, chondroitin sulfate, and the two in combination for painful knee osteoarthritis. Among many other results, they found that 188 of 313 randomly selected patients on placebo obtained at least 20% decrease in their WOMAC pain scores and 211 of 317 randomly selected patients on the combined supplements obtained a similar decrease. Do these results show a significant effect of the supplements versus the placebo?

Now, the sample proportions $\widehat{P}_1$ and $\widehat{P}_2$ of successful decreases are binomial (divided by n) with expected values $\widehat{p}_1 = \frac{188}{313} \approx 0.601$ and $\widehat{p}_2 = \frac{211}{317} \approx 0.666$. We take $H_0 : p_2 = p_1$ and $H_A : p_2 > p_1$. For the computation of $\widehat{\sigma}$ we use the pooled samples with $\widehat{p} = \frac{399}{630}$. Thus, under H_0, $\widehat{P}_2 - \widehat{P}_1$ is approximately normal with mean $\mu = 0$ and $\widehat{\sigma} = \sqrt{\widehat{p}(1 - \widehat{p})\left(\frac{1}{n_1} + \frac{1}{n_2}\right)} = \sqrt{\frac{399}{630}\left(1 - \frac{399}{630}\right)\left(\frac{1}{313} + \frac{1}{317}\right)} \approx 0.0384$. Hence, $P\left(\widehat{P}_2 - \widehat{P}_1 > 0.065\right) \approx 1 - \Phi\left(\frac{0.065}{0.0384}\right) \approx 1 - \Phi(1.693) \approx 0.045$. The effect seems to be just barely significant.

Note, however, that the authors reported an unexplained P-value of 0.09 and drew the conclusion that the result was not significant. The discrepancy is probably due to their use of a two-tailed test, but that seems to be unwarranted, since we want to test the efficacy of the supplements and not their absolute difference from the placebo. Among their other results, however, they reported a much more significant response to the combined therapy for patients with moderate-to-severe pain at baseline, than the numbers above for all patients. (See Exercise 8.6.4.)

The very high placebo effect can probably be explained by the patients' use of acetaminophen in addition to the experiment.   ♦

The t-test can also be generalized for two independent samples:

Definition 8.6.2. Two-Sample t-Test. In this test, we compare the unknown means μ_1 and μ_2 of two normal populations with unknown common $\sigma = \sigma_1 = \sigma_2$, using independent, small samples of sizes n_1 and n_2, respectively, from the two normal distributions. Again, the test hypotheses are $H_0 : \mu_1 = \mu_2$, or equivalently, $\mu = \mu_1 - \mu_2 = 0$, and one of the alternatives $H_A : \mu > 0, \mu < 0$, or $\mu \neq 0$. Under H_0, with

$$\widehat{S}_{\overline{X}} = \sqrt{\frac{\widehat{\Sigma}_1^2}{n_2} + \frac{\widehat{\Sigma}_2^2}{n_1}} \cdot \sqrt{\frac{n_1 + n_2}{n_1 + n_2 - 2}}, \tag{8.134}$$

---

[10] D.O. Clegg et al. Glucosamine, Chondroitin Sulfate, and the Two in Combination for Painful Knee Osteoarthritis. NEJM. Feb. 2006.

the test statistic

$$T = \frac{\overline{X}_1 - \overline{X}_2}{\widehat{S}_{\overline{X}}}, \tag{8.135}$$

has a t-distribution with $n_1 + n_2 - 2$ degrees of freedom.

We consider this test only under the assumption $\sigma_1 = \sigma_2$, which is very reasonable in many applications: For instance, if the two samples are taken from a treatment group and a control group, then the assumption underlying H_0, that the treatment has no effect, would imply that the two populations have the same characteristics, and so not only their means but also their variances are equal. The case $\sigma_1 \neq \sigma_2$ is discussed in more advanced texts.

We can verify the distribution of the T above as follows.

Under H_0, $\overline{X}_1$ and $\overline{X}_2$ are sample means of samples of sizes n_1 and n_2 from a normal population with mean μ and standard deviation σ. Thus, their means are the same μ and their variances are $\frac{\sigma^2}{n_1}$ and $\frac{\sigma^2}{n_2}$. Hence, $\overline{X}_1 - \overline{X}_2$ is normal with standard deviation $\sigma_{\overline{X}} = \sqrt{\frac{\sigma^2}{n_1} + \frac{\sigma^2}{n_2}}$, and $Z = \frac{\overline{X}_1 - \overline{X}_2}{\sigma_{\overline{X}}}$ is standard normal.

By Theorem 8.4.2,

$$\frac{S_1^2}{\sigma^2} = \frac{1}{\sigma^2} \sum_{i=1}^{m} \left(X_{1i} - \overline{X}_1 \right)^2 \text{ and } \frac{S_2^2}{\sigma^2} = \frac{1}{\sigma^2} \sum_{i=1}^{n} \left(X_{2i} - \overline{X}_2 \right)^2 \tag{8.136}$$

are chi-square random variables with $n_1 - 1$ and $n_2 - 1$ degrees of freedom, respectively. Thus,

$$V = \frac{S_1^2}{\sigma^2} + \frac{S_2^2}{\sigma^2} \tag{8.137}$$

is chi-square with $n_1 + n_2 - 2$ degrees of freedom.

Hence, by Theorem 8.4.5,

$$U = Z \sqrt{\frac{n_1 + n_2 - 2}{V}} \tag{8.138}$$

has a t-distribution with $n_1 + n_2 - 2$ degrees of freedom. Now, we show that this U is the same as the T in Equation 8.135:

Indeed, $\widehat{\Sigma}_1^2 = \frac{S_1^2}{n_1}$ and $\widehat{\Sigma}_2^2 = \frac{S_2^2}{n_2}$, and so,

$$V = \frac{n_1 \widehat{\Sigma}_1^2}{\sigma^2} + \frac{n_2 \widehat{\Sigma}_2^2}{\sigma^2}. \tag{8.139}$$

Thus,

$$
\begin{aligned}
U &= \frac{\overline{X}_1 - \overline{X}_2}{\sqrt{\frac{\sigma^2}{n_1} + \frac{\sigma^2}{n_2}}} \sqrt{\frac{n_1 + n_2 - 2}{\frac{n_1 \widehat{\Sigma}_1^2}{\sigma^2} + \frac{n_2 \widehat{\Sigma}_2^2}{\sigma^2}}} = \frac{\overline{X}_1 - \overline{X}_2}{\sqrt{\frac{1}{n_1} + \frac{1}{n_2}}} \sqrt{\frac{n_1 + n_2 - 2}{n_1 \widehat{\Sigma}_1^2 + n_2 \widehat{\Sigma}_2^2}} \\
&= \left(\overline{X}_1 - \overline{X}_2 \right) \sqrt{\frac{n_1 + n_2 - 2}{n_1 \widehat{\Sigma}_1^2 + n_2 \widehat{\Sigma}_2^2} \cdot \frac{n_1 n_2}{n_1 + n_2}} \\
&= \left(\overline{X}_1 - \overline{X}_2 \right) \sqrt{\frac{1}{\frac{n_1 \widehat{\Sigma}_1^2 + n_2 \widehat{\Sigma}_2^2}{n_1 n_2}} \cdot \frac{n_1 + n_2 - 2}{n_1 + n_2}} = \frac{\left(\overline{X}_1 - \overline{X}_2 \right)}{\sqrt{\frac{\widehat{\Sigma}_1^2}{n_2} + \frac{\widehat{\Sigma}_2^2}{n_1}} \cdot \sqrt{\frac{n_1 + n_2}{n_1 + n_2 - 2}}} = T,
\end{aligned}
\tag{8.140}
$$

as was to be shown.

Example 8.6.3. Cure for Stuttering in Children.

Mark Jones et al.[11] conducted an experiment in which they compared a treatment, called the Lidcombe Programme, with "no treatment" as control. Here we describe a much abbreviated version of the experiment and the results.

"The children allocated to the Lidcombe program arm received the treatment according to the program manual. Throughout the program, parents provide verbal contingencies for periods of stutter free speech and for moments of stuttering. This occurs in conversational exchanges with the child in the child's natural environment. The contingencies for stutter free speech are acknowledgment ("That was smooth"), praise ("That was good talking"), and request for self-evaluation ("Were there any bumpy words then?"). The contingencies for unambiguous stuttering are acknowledgment ("That was a bit bumpy") and request for self-correction ("Can you say that again?"). The program is conducted under the guidance of a speech pathologist. During the first stage of the program, a parent conducts the treatment for prescribed periods each day, and parent and child visit the speech pathologist once a week. The second stage starts when stuttering has been maintained at a frequency of less than 1.0% of syllables stuttered over three consecutive weeks inside and outside the clinic and is designed to maintain those low levels."

The authors measured the severity of stuttering (% of syllables stuttered) before randomization (that is, the random assignment of children to the two groups) and after nine months. They assumed that the hypothetical populations corresponding to the two groups were normal and independent, and consequently they used a two-sample t-test. They obtained the following means and SDs (the latter in parentheses):

	Treatment	Control
n	27	20
Before	6.4 (4.3)	6.8 (4.9)
At nine months	1.5 (1.4)	3.9 (3.5)

---

[11] Randomized controlled trial of the Lidcombe programme of early stuttering intervention. Mark Jones et al., BMJ Sep. 2005.

At nine months, from Equation 8.133, $\widehat{\sigma}_{\overline{X}_2 - \overline{X}_1} = \sqrt{\frac{1.4^2}{27} + \frac{3.5^2}{20}} \approx 0.8$. Thus, a 95% confidence interval for the difference $\delta = \mu_2 - \mu_1$ between the two populations in average % of syllables stuttered at nine months is approximately 2.4 ± 1.6.

Apparently, the authors made no use of the "before randomization" figures. They should have compared the *improvements* of the two groups: $6.4 - 1.5 = 4.9$ to $6.8 - 3.9 = 2.9$, rather than just the end results. We cannot do this comparison from the data presented, because we have no way of knowing the SDs of these differences. The "before" and "after" figures are not independent, for they refer to the same children, and so we cannot use Equation 8.133 to compute the SDs of the improvements. The only way these SDs could have been obtained would have been to note the improvement of each child and to compute the SDs from those.

To test the significance of the nine months results, the authors considered $H_0 : \delta = 0$ and $H_A : \delta > 0$. Under H_0 the t-value for the difference is about $2.4 / \left[0.8 \sqrt{(27 + 20)/(27 + 20 - 2)} \right] \approx 2.9$ with $n_1 + n_2 - 2 = 45$ degrees of freedom. By statistical software, $P(T > 2.9) \approx 0.003$. This result is highly significant: the treatment is effective.   ♦

Next, we present a test for comparing the *variances* of two independent normal populations. Since the normalized sample variances from Equation 8.136 are chi-square, with $n_1 - 1$ and $n_2 - 1$ degrees of freedom, respectively, it is customary to compare the *unbiased* sample variances

$$\widehat{V}_1 = \frac{1}{n_1 - 1} \sum_{i=1}^{m} \left(X_{1i} - \overline{X}_1 \right)^2 \text{ and } \widehat{V}_2 = \frac{1}{n_2 - 1} \sum_{i=1}^{n} \left(X_{2i} - \overline{X}_2 \right)^2 \quad (8.141)$$

to each other. For this comparison, we use their *ratio*, rather than their difference, because when $\sigma_1 = \sigma_2 = \sigma$, the sampling distribution of the difference depends on σ, but that of the ratio does not.

Such a ratio has a special, somewhat unfortunately named distribution, because it conflicts with the notation for d.f.'s. It was so named in honor of its discoverer, Ronald A. Fisher.

Definition 8.6.3. F-Distributions. *Let χ_m^2 and χ_n^2 be independent chi-square random variables with m and n degrees of freedom, respectively. Then*

$$F_{m,n} = \frac{\chi_m^2 / m}{\chi_n^2 / n} = \frac{n \chi_m^2}{m \chi_n^2} \qquad (8.142)$$

is said to have an F-distribution with m and n degrees of freedom.

Theorem 8.6.1. Density of F-Distributions. *The density of the $F_{m,n}$ above is given by*

$$f(x) = \begin{cases} 0 & \text{if } x \le 0 \\ c\dfrac{x^{(m/2)-1}}{(mx+n)^{(m+n)/2}} & \text{if } x > 0, \end{cases} \tag{8.143}$$

where

$$c = \frac{\Gamma\left(\frac{m+n}{2}\right) m^{m/2} n^{n/2}}{\Gamma\left(\frac{m}{2}\right) \Gamma\left(\frac{n}{2}\right)}. \tag{8.144}$$

Proof. Theorem 7.4.4 gives the density of a chi-square variable as

$$f_{\chi_n^2}(x) = \begin{cases} 0 & \text{if } x \le 0 \\ \dfrac{1}{2^{n/2}\,\Gamma(n/2)} x^{\frac{n}{2}-1} e^{-\frac{x}{2}} & \text{if } x > 0. \end{cases} \tag{8.145}$$

Hence, by Example 5.3.1, the density of χ_n^2/n is

$$f_{\chi_n^2/n}(x) = \begin{cases} 0 & \text{if } x \le 0 \\ \dfrac{n}{2^{n/2}\,\Gamma(n/2)} (nx)^{\frac{n}{2}-1} e^{-\frac{nx}{2}} & \text{if } x > 0. \end{cases} \tag{8.146}$$

Now, we apply Theorem 8.4.3 to the ratio of the two scaled chi-square random variables in Definition 8.6.3, with densities as given in Equation 8.146: For $x > 0$,

$$\begin{aligned} f(x) &= \int_0^\infty y f_{\chi_n^2/n}(y)\, f_{\chi_m^2/m}(xy)\, dy \\ &= \int_0^\infty y \frac{n}{2^{n/2}\,\Gamma(n/2)} (ny)^{\frac{n}{2}-1} e^{-\frac{ny}{2}} \frac{m}{2^{m/2}\,\Gamma(m/2)} (mxy)^{\frac{m}{2}-1} e^{-\frac{mxy}{2}}\, dy \\ &= \frac{m^{m/2} n^{n/2} x^{(m/2)-1}}{2^{(m+n)/2}\,\Gamma(m/2)\,\Gamma(n/2)} \int_0^\infty y^{\frac{m+n}{2}-1} e^{-\frac{(mx+n)y}{2}}\, dy. \end{aligned} \tag{8.147}$$

If we change the variable y to $u = \frac{(mx+n)y}{2}$ in the last integral, then we get

$$\begin{aligned} f(x) &= \frac{m^{m/2} n^{n/2} x^{(m/2)-1}}{2^{(m+n)/2}\,\Gamma(m/2)\,\Gamma(n/2)} \cdot \frac{2^{(m+n)/2}}{(mx+n)^{(m+n)/2}} \int_0^\infty y^{\frac{m+n}{2}-1} e^{-u}\, du \\ &= \frac{m^{m/2} n^{n/2} x^{(m/2)-1}}{\Gamma(m/2)\,\Gamma(n/2)} \cdot \frac{\Gamma\left(\frac{m+n}{2}\right)}{(mx+n)^{(m+n)/2}}. \end{aligned} \tag{8.148}$$

∎

Note that the explicit expression for the F-density is not very useful. We obtain associated probabilities from tables or by computer.

Definition 8.6.4. F-Test. *We use this test for comparing the standard deviations σ_1 and σ_2 of two normal populations with unknown σ_1 and σ_2 and arbitrary μ_1 and μ_2. We take two independent samples of arbitrary* *sizes n_1 and n_2, respectively, from the two populations. The null hypothesis is $H_0 : \sigma_1 = \sigma_2$, and we test against one of the alternative hypotheses $H_A : \sigma_1 < \sigma_2$, $\sigma_1 > \sigma_2$, or $\sigma_1 \neq \sigma_2$. We consider the test statistic $\widehat{V}_1/\widehat{V}_2$ (see Equation 8.141), which has an F-distribution with n_1 and n_2 degrees of freedom under H_0.*

Let us mention that this test is very sensitive to deviations from normality, and if the populations are not very close to normal, then other tests must be used.

Example 8.6.4. Oxygen in Wastewater.

Miller and Miller[12] discuss the following example. A proposed method for the determination of the chemical oxygen demand of wastewater is compared with the accepted mercury salt method. The measurements are assumed to come from independent normal populations. The new method is considered to be better than the old one, if its SD is smaller than the SD of the old method.

The following results were obtained:

	$\widehat{\mu}$ (mg/L)	$\widehat{V}$ (mg/L)	n
1. Standard Method	72	10.96	6
2. Proposed Method	72	2.28	8

Thus, we use the test statistic $F_{5,7}$, which now has the value $\frac{10.96}{2.28} \approx 4.8$. The null hypothesis is $H_0 : \sigma_1 = \sigma_2$, and the alternative is $H_A : \sigma_1 > \sigma_2$. Thus, the P-value is the probability of the right tail. Statistical software gives $P(F_{5,7} > 4.8) \approx 0.03$. This result is significant, that is, we accept H_A that the new method is better.   ♦

Exercises

Exercise 8.6.1.

St. John's wort extract (hypericum) is a popular herbal supplement for the treatment of depression. Researchers in Germany conducted an experiment, in which they showed that it compares favorably with a standard drug called paroxetine.[13] Among other results, they found the following mean decreases on the Montgomery-Åsberg depression rating scale, from baseline to day 42:

---

[12] Statistics for Analytical Chemistry, J.C. Miller and J. N. Miller.

[13] Acute treatment of moderate-to-severe depression with hypericum extract WS 5570 (St John's wort): randomised controlled double blind non-inferiority trial versus paroxetine

A Szegedi, R Kohnen, A Dienel, M Kieser, BMJ, March 2005.

	Hypericum	Paroxetine
n	122	122
mean (SD)	16.4 (10.7)	12.6 (10.6)

Find the P-value of a two-sample z-test, to show that the superior efficacy of St. John's wort extract is highly significant.

Exercise 8.6.2.

Show that a random variable X has a t-distribution with n degrees of freedom if and only if X^2 has an F distribution with 1 and n degrees of freedom.

Exercise 8.6.3.

Prove that $E(F_{m,n}) = \frac{n}{n-2}$ if $n > 2$. *Hint*: use the independence of the chi-square variables in the definition of $F_{m,n}$.

Exercise 8.6.4.

D. O. Clegg et al.(Footnote 10, page 328) reported a highly significant response to the combined therapy with glucosamine and chondroitin sulfate for patients with moderate-to-severe pain at baseline. They found that 38 of 70 such randomly selected patients on placebo obtained at least 20% decrease in their WOMAC pain scores and 57 of 72 such randomly selected patients on the combined supplements obtained a similar decrease. Find the P-value of the effect of the supplements versus the placebo.

8.7 Kolmogorov-Smirnov Tests

In the 1930s, two Russian mathematicians A. N. Kolmogorov and N. V. Smirnov developed several goodness of fit tests, two of which we are going to describe here. The first of these tests is designed to determine whether sample data come from a given distribution, and the second test, whether data of two samples come from the same distribution or not. These are instances of nonparametric tests.

These tests use a distribution function constructed from sample data:

Definition 8.7.1. *Empirical or Sample Distribution Function. Let $x_1, x_2, \ldots, x_n$ be arbitrary real numbers and assign probability $\frac{1}{n}$ to each of them. This discrete uniform distribution is called the empirical or sample distribution of the data. The corresponding distribution function is called the empirical or sample distribution function $F_n(x)$. In other words,*

$$F_n(x) = \frac{1}{n} \cdot (number\ of\ x_i \leq x). \tag{8.149}$$

Clearly, $F_n(x)$ is a step function, increasing from zero to one, and is continuous from the right. If the x_i values are distinct, then $F_n(x)$ has a jump of size $\frac{1}{n}$ at each of the x_i values. If the x_i are sample values from a population with continuous F, then they should be distinct, although in practice they may not be, because of rounding.

Example 8.7.1. An Empirical Distribution Function.

The graph in Figure 8.5 shows the empirical distribution function for a sample of size $n = 4$ with distinct x_i values. It has jumps of size $\frac{1}{4}$ at the sample values x_1, x_2, x_3, x_4.   ♦

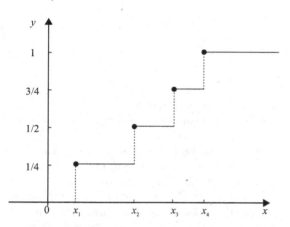

Fig. 8.5. The empirical distribution function of a sample

The tests of Kolmogorov and Smirnov use the following quantity as test statistic:

Definition 8.7.2. Kolmogorov-Smirnov Distance. *The Kolmogorov -Smirnov (K-S) distance of two distribution functions F and G is defined as the quantity*

$$d = \sup_x |F(x) - G(x)|. \tag{8.150}$$

(See Figure 8.6.)

Lemma 8.7.1. Alternative Expression for a K-S Distance. *If F is a continuous d. f. and F_n an empirical d.f. for distinct x_i values, then the K-S distance of F and F_n is given by*

$$d_n = \max_{1 \le i \le n} \left(\max\{F_n(x_i) - F(x_i), F(x_i) - F_n(x_i) + \frac{1}{n}\} \right). \tag{8.151}$$

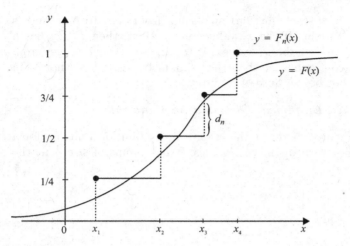

Fig. 8.6. The K_S distance

Proof. Assume that the x_i values are in increasing order and let $x_0 = -\infty$.

Clearly, $\sup_x |F(x) - F_n(x)|$ must be attained at one of the x_i values. For, on an interval $[x_{i-1}, x_i)$ or $(-\infty, x_1)$, where $F_n(x)$ is constant, $F(x) - F_n(x)$ is increasing (together with $F(x)$), and so its supremum is reached at the right endpoint x_i of the interval, and its minimum at the left endpoint x_{i-1} of the interval. Thus, $\sup_{x_{i-1} \leq x < x_i} |F(x) - F_n(x)|$ is the larger of the vertical distances $|F(x_{i-1}) - F_n(x_{i-1})|$ and $|F(x_i) - F_n(x_i)|$, and so $\sup_x |F(x) - F_n(x)|$ is the largest of the $2n$ such distances. This maximum distance can also be found by first finding the larger of the two distances at each x_i, that is, the larger of the vertical distances from the graph of F to the two corners of the graph of F_n at x_i and then finding the maximum of those as i varies from 1 to n. This procedure can be done without absolute values as stated in the theorem. ∎

Definition 8.7.3. *One-Sample Kolmogorov-Smirnov Test.* *Suppose we want to test whether a certain random variable X has a given continuous d. f. F (the null hypothesis) or not (the alternative hypothesis).*

Consider a random sample of X with distinct observed values $x_1, x_2, \ldots, x_n$. Construct the corresponding empirical d. f. F_n and find the K-S distance d_n between F and F_n. Tables and software are available for the null-distribution of the random variable

$$D_n = \max_{1 \leq i \leq n} \left(\max\{F_n(X_i) - F(X_i), F(X_i) - F_n(X_i) + \frac{1}{n}\} \right). \quad (8.152)$$

For small samples, use one of those to find the P-value $P(D_n \geq d_n)$. For large n, use the formula

$$P\left(D_n \geq \sqrt{\frac{2}{n}} c\right) \approx 2 \sum_{k=1}^{\infty} (-1)^{k-1} e^{-2k^2 c^2}. \quad (8.153)$$

Reject the null hypothesis if P is small and accept it otherwise.

Remarks.

1. The null-distribution of D_n does not depend on F, that is, a single table works for any continuous F.
2. For discrete random variables, the P-values given in the table are only upper bounds, that is, the true P-value can be much smaller than the one obtained from the table. Thus, the K-S test can be used also for discrete random variables if it leads to the rejection of H_0.
3. F must be fully specified. If parameters are estimated from the data, then the test is only approximate. Separate tables have been obtained by simulation for the most important parametric families of distributions to deal with this problem; we do not discuss them.
4. The test is more sensitive to data at the center of the distribution than at the tails. Various modifications have been developed to correct for this problem; we do not discuss them.
5. For small samples, the test has low power for Type 2 errors, that is, it accepts the null hypothesis too easily when it should not.

Example 8.7.2. Are Grades Normal?

In a small class, the grades on a calculus exam were 12, 19, 22, 43, 52, 56, 68, 76, 88, and 95. Do they come from a normal distribution?

First, we compute μ and σ of the data, and then we use a K-S test as follows: We find $\mu \approx 53$ and $\sigma \approx 29$ and we take F to be the normal d.f. with these parameters. (Note that, by Remark 3 above, the test is only approximate, because the parameters are estimated from the data. However, the K-S distance of the sample d.f. from the normal d.f. with these parameters is correct, and the test is acceptable.) Also, $n = 10$, and so F_n is a step function with jumps of size $1/10$. Below, we tabulate the values of $F(x_i)$, $F_n(x_i)$, $d_i^+ = F_n(x_i) - F(x_i)$, and $d_i^- = F(x_i) - F_n(x_i) + \frac{1}{n} = \frac{1}{n} - d_i^+$, for each grade x_i:

x_i	12	19	22	43	52	56	68	76	88	95
$F(x_i)$	.079	.121	.143	.365	.486	.541	.698	.786	.886	.926
$F_n(x_i)$	.1	.2	.3	.4	.5	.6	.7	.8	.9	1
d_i^+	.021	.079	.157	.035	.014	.059	.002	.014	.014	.074
d_i^-	.079	.021	−.057	.065	.086	.041	.098	.086	.086	.026

Hence $d_n \approx 0.157$. In the table, the entry for $n = 10$ under P $= 0.20$ is 0.322. A smaller d_n supports H_0 more strongly. Thus, 0.157 would fall under a P-value considerably higher than 0.20. So, we accept the null hypothesis: the grades may well come from a normal distribution.   ◆

Definition 8.7.4. *Two-Sample Kolmogorov-Smirnov Test*. *Suppose we want to test whether two independent random samples of sizes m and n, respectively, have the same continuous d. f. (the null hypothesis) or not (the alternative hypothesis).*

Let $F_m(x)$ and $G_n(x)$ denote the empirical distribution functions of the two samples. Compute their K-S distance

$$d_{mn} = \sup_x |F_m(x) - G_n(x)|. \tag{8.154}$$

We have tables and software for the null-distribution of the corresponding test statistic D_{mn}. For small samples, use one of those to find the P-value $P(D_{mn} \geq d_{mn})$. For large samples, use the formula

$$P\left(D_{mn} \geq c\sqrt{\frac{m+n}{mn}}\right) \approx 2\sum_{k=1}^{\infty} (-1)^{k-1} e^{-2k^2 c^2}. \tag{8.155}$$

Reject the null hypothesis if P is small and accept it otherwise.

Example 8.7.3. Grades of Men and Women.

Suppose that on an exam in a large statistics class, the grades of $m = 5$ randomly selected men were 25, 36, 58, 79, and 96, and the grades of $n = 6$ randomly selected women were 32, 44, 51, 66, 89, and 93. Use the two-sample K-S test to determine whether the two sets come from the same distribution.

We want to use the result of Exercise 8.7.3 to compute d_{mn}. We list the necessary quantities in the following table:

z_i	25	36	58	79	96
$F_m(z_i)$	1/5	2/5	3/5	4/5	1
$G_n(z_i)$	0	1/6	3/6	4/6	1
$|F_m(z_i) - G_n(z_i)|$	6/30	7/30	3/30	4/30	0/30

z_i	32	44	51	66	89	93
$F_m(z_i)$	1/5	2/5	2/5	3/5	4/5	4/5
$G_n(z_i)$	1/6	2/6	3/6	4/6	5/6	1
$|F_m(z_i) - G_n(z_i)|$	1/30	2/30	3/30	2/30	1/30	6/30

Hence, $d_{mn} = 7/30$. The critical value at $m = 5$ and $n = 6$ in the two-sample K-S table for $\alpha = 0.05$ is 20/30. Since d_{mn} is less than this, we accept the null hypothesis, that the men and women have the same grade distribution, at the 5% level. In fact, the P-value is apparently much higher than 0.05. (In general, a small d_{mn} value supports the null-hypothesis, while a high one supports the alternative.) ♦

Exercises

Exercise 8.7.1.

Use the one-sample K-S test to determine whether a sample of size $n = 300$ comes from a population with a given continuous d.f. F, if $d_n = 0.06$.

Exercise 8.7.2.

Suppose the grades in a class were 20, 70, 20, 40, 70, 50, 50, 70, 80, and 80. Find and plot the empirical d.f. of this sample.

Exercise 8.7.3.

Let $x_1, x_2, \ldots, x_m$ and $y_1, y_2, \ldots, y_n$ be the observed values of two samples. Let $\{z_1, z_2, \ldots, z_l\} = \{x_1, x_2, \ldots, x_m\} \cup \{y_1, y_2, \ldots, y_n\}$. Prove that $d_{mn} = \max_i |F_m(z_i) - G_n(z_i)|$.

Exercise 8.7.4.

In Exercise 8.5.5, 20 random numbers from a calculator were given, and the chi-square test was used to decide whether the calculator generates random numbers from the uniform distribution (apart from rounding) over the interval $[0, 1]$. Answer the same question using the K-S test.

Exercise 8.7.5.

Suppose we have two samples of sizes $m = 200$ and $n = 300$, respectively, and we find $d_{mn} = 0.08$. Use the two-sample K-S test to decide whether to accept H_0 that they come from the same population.

Exercise 8.7.6.

Suppose the grades in other samples than in Example 8.7.3 were found to be 25, 28, 39, 52, 75, and 96 for the men and 38, 44, 51, 66, 89, 93, and 98 for the women. Use the two-sample K-S test to determine whether the two sets come from the same distribution.

8.8 Regression

In Section 7.5, we saw that the conditional density $E(X_2|X_1 = x_1)$ for a bivariate normal pair (X_1, X_2) is a linear function of x_1 whose graph in the x_1x_2-plane is called the regression line or least squares line of X_2 on X_1. Changing the notation from (x_1, x_2) to (x, y), we can write its equation as (see Equation 7.142)

$$y = \mu_2 + \rho \frac{\sigma_2}{\sigma_1}(x - \mu_1). \tag{8.156}$$

where the parameters are those of the underlying bivariate normal distribution.

In most applications, the parameters are unknown and we estimate them from the empirical distribution obtained from observed sample data, that is, from the distribution that assigns probability $1/n$ to each data point (x_1, x_2) (as in the one-dimensional case in Definition 8.7.1). In Theorem 6.4.5 we observed that the least squares line for the empirical distribution is given by Equation 8.156, with the parameters replaced by their estimates from the data. So, the empirical least squares line is given by

$$y = \widehat{\mu}_2 + \widehat{\rho}\frac{\widehat{\sigma}_2}{\widehat{\sigma}_1}(x - \widehat{\mu}_1), \tag{8.157}$$

where the parameters are given by Equations 6.8, 6.62, and 6.145.

We can show that these estimated parameters are maximum likelihood estimates for the likelihood function built from the conditional density given in Equation 7.144. Thus, changing (x_1, x_2) to (x, y), and assuming that the Y_i are independent under the condition $X_i = x_i$ for all i, we write

$$L = \prod_{i=1}^{n}\left(\frac{1}{\sqrt{2\pi(1-\rho^2)}\sigma_2}\exp\frac{-\left(y_i - \mu_2 - \rho\sigma_2\frac{x_i - \mu_1}{\sigma_1}\right)^2}{2(1-\rho^2)\sigma_2^2}\right)$$

$$= \left(\frac{1}{\sqrt{2\pi(1-\rho^2)}\sigma_2}\right)^n\exp\frac{-\sum\left(y_i - \mu_2 - \rho\sigma_2\frac{x_i - \mu_1}{\sigma_1}\right)^2}{2(1-\rho^2)\sigma_2^2}. \tag{8.158}$$

Clearly, for any values of σ_2 and ρ, this function is a maximum when the sum in the exponent is a minimum. Writing $a = \mu_2$, $b = \rho\sigma_2/\sigma_1$ and $\mu_1 = \overline{x}$, we want to minimize

$$Q(a, b) = \sum(y_i - a - b(x_i - \overline{x}))^2, \tag{8.159}$$

that is, we want to find the least squares line $y = \widehat{a} + \widehat{b}(x - \overline{x})$ for the given points (x_i, y_i). We can do this algebraically by expanding and completing the squares of a and b as follows:

$$Q\left(a,b\right) = \sum y_i^2 + na^2 - 2a \sum y_i + b^2 \sum \left(x_i - \overline{x}\right)^2 - 2b \sum \left(x_i - \overline{x}\right) y_i$$

$$= \sum y_i^2 + n \left(a - \frac{1}{n} \sum y_i\right)^2 - \frac{1}{n} \left(\sum y_i\right)^2$$

$$+ \sum \left(x_i - \overline{x}\right)^2 \left(b - \frac{\sum \left(x_i - \overline{x}\right) y_i}{\sum \left(x_i - \overline{x}\right)^2}\right)^2 - \frac{\left(\sum \left(x_i - \overline{x}\right) y_i\right)^2}{\sum \left(x_i - \overline{x}\right)^2}.$$

$$(8.160)$$

Hence $Q\left(a,b\right)$ is a minimum for the a,b values

$$\widehat{a} = \frac{1}{n} \sum y_i = \overline{y} \qquad\qquad\qquad\qquad (8.161)$$

and

$$\widehat{b} = \frac{\sum \left(x_i - \overline{x}\right) y_i}{\sum \left(x_i - \overline{x}\right)^2}$$

$$= \frac{\frac{1}{n} \sum x_i y_i - \overline{x}\,\overline{y}}{\left(\frac{1}{n} \sum \left(x_i - \overline{x}\right)^2 \frac{1}{n} \sum \left(y_i - \overline{y}\right)^2\right)^{1/2}} \cdot \frac{\left(\frac{1}{n} \sum \left(y_i - \overline{y}\right)^2\right)^{1/2}}{\left(\frac{1}{n} \sum \left(x_i - \overline{x}\right)^2\right)^{1/2}} = \widehat{\rho} \frac{\widehat{\sigma}_2}{\widehat{\sigma}_1},$$

$$(8.162)$$

where $\overline{y}, \widehat{\rho}, \widehat{\sigma}_1, \widehat{\sigma}_2$ are the parameters of the empirical distribution estimating the corresponding parameters of the bivariate normal distribution.

The minimum value of $Q\left(a,b\right)$ is then

$$Q\left(\widehat{a}, \widehat{b}\right) = \sum \left(y_i - \widehat{a} - \widehat{b}\left(x_i - \overline{x}\right)\right)^2. \qquad\qquad (8.163)$$

In order to find the MLE for σ_2 and ρ, we take the logarithm of L from Equation 8.158 after substituting $\widehat{a}$ and $\widehat{b}$ and set the partial derivatives of

$$\log L\left(a, b, \sigma_2, \rho\right) = -n \log \left(\sqrt{2\pi \left(1 - \rho^2\right)}\right) - n \log \sigma_2 - \frac{\sum \left(y_i - \widehat{a} - \widehat{b}\left(x_i - \overline{x}\right)\right)^2}{2\left(1 - \rho^2\right) \sigma_2^2},$$

$$(8.164)$$

with respect to σ_2 and ρ equal to zero:

$$\frac{\partial}{\partial \sigma_2} \log L\left(a, b, \sigma_2, \rho\right) = -\frac{n}{\sigma_2} + \frac{\sum \left(y_i - \widehat{a} - \widehat{b}\left(x_i - \overline{x}\right)\right)^2}{\left(1 - \rho^2\right) \sigma_2^3} = 0, \qquad (8.165)$$

and

$$\frac{\partial}{\partial \rho} \log L\left(a, b, \sigma_2, \rho\right) = \frac{n\rho}{1 - \rho^2} - \frac{\rho \sum \left(y_i - \widehat{a} - \widehat{b}\left(x_i - \overline{x}\right)\right)^2}{\left(1 - \rho^2\right)^2 \sigma_2^2} = 0. \quad (8.166)$$

Both of the preceding equations result in

$$(1 - \widehat{\rho}^2) \widehat{\sigma}_2^2 = \frac{1}{n} \sum \left(y_i - \widehat{a} - \widehat{b}(x_i - \overline{x}) \right)^2 = \frac{1}{n} Q\left(\widehat{a}, \widehat{b}\right), \qquad (8.167)$$

which shows that the MLE of the variance $(1 - \rho^2) \sigma_2^2$ of the conditional density, given in Equation 7.144, equals the average least squares deviation of the data about the regression line.

In some applications, x is a controlled variable and not random, and for each value x_i chosen by the experimenter, we assume that

$$Y_i = g(x_i) + \epsilon_i, \qquad (8.168)$$

for all i, where the ϵ_i are i.i.d., normal r.v.'s with mean 0 and variance σ^2. The name regression is retained for this model, too, although there is no regression effect in this case. The function g is called the *regression function*. It is usually assumed to have a given form, and if $g(x) = a + b(x - \overline{x})$, then we speak of a *simple linear regression*, where the adjective "simple" indicates that x is one-dimensional and "linear" refers to linearity in the parameters a and b. The discussion above applies, with minor modifications, in this case, too. If g is a function of several variables, then we speak of *multiple regression*, but we shall only discuss the simple linear case with $g(x) = a + b(x - \overline{x})$, that is, with

$$Y_i = a + b(x_i - \overline{x}) + \epsilon_i, \qquad (8.169)$$

except for two exercises on multiple linear regression.

Thus in this case

$$E(Y_i) = a + b(x_i - \overline{x}) + E(\epsilon_i) = a + b(x_i - \overline{x}) \qquad (8.170)$$

and the likelihood function is

$$\begin{aligned}
L &= \prod_{i=1}^{n} \left(\frac{1}{\sqrt{2\pi}\sigma} \exp \frac{-(y_i - a - b(x_i - \overline{x}))^2}{2\sigma^2} \right) \\
&= \left(\frac{1}{\sqrt{2\pi}\sigma} \right)^n \exp \frac{-\sum (y_i - a - b(x_i - \overline{x}))^2}{2\sigma^2}. \qquad (8.171)
\end{aligned}$$

Similarly as above, we get the same MLE estimates as in the bivariate normal case and the corresponding estimators

$$\widehat{A} = \frac{1}{n} \sum Y_i \qquad (8.172)$$

and

$$\widehat{B} = \frac{\sum (x_i - \overline{x}) Y_i}{\sum (x_i - \overline{x})^2}. \tag{8.173}$$

These estimators are unbiased (conditionally in the bivariate normal case, which we do not write):

$$E\left(\widehat{A}\right) = \frac{1}{n} \sum E\left(Y_i\right) = \frac{1}{n} \sum \left(a + b\left(x_i - \overline{x}\right)\right)$$

$$= \frac{1}{n} \left(na + b \sum x_i - bn\overline{x}\right) = a + b\frac{1}{n} \sum x_i - b\overline{x} = a, \tag{8.174}$$

and

$$E\left(\widehat{B}\right) = \frac{\sum (x_i - \overline{x}) E\left(Y_i\right)}{\sum (x_i - \overline{x})^2} = \frac{\sum (x_i - \overline{x}) (a + b(x_i - \overline{x}))}{\sum (x_i - \overline{x})^2}$$

$$= \frac{a \sum (x_i - \overline{x})}{\sum (x_i - \overline{x})^2} + \frac{b \sum (x_i - \overline{x})^2}{\sum (x_i - \overline{x})^2} = b. \tag{8.175}$$

Since the Y_i are i.i.d. normal with common variance σ^2 (conditionally in the bivariate normal case, with variance $(1 - \rho^2) \sigma_2^2$, by Equation 7.143), and we have

$$Var\left(\widehat{A}\right) = \left(\frac{1}{n}\right)^2 \sum Var\left(Y_i\right) = \frac{\sigma^2}{n}, \tag{8.176}$$

$$Var\left(\widehat{B}\right) = \frac{\sum (x_i - \overline{x})^2}{\left(\sum (x_i - \overline{x})^2\right)^2} \sigma^2 = \frac{n\widehat{\sigma}_1^2}{(n\widehat{\sigma}_1^2)^2} \sigma^2 = \frac{\sigma^2}{n\widehat{\sigma}_1^2}, \tag{8.177}$$

and, by Exercise 6.4.10,

$$Cov\left(\widehat{A}, \widehat{B}\right) = Cov\left(\frac{1}{n} \sum Y_j, \frac{\sum (x_i - \overline{x}) Y_i}{\sum (x_i - \overline{x})^2}\right)$$

$$= \frac{\sum_j \sum_i (x_i - \overline{x}) Cov\left(Y_i, Y_j\right)}{n \sum (x_i - \overline{x})^2} = 0, \tag{8.178}$$

since, Y_i, Y_j being independent, $Cov\left(Y_i, Y_j\right) = 0$ for $i \neq j$, $Cov\left(Y_i, Y_i\right) = \sigma^2$, and $\sum_i (x_i - \overline{x}) = 0$.

In the simple linear case we estimate the variance of each Y_i as

$$\widehat{\sigma}^2 = \frac{1}{n} \sum \left(y_i - \widehat{a} - \widehat{b}(x_i - \overline{x})\right)^2. \tag{8.179}$$

Hence

$$\widehat{\sigma}^2 = \sum \left(y_i - \overline{y} - \widehat{\rho}\frac{\widehat{\sigma}_2}{\widehat{\sigma}_1}\left(x_i - \overline{x}\right)\right)^2$$

$$= \frac{1}{n}\sum\left(y_i - \overline{y}\right)^2 + \frac{1}{n}\sum\left(\widehat{\rho}\frac{\widehat{\sigma}_2}{\widehat{\sigma}_1}\left(x_i - \overline{x}\right)\right)^2 - \frac{2}{n}\sum\left(\widehat{\rho}\frac{\widehat{\sigma}_2}{\widehat{\sigma}_1}\left(x_i - \overline{x}\right)\left(y_i - \overline{y}\right)\right)$$

$$= \widehat{\sigma}_2^2 + \widehat{\rho}^2\frac{\widehat{\sigma}_2^2}{\widehat{\sigma}_1^2}\widehat{\sigma}_1^2 - \widehat{\rho}^2\frac{\widehat{\sigma}_2^2}{\widehat{\sigma}_1^2}\frac{2}{n}\sum\left(\left(x_i - \overline{x}\right)\left(y_i - \overline{y}\right)\right)^2$$

$$= \widehat{\sigma}_2^2 + \widehat{\rho}^2\frac{\widehat{\sigma}_2^2}{\widehat{\sigma}_1^2}\widehat{\sigma}_1^2 - 2\widehat{\rho}\frac{\widehat{\sigma}_2}{\widehat{\sigma}_1}\widehat{\rho}\widehat{\sigma}_1\widehat{\sigma}_2 = \widehat{\sigma}_2^2\left(1 - \widehat{\rho}^2\right), \tag{8.180}$$

just as in the bivariate normal case.

Now, $\widehat{A}$ and $\widehat{B}$ are both normal, because they are linear combinations of the i.i.d. normal variables Y_i, and, by Theorem 7.5.5, they form a bivariate normal pair. Thus, by Theorem 7.5.1, they are independent (conditionally in the bivariate normal case).

One of the main uses of regression analysis is the prediction of y-values for new unmeasured x-values. The theory above can be used to find confidence intervals for the prediction error. So, if we write in the bivariate normal case

$$E\left(Y_0|X = x_0\right) = \widehat{a} + \widehat{b}\left(x_0 - \overline{x}\right) \tag{8.181}$$

for the best prediction at $x = x_0$, then the variance of the prediction error is

$$Var\left(Y_0 - \widehat{A} - \widehat{B}\left(x_0 - \overline{x}\right)|X = x_0\right)$$

$$= Var\left(Y_0|X = x_0\right) + Var\left(\widehat{A}|X = x_0\right) + \left(x_0 - \overline{x}\right)^2 Var\left(\widehat{B}|X = x_0\right), \tag{8.182}$$

where the first term is the variance of the y-score around the best prediction and the second and third terms make up the variance of the prediction. The MLE of this variance is

$$\widehat{\sigma}^2 = \left(1 - \widehat{\rho}^2\right)\left(\widehat{\sigma}_2^2 + \frac{\widehat{\sigma}_2^2}{n} + \left(x_0 - \overline{x}\right)^2\frac{\widehat{\sigma}_2^2}{n\widehat{\sigma}_1^2}\right)$$

$$= \widehat{\sigma}_2^2\left(1 - \widehat{\rho}^2\right)\left(1 + \frac{1}{n} + \frac{\left(x_0 - \overline{x}\right)^2}{n\widehat{\sigma}_1^2}\right). \tag{8.183}$$

This result shows that the prediction is best at $x_0 = \overline{x}$, and the farther x_0 is from $\overline{x}$ the worse it becomes.

Example 8.8.1. Prediction of Exam Score.

In Example 6.4.4, we obtained the empirical least squares line for five data points (x_i, y_i) and found, with our current notation, $\widehat{\mu}_1 = \widehat{\mu}_2 = 70, \widehat{\sigma}_1 = \sqrt{5220 - 70^2} \approx 17.889, \widehat{\sigma}_2 = \sqrt{5130 - 70^2} \approx 15.166, \widehat{\rho} \approx \frac{5140 - 70^2}{17.889 \cdot 15.166} \approx$

0.88, $\widehat{a} = 70$, and $\widehat{b} = 0.75$. Assuming an underlying bivariate normal distribution, find a 90% confidence interval for the y-score of a hypothetical student whose x-score was $x_0 = 100$.

The prediction for the y-score is

$$y_0 = 70 + 0.75\,(100 - 70) = 92.5. \tag{8.184}$$

This result illustrates the regression effect; the first score of 100 was followed on average by the lower score of 92.5.

A 90% confidence interval for this y-score is the interval given by $P(|Y - y_0| < c) = 0.95$ for a t-distribution with mean y_0 and variance $\widehat{\sigma}^2$ given by Equation 8.183. We must use the t-distribution, since $\widehat{\sigma}^2$ is just an estimate of the variance and n is small. We need the t-distribution with $n - 2 = 3$ degrees of freedom because the two means are fixed. (This issue is explained in more advanced texts.) Thus we must standardize with $\sigma^+ = \widehat{\sigma}\sqrt{5/3}$ rather than with $\widehat{\sigma}$:

$$P\left(|Y - y_0| < c\right) = P\left(|T(3)| < \frac{c}{\widehat{\sigma}\sqrt{5/3}}\right) = 0.90, \tag{8.185}$$

and Equation 8.183 gives

$$\widehat{\sigma}$$
$$= \left[(5130 - 70^2)\left(1 - \frac{\left(5140 - 70^2\right)^2}{(5130 - 70^2)\cdot(5220 - 70^2)}\right)\left(1.2 + \frac{30^2}{5\cdot(5220 - 70^2)}\right)\right]^{1/2}$$
$$\approx 9.4. \tag{8.186}$$

So $c \approx 9.4 \cdot \sqrt{5/3} \cdot t_{.95}(3) \approx 9.4 \cdot \sqrt{5/3} \cdot 2.35 \approx 28.5$. Thus an approximate 90% confidence interval for the predicted y-score at $x_0 = 100$ is the interval 92.5 ± 28.5. Unfortunately, this interval is very wide, but the prediction is based on a very small sample. (Also, it may seem odd that the interval goes beyond a score of 100, but if the scores were capped at 100, then the normal model would not apply. Often this is not a problem or the limits of applicability of a normal model are fudged.)   ♦

Example 8.8.2. Comparison of Two Fertilizers. **This example is taken from a classic work of R. A. Fisher.**[14].

The yields of grain in bushels per acre were obtained from two plots during thirty years. The only difference in treatment was that one plot received nitrate of soda while the other received an equivalent quantity of nitrogen as sulfate of ammonia. In the course of the experiment, the first plot appears

---

[14] R. A. Fisher, Statistical Methods for Research Workers. Oliver & Boyd, 1925. Also available at http://psychclassics.yorku.ca/Fisher/Methods/index.htm

to be gaining in yield on the second one. Is this apparent gain significant? In other words, we want to test the null hypothesis that the slope b of the regression line is 0 versus the alternative that it is positive.

The harvest years are taken as the x_i values and the corresponding differences in yields as the y_i values (Table 8.1). Thus, we are using the regression model of Equation 8.169 with prescribed nonrandom x values.

x_i	y_i
1	−3.38
2	−4.53
3	−1.09
4	−1.38
5	−4.66
6	+4.90
7	−1.19
8	+7.56
9	+1.90
10	+5.28
11	+3.84
12	+2.59
13	+6.97
14	+8.62
15	+10.75
16	+4.13
17	+12.13
18	+11.63
19	+13.06
20	−1.37
21	+3.87
22	+7.81
23	+21.00
24	+5.00
25	+4.69
26	−0.25
27	+9.31
28	−2.94
29	+7.07
30	+2.69

The difference of the yields of two plots vs. years.   (Table 8.1)

Using the mathematical software *Maple* we find $\widehat{\sigma}_1 = 8.6554$, $\widehat{\sigma}_2 = 5.8326$, $\widehat{\rho} = 0.39592$, and so $\widehat{b} = \widehat{\rho}\frac{\widehat{\sigma}_2}{\widehat{\sigma}_1} = 0.39592 \cdot \frac{5.8326}{8.6554} = 0.266\,80$. Furthermore, $Var\left(\widehat{B}\right) = \frac{\sigma_2^2(1-\widehat{\rho}^2)}{n\widehat{\sigma}_1^2} = \frac{5.8326^2(1-0.395\,92^2)}{30 \cdot 8.6554^2} = 0.01276$ and so $SD\left(\widehat{B}\right) = \sqrt{0.01276} = 0.11296$. We use the t-test with 28 degrees of freedom, since the variance was estimated from the sample with two means already calculated. Hence, we standardize $\widehat{b}$ as $t = (0.26679/0.11296)\sqrt{28/30} = 2.282$. *Maple* gives the P-value $1 - \text{TDist}(2.282, 28) = 0.015145$.

Let us remark that Fisher obtained the same t-value, but he concluded, with the less accurate tables of his time, that P is between 0.02 and 0.05. He inferred that "The result must be judged significant, though barely so; in view of the data we cannot ignore the possibility that on this field, and in conjunction with the other manures used, nitrate of soda has conserved the fertility better than sulphate of ammonia ; these data do not, however, demonstrate the point beyond possibility of doubt." This conclusion can be drawn from our P-value as well.   ◆

Example 8.8.3. Exercise Capacity and Age.

M. Gulati & al.[15] have studied the dependence of exercise capacity on age for 5721 asymptomatic women and of 4471 symptomatic women. They measured exercise capacity in MET (metabolic equivalent) units obtained from treadmill stress tests. They found that the average exercise capacity (y) for each age (x) depended linearly on age for each group. For the asymptomatic group they calculated the parameter values $\overline{x} = 52.4$, $\widehat{\sigma}_1 = 10.8$, $\overline{y} = 8.0$, $\widehat{\sigma}_2 = 2.7$, and $\widehat{\rho} = -0.51$ and obtained the regression equation

$$y = 14.7 - 0.13x. \tag{8.187}$$

Indeed with these parameter values our Equation 8.157 gives

$$y = \widehat{\mu}_2 + \widehat{\rho}\frac{\widehat{\sigma}_2}{\widehat{\sigma}_1}(x - \widehat{\mu}_1) = 8 - 0.51\frac{2.7}{10.8}(x - 52.4) = 14.681 - 0.1275x, \tag{8.188}$$

which can be rounded to Equation 8.187.

The authors then used the regression equation to show that the amount of deviation from one's age-predicted exercise capacity was correlated with the risk of both death from any cause and death from cardiac causes. They did this also for the symptomatic group and for some subgroups as well. (The nomogram of the paper's title is a graphic device for representing functions of two variables, in this case the relative deviation of the observed exercise capacity from the predicted average for that age as a function of age and the observed exercise capacity.)   ◆

---

[15] M Gulati & al. The Prognostic Value of a Nomogram for Exercise Capacity in Women. N Engl J Med 2005; 353:468-475

Exercises

Exercise 8.8.1.

In our discussion, we found the equation of the least squares line in the form $y = \hat{a} + \hat{b}(x - \bar{x})$ for the given points (x_i, y_i). Modify the formulas to obtain it in the form $y = \hat{c} + \hat{b}x$ and find the mean and the variance of $\hat{C}$ and the covariance of $\hat{B}$ and $\hat{C}$. Why is this form less desirable?

Exercise 8.8.2.

Show that $\hat{b}$ in Equation 8.162 can also be written as

$$\hat{b} = \frac{\sum (x_i - \bar{x})(y_i - \bar{y})}{\sum (x_i - \bar{x})^2}.$$

Exercise 8.8.3.

Use the normal model of Equation 8.169 and the result of Exercise 8.8.1 to find the standard deviations of the coefficients of Equation 8.187.

Exercise 8.8.4.

Using the data in Exercise 6.4.11 and assuming an underlying bivariate normal distribution, find an 80% confidence interval for the y-score of a hypothetical student whose x-score was $x_0 = 100$.

Exercise 8.8.5.

Using the data of Example 8.8, find the regression equation of X on Y, and assuming the same underlying bivariate normal distribution, find a 90% confidence interval for the x-score of a hypothetical student whose y-score was $y_0 = 100$. Why is there apparently no regression effect in this case?

Exercise 8.8.6.

Using the data of Example 8.8.4, find the regression equation of X on Y, and assuming the same underlying bivariate normal distribution, find an 80% confidence interval for the x-score of a hypothetical student whose y-score was $y_0 = 100$.

Exercise 8.8.7.

Show that in the simple linear model, $\widehat{B}$ is not necessarily a consistent estimator of b, that is, that $\lim_{n\to\infty} \mathrm{P}\left(\left|\widehat{B} - b\right| < \varepsilon\right)$ can be less than 1. (*Hint:* Choose $x_i = 0$ for $i = 1, 2, \ldots, n-1$ and $x_n = 1$. Compute $Var\left(\widehat{B}\right)$ and use the fact that $\widehat{B}$ is normal to compute the limit.)

Exercise 8.8.8.

In the simple linear model, find a sequence of x_i values such that $\widehat{B}$ is a consistent estimator of b, that is, that $\lim_{n\to\infty} \mathrm{P}\left(\left|\widehat{B} - b\right| < \varepsilon\right) = 1$. (*Hint:* Compute $Var\left(\widehat{B}\right)$ and use the fact that $\widehat{B}$ is normal to compute the limit.)

Appendix 1: Tables

Table 1. Standard normal d.f.

$$\Phi(z) = \int_{-\infty}^{z} \frac{1}{\sqrt{2\pi}} e^{-u^2/2} du = P(Z \le z)$$

z	0	1	2	3	4	5	6	7	8	9
0.0	0.5000	0.5040	0.5080	05120	05160	0.5199	0.5239	0.5279	05319	05359
0.1	0.5398	0.5438	0.5478	05517	05557	0.5596	0.5636	0.5675	05714	05753
0.2	0.5793	0.5832	0.5871	05910	05948	0.5987	0.6026	0.6064	0.6103	0.6141
0.3	0.6179	0.6217	0.6255	0.6293	0.6331	0.6368	0.6406	0.6443	0.6480	0.6517
0.4	0.6554	0.6591	0.6628	0.6664	0.6700	0.6736	0.6772	0.6808	0.6844	0.6879
0.5	0.6915	0.6950	0.6985	0.7019	0.7054	0.7088	0.7123	0.7157	0.7190	0.7224
0.6	0.7257	0.7291	0.7324	0.7357	0.7389	0.7422	0.7454	0.7486	0.7517	0.7549
0.7	0.7580	0.7611	0.7642	0.7673	0.7703	0.7734	0.7764	0.7794	0.7823	0.7852
0.8	0.7881	0.7910	0.7939	0.7967	0.7995	0.8023	0.8051	0.8078	0.8106	0.8133
0.9	0.8159	0.8186	0.8212	0.8238	0.8264	0.8289	0.8315	0.8340	0.8365	0.8389
1.0	0.8413	0.8438	0.8461	0.8485	0.8508	0.8531	0.8554	0.8577	0.8599	0.8621
1.1	0.8643	0.8665	0.8686	0.8708	0.8729	0.8749	0.8770	0.8790	0.8810	0.8830
1.2	0.8849	0.8869	0.8888	0.8907	0.8925	0.8944	0.8962	0.8980	0.8997	0.9015
1.3	0.9032	0.9049	0.9066	0.9082	0.9099	0.9115	0.9131	0.9147	0.9162	0.9177
1.4	0.9192	0.9207	0.9222	0.9236	0.9251	0.9265	0.9278	0.9292	0.9306	0.9319
1.5	0.9332	0.9345	0.9357	0.9370	0.9382	0.9394	0.9406	0.9418	0.9430	0.9441
1.6	0.9452	0.9463	0.9474	0.9484	0.9495	0.9505	0.9515	0.9525	0.9535	0.9545
1.7	0.9554	0.9564	0.9573	0.9582	0.9591	0.9599	0.9608	0.9616	0.9625	0.9633
1.8	0.9641	0.9648	0.9656	0.9664	0.9671	0.9678	0.9686	0.9693	0.9700	0.9706
1.9	0.9713	0.9719	0.9726	0.9732	0.9738	0.9744	0.9750	0.9756	0.9762	0.9767
2.0	0.9772	0.9778	0.9783	0.9788	0.9793	0.9798	0.9803	0.9808	0.9812	0.9817
2.1	0.9821	0.9826	0.9830	0.9834	0.9838	0.9842	0.9846	0.9850	0.9854	0.9857
2.2	0.9861	0.9864	0.9868	0.9871	0.9874	0.9878	0.9881	0.9884	0.9887	0.9890
2.3	0.9893	0.9896	0.9898	0.9901	0.9904	0.9906	0.9909	0.9911	0.9913	0.9916
2.4	0.9918	0.9920	0.9922	0.9925	0.9927	0.9929	0.9931	0.9932	0.9934	0.9936
2.5	0.9938	0.9940	0.9941	0.9943	0.9945	0.9946	0.9948	0.9949	0.9951	0.9952
2.6	0.9953	0.9955	0.9956	0.9957	0.9959	0.9960	0.9961	0.9962	0.9963	0.9964
2.7	0.9965	0.9966	0.9967	0.9968	0.9969	0.9970	0.9971	0.9972	0.9973	0.9974
2.8	0.9974	0.9975	0.9976	0.9977	0.9977	0.9978	0.9979	0.9979	0.9980	0.9981
2.9	0.9981	0.9982	0.9982	0.9983	0.9984	0.9984	0.9985	0.9985	0.9986	0.9986
3.0	0.9987	0.9990	0.9993	0.9995	0.9997	0.9998	0.9998	0.9999	0.9999	1.0000

© Springer International Publishing Switzerland 2016
G. Schay, *Introduction to Probability with Statistical Applications*,
DOI 10.1007/978-3-319-30620-9

Table 2. Percentiles of the t distribution

df	$t_{.60}$	$t_{.70}$	$t_{.80}$	$t_{.90}$	$t_{.95}$	$t_{.975}$	$t_{.99}$	$t_{.995}$
1	.325	.727	1.376	3.078	6.314	12.706	31.821	63.657
2	.289	.617	1.061	1.886	2.920	4.303	6.965	9.925
3	.277	.584	.978	1.638	2.353	3.182	4.541	5.841
4	.271	.569	.941	1.533	2.132	2.776	3.747	4.604
5	.267	.559	.920	1.476	2.015	2.571	3.365	4.032
6	.265	.553	.906	1.440	1.943	2.447	3.143	3.707
7	.263	.549	.896	1.415	1.895	2.365	2.998	3.499
8	.262	.546	.889	1.397	1.860	2.306	2.896	3.355
9	.261	.543	.883	1.383	1.833	2.262	2.821	3.250
10	.260	.542	.879	1.372	1.812	2.228	2.764	3.169
11	.260	.540	.876	1.363	1.796	2.201	2.718	3.106
12	.259	.539	.873	1.356	1.782	2.179	2.681	3.055
13	.259	.538	.870	1.350	1.771	2.160	2.650	3.012
14	.258	.537	.868	1.345	1.761	2.145	2.624	2.977
15	.258	.536	.866	1.341	1.753	2.131	2.602	2.947
16	.258	.535	.865	1.337	1.746	2.120	2.583	2.921
17	.257	.534	.863	1.333	1.740	2.110	2.567	2.898
18	.257	.534	.862	1.330	1.734	2.101	2.552	2.878
19	.257	.533	.861	1.328	1.729	2.093	2.539	2.861
20	.257	.533	.860	1.325	1.725	2.086	2.528	2.845
21	.257	.532	.859	1.323	1.721	2.080	2.518	2.831
22	.256	.532	.858	1.321	1.717	2.074	2.508	2.819
23	.256	.532	.858	1.319	1.714	2.069	2.500	2.807
24	.256	.531	.857	1.318	1.711	2.064	2.492	2.797
25	.256	.531	.856	1.316	1.708	2.060	2.485	2.787
26	.256	.531	.856	1.315	1.706	2.056	2.479	2.779
27	.256	.531	.855	1.314	1.703	2.052	2.473	2.771
28	.256	.530	.855	1.313	1.701	2.048	2.467	2.763
29	.256	.530	.854	1.311	1.699	2.045	2.462	2.756
30	.256	.530	.854	1.310	1.697	2.042	2.457	2.750
40	.255	.529	.851	1.303	1.684	2.021	2.423	2.704
60	.254	.527	.848	1.296	1.671	2.000	2.390	2.660
120	.254	.526	.845	1.289	1.658	1.980	2.358	2.617
∞	.253	.524	.842	1.282	1.645	1.960	2.326	2.576

Table 3. Percentiles of the x^2 distribution

df	$\chi^2_{.005}$	$\chi^2_{.01}$	$\chi^2_{.025}$	$\chi^2_{.05}$	$\chi^2_{.10}$	$\chi^2_{.90}$	$\chi^2_{.95}$	$\chi^2_{.975}$	$\chi^2_{.99}$	$\chi^2_{.995}$
1	.000039	.00016	.00098	.0039	.0158	2.71	3.84	5.02	6.63	7.88
2	.0100	.0201	.0506	.1026	.2107	4.61	5.99	7.38	9.21	10.60
3	.0717	.115	.216	.352	.584	6.25	7.81	9.35	11.34	12.84
4	.207	.297	.484	.711	1.064	7.78	9.49	11.14	13.28	14.86
5	.412	.554	.831	1.15	1.61	9.24	11.07	12.83	15.09	16.75
6	.676	.872	1.24	1.64	2.20	10.64	12.59	14.45	16.81	18.55
7	.989	1.24	1.69	2.17	2.83	12.02	14.07	16.01	18.48	20.28
8	1.34	1.65	2.18	2.73	3.49	13.36	15.51	17.53	20.09	21.96
9	1.73	2.09	2.70	3.33	4.17	14.68	16.92	19.02	21.67	23.59
10	2.16	2.56	3.25	3.94	4.87	15.99	18.31	20.48	23.21	25.19
11	2.60	3.05	3.82	4.57	5.58	17.28	19.68	21.92	24.73	26.76
12	3.07	3.57	4.40	5.23	6.30	18.55	21.03	23.34	26.22	28.30
13	3.57	4.11	5.01	5.89	7.04	19.81	22.36	24.74	27.69	29.82
14	4.07	4.66	5.63	6.57	7.79	21.06	23.68	26.12	29.14	31.32
15	4.60	5.23	6.26	7.26	8.55	22.31	25.00	27.49	30.58	32.80
16	5.14	5.81	6.91	7.96	9.31	23.54	26.30	28.85	32.00	34.27
18	6.26	7.01	8.23	9.39	10.86	25.99	28.87	31.53	34.81	37.16
20	7.43	8.26	9.59	10.85	12.44	28.41	31.41	34.17	37.57	40.00
24	9.89	10.86	12.40	13.85	15.66	33.20	36.42	39.36	42.98	45.56
30	13.79	14.95	16.79	18.49	20.60	40.26	43.77	46.98	50.89	53.67
40	20.71	22.16	24.43	26.51	29.05	51.81	55.76	59.34	63.69	66.77
60	35.53	37.48	40.48	43.19	46.46	74.40	79.08	83.30	88.38	91.95
120	83.85	86.92	91.58	95.70	100.62	140.23	146.57	152.21	158.95	163.64

For large degrees of freedom,

$$\chi^2_P = \frac{1}{2}(z_P + \sqrt{2v - 1}^2) \text{ approximately,}$$

where v = degrees of freedom and zp is given by Table 1.

Table 4. One-sample Kolmogorov-Smirnov test *(If calculated d_n is greater than the value shown, then reject the null hypothesis at the chosen level of significance)*

Sample Size n	Level of significance for d_n				
	.20	.15	.10	.05	.01
1	.900	.925	.950	.975	.995
2	.684	.726	.776	.842	.929
3	.565	.597	.642	.708	.828
4	.494	.525	.564	.624	.733
5	.446	.474	.510	.565	.669
6	.410	.436	.470	.521	.618
7	.381	.405	.438	.486	.577
8	.358	.381	.411	.457	.543
9	.339	.360	.388	.432	.514
10	.322	.342	.368	.410	.490
11	.307	.326	.352	.391	.468
12	.295	.313	.338	.375	.450
13	.284	.302	.325	.361	.433
14	.274	292	.314	.349	.418
15	.266	283	.304	.338	.404
16	.258	274	.295	.328	.392
17	.250	266	.286	.318	.381
18	.244	259	.278	.309	.371
19	.237	252	.272	.301	.363
20	.231	246	.264	294	.356
25	.210	220	.240	270	.320
30	.190	200	.220	240	.290
35	.180	.190	.210	230	.270
Over35	$\frac{1.07}{\sqrt{n}}$	$\frac{1.14}{\sqrt{n}}$	$\frac{1.22}{\sqrt{n}}$	$\frac{1.36}{\sqrt{n}}$	$\frac{1.63}{\sqrt{n}}$

Table 5. Critical values for the two-sample Kolmogorov-Smirnov statistic

Sample size n_2		n_1=1	2	3	4	5	6	7	8	9	10	12	15
1		*	*	*	*	*	*	*	*	*	*	*	*
		*	*	*	*	*	*	*	*	*	*	*	*
2			*	*	*	*	*	*	7/8	16/18	9/10	11/12	26/30
			*	*	*	*	*	*	*	*	*	*	*
3				*	*	12/15	5/6	18/21	18/24	7/9	24/30	9/12	11/15
				*	*	*	*	*	*	8/9	27/30	11/12	13/15
4					3/4	16/20	9/12	21/28	6/8	27/36	14/20	8/12	41/60
					*	*	10/12	24/28	7/8	32/36	16/20	10/12	48/60
5						4/5	20/30	25/35	27/40	31/45	7/10	40/60	10/15
						4/5	25/30	30/35	32/40	36/45	8/10	48/60	11/15
6							4/6	29/42	16/24	12/18	19/30	7/12	18/30
							5/6	35/42	18/24	14/18	22/30	9/12	22/30
7								5/7	35/56	40/63	43/70	51/84	61/105
								5/7	42/56	47/63	53/70	58/84	70/105
8									5/8	45/72	23/40	14/24	66/120
									6/8	54/72	28/40	16/24	80/120
9										5/9	52/90	20/36	24/45
										6/9	62/90	24/36	29/45
10											6/10	32/60	15/30
											7/10	39/60	19/30
12												6/12	30/60
												7/12	35/60
15													7/15
													8/15

Notes: 1. Reject H_0 at the 5% or 1% level if $d = \sup |F_{n_2}(x) - F_{n_1}(x)|$ equals or exceeds the tabulated value. The upper value corresponds to $\alpha = .05$ and the lower to $\alpha = .01$.

2. Where * appears, do not reject H_0 at the given level.

3. For large values of n_1 and n_2, the following approximate formulas may be used:

$\alpha = .05 :\ 1.36\sqrt{\frac{n_1+n_2}{n_1 n_2}}$.

$\alpha = .01 :\ 1.63\sqrt{\frac{n_1+n_2}{n_1 n_2}}$.

Appendix 2: Answers and Hints to Selected Odd-Numbered Exercises

2.1.1.

a) The sample points are HH, HT, TH, TT, and the elementary events are $\{HH\}, \{HT\}, \{TH\}, \{TT\}$.

2.1.3.

a) Two possible sample spaces to describe three tosses of a coin are:
$S_1 = \{$an even # of H's, an odd # of H's$\}$,
$S_2 = \{HHHH, HHHT, HHTH, HHTT, HTHH, HTHT, HTTH,$
$HTTT, THHH, THHT, THTH, THTT, TTHH, TTHT, TTTH,$
$TTTT\}$, where the fourth letter is to be ignored in each sample point.

c) It is not possible to find an event corresponding to the statement $p =$ "at most one tail is obtained in three tosses" in every conceivable sample space for the tossing of three coins, because some sample spaces are too coarse, that is, the sample points that contain this outcome also contain opposite outcomes. For instance, in S_1 above, the sample point "an even # of H's" contains the outcomes HHT, HTH, THH for which our p is true, and it also contains the outcome TTT, for which it is not true. Thus, p has no truth set in S_1.

2.1.5.

In the 52-element sample space for the drawing of a card:

a) The event corresponding to the statement $p =$ "an Ace or a red King is drawn" is $P = \{AS, AH, AD, AC, KH, KD\}$.

b) A statement corresponding to the event $U = \{AH, KH, QH, JH\}$ is $u =$ "the Ace of hearts or a heart face card is drawn."

2.1.7.

One possible sample space is $S = \{$January, February,$\ldots$, December$\}$.

2.2.1.

a) $\{1, 3, 5, 7, 9\}$ or $\{k : k = 2n + 1, n = 0, 1, 2, 3, 4\}$.

© Springer International Publishing Switzerland 2016
G. Schay, *Introduction to Probability with Statistical Applications*,
DOI 10.1007/978-3-319-30620-9

2.2.5.
$A \cap B \cap C = \{1\}, (A \cap B) \cap C = \{1,4\} \cap \{1,2,3,7\} = \{1\}$, etc.

2.2.7.

a) $A \cap (B \cup C) = \{1,3,4,5\} \cap \{1,2,3,4,6,7\} = \{1,3,4\}$,

but $(A \cap B) \cup C = \{1,4\} \cup \{1,2,3,7\} = \{1,2,3,4,7\}$.

2.2.9.
Draw a Venn diagram with non-overlapping sets A and B and number the three regions.

2.2.11.

1. First, assume that $A \cup B = B$, that is, that $\{x : x \in A \text{ or } x \in B\} = B$. Hence, if $x \in A$, then x must also belong to B, which means that $A \subset B$. Alternatively, by the definition of unions, $A \subset A \cup B$, and so, if $A \cup B = B$, then substituting B for $A \cup B$ in the previous relation, we obtain that $A \cup B = B$ implies $A \subset B$.
2. Conversely, assume that $A \subset B$, and proceed similarly as above.

2.2.13.

a) $(A - B) - C = \{3,5\} - \{1,2,3,7\} = \{5\}$ and $(A - C) - (B - C) = \{4,5\} - \{4,6\} = \{5\}$.

2.3.1.

a) The event R corresponding to $r =$ "b is 4 or 5" is the region consisting of the fourth and fifth columns in Figure 2.4, that is, $R = \{(b,w) : b = 4,5 \text{ and } w = 1,2,\ldots,6\}$.

2.3.3.
$P_4 = \{2,3,4\} = \overline{A}BC \cup A\overline{B}C \cup AB\overline{C}$.

2.3.7.
$(A \triangle B) \triangle C = \{1,5,6,7\}$.

3.1.1.
Let $A =$ set of drinkers and $B =$ set of smokers. Then $n(AB) = 23$.

3.1.5.
$n(A) + n(B) + n(C) - n(A \cap B) - n(A \cap C) - n(B \cap C) + n(A \cap B \cap C)$
$= n(1,3,4,5) + n(1,2,4,6) + n(1,2,3,7) - n(1,4) - n(1,3) - n(1,2) + n(1)$, etc.

3.1.7.

1. By the definition of indicators, $I_{AB}(s) = 1 \Leftrightarrow s \in AB$. By the definition of intersection, $s \in AB \Leftrightarrow (s \in A \text{ and } s \in B)$, and, by the definition of indicators, $(s \in A \text{ and } s \in B) \Leftrightarrow (I_A(s) = 1 \text{ and } I_B(s) = 1)$. Since

$1 \cdot 1 = 1$ and $1 \cdot 0 = 0 \cdot 0 = 0$, clearly, $(I_A(s) = 1$ and $I_B(s) = 1) \Leftrightarrow$ $I_A(s) I_B(s) = 1$. Now, by the transitivity of equivalence relations, $I_{AB}(s) = 1 \Leftrightarrow I_A(s) I_B(s) = 1$, which is equivalent to $I_{AB} = I_A I_B$.

3.2.1.

a) $S = \{ASAH, ASAD, ASAC, AHAS, AHAD, AHAC, ADAS, ADAH,$
$ADAC, ACAS, ACAH, ACAD\}$.

3.2.5.
a) 24360, b) 27000.

3.2.7.
a) 14, b) 30.

3.3.1.
20, 120, 8, 1, 1.

3.3.5.
$\{ABC\ ACB\ BAC\ BCA\ CAB\ CBA\}$, etc.

3.3.7.
360.

3.3.9.

a) 24, b) 12.

3.3.11.

a) 1666,
b) 1249,
c) 416,
d) 2500.

3.4.1.
The tenth row is

1	10	45	120	210	252	210	120	45	10	1.

3.4.5.
45.

3.4.7.

a) 5^n

3.4.9.

a) $2^n - 1 - n,$

3.5.1.

a) 64,
b) 15,
c) 60,
d) 240.

3.5.3.

a) 420,
b) 60,
c) 300,
d) 240.

3.5.5.

a) 210,
b) $-22,680$.

3.5.7.

a) 66,
b) 36.

4.1.1.

g) $P(\overline{B} \cap C) = 0,$
h) $P(\overline{B} \cup \overline{C}) = \frac{48}{52}.$

4.1.3.
$A = A\overline{B} \cup AB$ and $A\overline{B} \cap AB = \emptyset$. Thus, by Axiom 3, $P(A) = P(A\overline{B}) + P(AB)$. Similarly, $P(B) = P(\overline{A}B) + P(AB)$. Add, use Axiom 3 again, and rearrange.

4.1.5.
The given relation is true if and only if $P(\overline{A}B) = 0$.
4.1.7.

a) This result follows at once from Theorem 4.1.2 because we are subtracting the (by Axiom 1) nonnegative quantity $P(AB)$ from $P(A) + P(B)$ on the right of Equation 4.1 to get $P(A \cup B)$.
c) Use induction.

4.2.1.

b) $P(A \text{ and } K) = \frac{8}{12} = \frac{2}{3}.$
d) Here, each unordered pair corresponds to two ordered pairs, and therefore each one has probability $2 \cdot \frac{1}{12} = \frac{1}{6}$. In Example 4.2.2, some unordered pairs correspond to two ordered pairs and some to one.

4.2.3.

We did not get P(at least one six) $= 1$, in spite of the fact that on each throw the probability of getting a six is $\frac{1}{6}$, and six times $\frac{1}{6}$ is 1, for two reasons: First, we would be justified in taking the $\frac{1}{6}$ six times here only if the events of getting a six on the different throws were mutually exclusive; then the probability of getting a six on one of the throws could be computed by Axiom 3 as $6 \cdot \frac{1}{6}$, but these are not mutually exclusive events. Second, the event of getting at least one six is not the same as the event of getting a six on the first throw, on the second, etc.

4.2.5.

$\frac{5}{9}$.

4.2.7.

$\frac{m!n!}{(m+n-1)!}$.

4.2.9.

$P(\text{jackpot}) = \frac{1}{5,245,786} \approx 2 \cdot 10^{-7}$ and $P(\text{match } 5) = \frac{108}{2,622,893} \approx 4 \cdot 10^{-5}$.

4.2.11.

a) $\frac{3}{8}$,
b) 0.441,
c) 0.189.

4.2.15.

P(all different) ≈ 0.507. (Note that we have included "straights" and "flushes" in the count, that is, cards with five consecutive denominations or five cards of the same suit, which are very valuable hands, while the other cases of different denominations are poor hands.)

4.2.17.

P(full house in poker) ≈ 0.0014.

4.2.19.

P(full house in poker dice) ≈ 0.0386.

4.3.1.

Let $E = $ "even" and $O = $ "odd," and consider the sample space $S = \{EEE, EEO, EOE,$
$EOO, OEE, OEO, OOE, OOO\}$ for throwing three dice. Compute $P(A)$, $P(B)$, and $P(AB)$.

4.3.5.

a) Let A and B be independent. Then $P(A\overline{B}) = P(A) - P(AB) = P(A) - P(A)P(B) = P(A)[1 - P(B)] = P(A)P(\overline{B})$.

4.3.7.
$p(0) = \left(\frac{1}{2}\right)^5$, $p(1) = 5 \cdot \left(\frac{1}{2}\right)^5$, $p(2) = 10 \cdot \left(\frac{1}{2}\right)^5$, etc.

4.3.9.
P(two of each color) ≈ 0.123.

4.3.11.
Expand both sides of $P(A(B \cup C)) = P(AB \cup AC)$.

4.4.3.
If $A = \{K \text{ or } 2\}$ and $B = \{J, Q, K\}$, then $P(A|B) = \frac{1}{3}$.

4.4.5.
Apply Theorem 4.4.1, part 3.

4.4.7.
P(two girls and one boy | one child is a girl) $= \frac{1}{2}$.

4.4.9.
Thus, P(two Kings | two face cards) $= \frac{1}{11}$.

4.4.11.
P(exactly one King |

4.4.13.

a)

$$P(A|B) = \frac{\binom{26}{8}}{\binom{39}{8}} = \frac{575}{22\,644} \approx 0.02539.$$

b)

$$P(\overline{A}|B) = 1 - \frac{\binom{26}{8}}{\binom{39}{8}} \approx 0.9746.$$

4.5.1.

c) $\frac{17}{40}$.

4.5.3.

a) $\frac{1}{33}$,
b) $\frac{5}{101}$,
c) $\frac{1}{17}$,
d) $\frac{1}{2}$.

4.5.5.

Equation 4.52 becomes $P(A_m) = P(A_{m+1}) \cdot p + P(A_{m-1}) \cdot q$ for $0 < m < n$, where $q = 1-p$ and A_m denotes the event that the gambler with initial capital m is ruined. Try to find constants λ such that $P(A_m) = \lambda^m$ for $0 < m < n$, just as in the analogous, but more familiar, case of linear homogeneous differential equations with constant coefficients. Solve the resulting quadratic equation $p\lambda^2 - \lambda + q = 0$. The general solution of the difference equation is then of the form $P(A_m) = a\lambda_1^m + b\lambda_2^m = a + b\left(\frac{q}{p}\right)^m$. As in Example 4.5.5, use the boundary conditions $P(A_0) = 1$ and $P(A_n) = 0$ to determine the constants a and b. Thus, the probability of the gambler's ruin is $P(A_m) = \frac{(q/p)^m - (q/p)^n}{1 - (q/p)^n}$, if he starts with m dollars and stops if he reaches n dollars. If $q < p$, that is, the game is favorable for our gambler, then $\lim_{m \to \infty} (q/p)^n = 0$, and so the gambler may play forever without getting ruined, and the probability that he does not get ruined is $1 - \left(\frac{q}{p}\right)^m$.

4.5.7.
$\frac{5}{17}$.

4.5.9.

$P(GG|G)$

$$= \frac{P(G|GG)P(GG)}{P(G|GG)P(GG) + P(G|BG)P(BG) + P(G|GB)P(GB) + P(G|BB)P(BB)}$$

$$= \frac{1}{2}.$$

4.5.11.
$P(WB|BW \cup WB) = 2/11$.

4.5.13.

Let $A =$ "the witness says the hit-and-run taxi was blue," $B_1 =$ "the hit-and-run taxi was blue," and $B_2 =$ "the hit-and-run taxi was black." Then $P(B_1|A) \approx 0.41$. Thus, the evidence against the blue taxi company is very weak.

4.5.15.

1. P(car is behind 2|3 is opened) $= 1/2$.
2. P(car is behind 2|3 is opened) $= 1/(p+1)$

5.1.1.

The p.f. of X is given by $f(x) = \binom{13}{x}\binom{39}{5-x}/\binom{52}{5}$ for $x = 0, 1, \ldots, 5$, and the d.f. of X is given by

$$F(x) \approx \begin{cases} 0 & \text{if } x < 0 \\ .222 & \text{if } 0 \leq x < 1 \\ .633 & \text{if } 1 \leq x < 2 \\ .907 & \text{if } 2 \leq x < 3 \\ .989 & \text{if } 3 \leq x < 4 \\ .999 & \text{if } 4 \leq x < 5 \\ 1 & \text{if } x \geq 5. \end{cases}$$

5.1.3.

The possible values of x are $0, \pm 2, and \pm 4$ and $f(0) = \frac{6}{16}$, $f(\pm 2) = \frac{4}{16}$, and $f(4) = \frac{1}{16}$. The histogram is

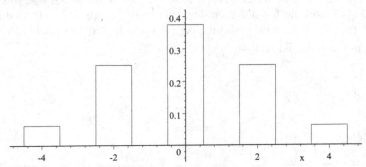

5.1.5.

The possible values of x are $3, 4, and 5$ and $f(3) = \frac{5}{8}$, $f(4) = \frac{5}{16}$, and $f(5) = \frac{1}{16}$.

5.1.7.

The p.f. is given by $f(2) = P(X = 2) = p^2 + q^2$, $f(3) = pq^2 + qp^2 = pq(q + p) = pq$, $f(4) = pq(p^2 + q^2)$, and $f(5) = p^2q^3 + q^2p^3 = (pq)^2(q + p) = (pq)^2$. Thus, in general, $f(2n) = (pq)^{n-1}(p^2 + q^2)$ and $f(2n + 1) = (pq)^n$ for $n = 1, 2, 3, \ldots$.

5.1.9.

The d.f is

$$F(x) = \begin{cases} 0 & \text{if } x < 1 \\ \lfloor x \rfloor / 6 & \text{if } 1 \leq x < 6 \\ 1 & \text{if } x \geq 6. \end{cases}$$

5.1.11.

First, we display the possible values of X in a table as a function of the outcomes on the two dice:

$w \backslash b$	1	2	3	4	5	6
1	0	1	2	3	4	5
2	1	0	1	2	3	4
3	2	1	0	1	2	3
4	3	2	1	0	1	2
5	4	3	2	1	0	1
6	5	4	3	2	1	0

Since each box has probability $1/36$, from here we can read off the values of the p.f. as

$$f(x) = \begin{cases} 6/36 & \text{if } x = 0 \\ 10/36 & \text{if } x = 1 \\ 8/36 & \text{if } x = 2 \\ 6/36 & \text{if } x = 3 \\ 4/36 & \text{if } x = 4 \\ 2/36 & \text{if } x = 5 \end{cases}$$

5.1.13.

Since $A_1, A_2, \ldots$ is a nondecreasing sequence of events, $A = A_1 \cup [\cup_{k=2}^{\infty}(A_k - A_{k-1})]$ and the terms of the union are disjoint, Axiom 2 gives $P(A) = P(A_1) + \sum_{k=2}^{\infty} P(A_k - A_{k-1})$. By the definition of infinite sums, the expression on the right is the limit of the partial sums, that is, $P(A) = \lim_{n \to \infty} [P(A_1) + \sum_{k=2}^{n} P(A_k - A_{k-1})]$. Apply Axiom 2 again.

5.1.15.

Let $\langle x_n \rangle$ be a sequence of real numbers decreasing to $-\infty$, and let $A_n = \{s : X(s) \leq x_n\}$ for every n. Then $F(x_n) = P(A_n)$ and $A_n \supset A_{n+1}$ for $n = 1, 2, \ldots$. Furthermore, $A = \cap_{k=1}^{\infty} A_k = \emptyset$, because there is no $s \in S$ for which the real number $X(s)$ can be $\leq x_n$ for every n, considering that $x_n \to -\infty$. Apply the result of Exercise 5.1.14 and the theorem from real analysis quoted in the hint.

5.2.1.

1. $C = \frac{1}{8}$.
4. $P(X < 1) = \frac{1}{16}$,
5. $P(2 < X) = \frac{3}{4}$.

5.2.3.

1. $C = 1$.
4. $P(X < 2) = \frac{1}{2}$.
5. $P(2 < |X|) = \frac{1}{2}$.

5.2.5.

1. Roll a die. If the number six comes up, then also spin a needle that can point with uniform probability density to any point on a scale from 0 to 1, and let X be the number the needle points to. If the die shows 1, then let $X = 1$, and if the die shows any number other than 1 or 6, then let $X = 2$.
2. $P(X < 1/2) = \frac{1}{12}$,
3. $P(X < 3/2) = \frac{1}{3}$,
4. $P(1/2 < X < 2) = \frac{1}{4}$,
5. $P(X = 1) = \frac{1}{6}$,
6. $P(X > 1) = \frac{2}{3}$,
7. $P(X = 2) = \frac{2}{3}$.

5.2.7.

2. $P(X < 1/2) = \frac{1}{10}$.
3. $P(X < 3/2) = \frac{3}{5}$.
4. $P(1/2 < X < 2) = \frac{7}{10}$.
5. $P(X = 1) = \frac{1}{2}$.
6. $P(X > 1) = \frac{3}{5}$.
7. $P(X = 2) = \frac{1}{5}$.

5.3.1.
First, make a table whose first row contains the possible values of x, the second row the corresponding values of $f_X(x)$, and the third row the values of $y = x^2 - 3x$. From this table extract a new table for the p.f. of Y,

5.3.3.
$$f_Y(y) = \begin{cases} 1/2 \text{ if } y = 0 \\ 1/2 \text{ if } y = \pi/4. \end{cases}$$

5.3.5.
$$F_Y(y) = \begin{cases} e^y \text{ if } y < 0 \\ 1 \text{ if } y \geq 0. \end{cases}$$

5.3.7.
$$F_Y(y) = \begin{cases} 0 & \text{if } y < 0 \\ \int_{-y}^{y} f_X(x)\, dx & \text{if } y \geq 0. \end{cases}$$

5.3.9.
$$F_Y(y) = P(Y \leq y) = \begin{cases} 0 & \text{if } y < -r \\ \frac{1}{2} + \frac{1}{\pi} \arcsin \frac{y}{r} & \text{if } -r \leq y < r \\ 1 & \text{if } r \leq y. \end{cases}$$

5.4.1.
First, make a 6×6 table with the possible values of X and Y on the margins and the corresponding values of $U = X + Y$ and $V = X - Y$ in the body of the table. Next, make an 11×11 table with the possible values of

U and V on the margins and the corresponding values of $f_{U,V}(u,v)$ in the body of the table, which are obtained from the first table, considering that each box there has probability $1/36$.

5.4.3.

1. $\frac{5}{324}$,
2. $\frac{5}{3888}$.

5.4.5.

$$F_Z(z) = \begin{cases} 0 & \text{if } z < 0 \\ z^2 & \text{if } 0 \le z < 1 \\ 1 & \text{if } 1 \le z. \end{cases}$$

5.4.7.

1. $C = 10$.
2. $f_X(x) = 10\left(x - x^4\right)/3$ if $0 \le x \le 1$, and $f_X(x) = 0$ otherwise. Similarly, $f_Y(y) = 5y^4$ for $0 \le y \le 1$, and $f_Y(y) = 0$ otherwise.
3. If $(x,y) \in D$, then $F(x,y) = \frac{5}{3}y^3x^2 - \frac{2}{3}x^5$. If $0 < x < 1$ and $y \ge 1$, then $F(x,y) = F(x,1) = \frac{5}{3}x^2 - \frac{2}{3}x^5$. If $0 < y < 1$ and $x \ge y$, then $F(x,y) = F(y,y) = y^5$. If $x \ge 1$ and $y \ge 1$, then $F(x,y) = 1$, and $F(x,y) = 0$ otherwise.
4. $P(X > Y^2) = 2/7$

5.4.9.

In the xy-plane, draw the four points (x_1, y_1), (x_1, y_2), (x_2, y_1), *and* (x_2, y_2) and the quarter planes to the left and below each of these points. Number the regions. The probabilities of the quarter planes are the values of F in the given points. Use this fact and the additivity axiom for the numbered regions to prove the formula.

5.4.11.

1. In the discrete case,

$$f_Z(z) = \sum_{x-y=z} f(x,y) = \sum_{x=-\infty}^{\infty} f(x, x-z) = \sum_{y=-\infty}^{\infty} f(y+z, y).$$

2. In the discrete case,

$$f_Z(z) = \sum_{2x-y=z} f(x,y) = \sum_{x=-\infty}^{\infty} f(x, 2x-z) = \sum_{y=-\infty}^{\infty} f\left(\frac{y+z}{2}, y\right).$$

5.5.1.

Compute some joint and marginal probabilities, and test whether the product rule for independence holds.

For instance, $f(0,1) = P(X = 0, Y = 1) = \frac{\binom{13}{0}\binom{13}{1}\binom{26}{1}}{\binom{52}{2}} = \frac{13}{51}$.

5.5.3.
Compare to Examples 5.4.5 and 5.5.2.

5.5.5.

1. By the definition of indicators, $I_{AB}(s) = 1 \Leftrightarrow s \in AB$. By the definition of intersection, $s \in AB \Leftrightarrow (s \in A \text{ and } s \in B)$, and, by the definition of indicators, $(s \in A \text{ and } s \in B) \Leftrightarrow (I_A(s) = 1 \text{ and } I_B(s) = 1)$. Since $1 \cdot 1 = 1$ and $1 \cdot 0 = 0 \cdot 0 = 0$, clearly, $(I_A(s) = 1 \text{ and } I_B(s) = 1) \Leftrightarrow I_A(s) I_B(s) = 1$. Now, by the transitivity of equivalence relations, $I_{AB}(s) = 1 \Leftrightarrow I_A(s) I_B(s) = 1$, which is equivalent to $I_{AB} = I_A I_B$.

5.5.7. $\frac{7}{16}$.

5.5.9. $\frac{2}{3} - \frac{\sqrt{3}}{2\pi}$.

5.5.11.

1. $F_Z(z) = \int_0^\infty f_Y(y) \left[\int_0^{z/y} f_X(x) dx \right] dy$, and $f_Z(z) = \int_0^\infty f_Y(y) f_X(\frac{z}{y}) \frac{1}{y} dy$.
2. $F_Z(z) = \int_0^\infty f_Y(y) \left[\int_0^{zy} f_X(x) dx \right] dy$, and $f_Z(z) = \int_0^\infty f_Y(y) f_X(zy) y \, dy$.
3. $f_Z(z) = -\ln z$ if $0 \le z < 1$, and $f_Z(z) = 0$ otherwise.

5.5.13.

1. $P(T > 200) \approx 0.135$.
2. $P(T < 400) \approx 0.330$.
3. $P(\max T_i < 200) \approx 2 \cdot 10^{-9}$.
4. $P(\min T_i < 40) \approx 0.999985$.

5.5.15.
Use Definition 5.2.3 and Corollary 5.5.1.

5.5.17.
Find the d.f. of $X = -T_2$, and use the sum formula for $Z = T_1 - T_2 = X + T_1$, separately, for $z < 0$ and $z \ge 0$.

5.5.19.

$$f_Z(z) = \begin{cases} 0 & \text{if } z < -1 \\ z + 1 & \text{if } -1 \le z < 0 \\ 1 - z & \text{if } 0 \le z < 1 \\ 0 & \text{if } 1 \le z. \end{cases}$$

5.6.1.
The joint distribution of X and Y is trinomial (with the third possibility being that we get any number other than 1 or 6), and so $f_{X,Y}(i,j) =$

$P(X = i, Y = j) = \binom{4}{i \ j \ k} \left(\frac{1}{6}\right)^i \left(\frac{1}{6}\right)^j \left(\frac{4}{6}\right)^k$ for $i, j, k = 0, 1, \ldots, 4$, $i+j+k = 4$. Use this formula to compute the marginals and the conditional p.f., which is given by $f_{X|Y}(i|j) = \frac{f_{X,Y}(i,j)}{f_Y(j)}$.

5.6.3.

By Example 5.5.4, $P(A|X = x) = \begin{cases} P\left(\frac{1}{2} < Y < x + \frac{1}{2}\right) = x & \text{if } 0 \le x < \frac{1}{2} \\ P\left(x - \frac{1}{2} < Y < \frac{1}{2}\right) = 1 - x & \text{if } \frac{1}{2} < x < 1, \end{cases}$

and $P(A) = \frac{1}{4}$, and, by Equation 5.141, $f_{X|A}(x) = \frac{P(A|X=x)f_X(x)}{P(A)}$. Substitute into the latter.

5.6.5.

First, compute the marginal densities. By definition, $f(x, y) = \begin{cases} 2 & \text{if } (x, y) \in D \\ 0 & \text{otherwise} \end{cases}$

and so $f_X(x) = \begin{cases} \int_0^{1-x} 2dy = 2(1-x) & \text{if } 0 < x < 1 \\ 0 & \text{otherwise} \end{cases}$ and

$f_Y(y) = \begin{cases} \int_0^{1-y} 2dx = 2(1-y) & \text{if } 0 < y < 1 \\ 0 & \text{otherwise.} \end{cases}$ Now use Equation 5.128 to

obtain the required conditional densities.

5.6.7.

1. First, compute $P(Z \le z|X = x) = F_{Z|X}(z|x) = P(x + Y \le z) = P(Y \le z - x)$. Next, find $f_{Z|X}(z|x) = \frac{\partial}{\partial z} F_{Z|X}(z|x)$. Then use Equation 5.144, to obtain $f_{X|Z}(x|z)$.

2. Draw a diagram to show that $P(X \le x, Z \le 1) = \begin{cases} 0 & \text{if } x < 0 \\ \frac{1}{2} - \frac{1}{2}(1-x)^2 & \text{if } 0 \le x \le 1 \\ \frac{1}{2} & \text{if } x > 1 \end{cases}$

and so that $F_{X|A}(x) = \frac{P(X \le x, Z \le 1)}{1/2} = \begin{cases} 0 & \text{if } x < 0 \\ 1 - (1-x)^2 & \text{if } 0 \le x < 1 \\ 1 & \text{if } x \ge 1. \end{cases}$ Differentiate

to obtain $f_{X|A}(x)$.

6.1.1.
$E(X) = \frac{85}{13} \approx 6.54$.

6.1.3.
$E(T) = \int_0^\infty t \cdot \lambda^2 t e^{-\lambda t} dt$. Integrate by parts twice to obtain $E(T) = \frac{2}{\lambda}$.

6.1.7.

The distribution of a discrete X is symmetric about a number α if all possible values of X are paired so that for each $x_i < \alpha$, there is a possible value $x_j > \alpha$ and vice versa, such that $\alpha - x_i = x_j - \alpha$ and $f(x_i) = f(x_j)$. For such X, $E(X) = \sum_{\text{all } i} x_i f(x_i) = \sum_{x_i < \alpha} x_i f(x_i) + \alpha f(\alpha) + \sum_{x_j > \alpha} x_j f(x_j)$. (Here $f(\alpha) = 0$, if α is not a possible value of X.) In the last term, apply the symmetry conditions and simplify.

6.1.9.
Use Theorem 6.1.3 and the integral from the solution of Exercise 6.1.3.

6.1.11.
Use Theorem 6.1.3.

6.1.13.

Example 5.3.1 gives, for continuous X and $Y = aX + b$, where $a \neq 0$, $f_Y(y) = \frac{1}{|a|} f_X\left(\frac{y-b}{a}\right)$. Use this expression to compute $E(aX + b) = E(Y)$, and in the integral change, the variable y to $x = \frac{y-b}{a}$ separately when $a < 0$ and when $a > 0$.

6.1.15.

In the expression for $E(X)$, factor out np and then use the binomial theorem.

6.1.17.

In this case, Theorem 6.1.6 does not apply, because X and Y are not independent. Nevertheless, Equation 6.53 is still true, and you have to check it directly.

6.1.19. $E(Z) = \frac{2}{3}$.

6.1.21.

Use the hint and the formula for the sum of a geometric series.

6.2.1.

For instance, if X is a continuous r.v. with density $f(x) = \begin{cases} 2/x^3 & \text{if } 1 < x < \infty \\ 0 & \text{otherwise} \end{cases}$ and $Y = -X$, then show that X and Y are as required.

6.2.3.

Prove $Var(aX + bY + c) = a^2 Var(X) + b^2 Var(Y)$.

6.2.7.

1. $E(X + 2Y) = \frac{3}{\lambda}$, and $Var(X + 2Y) = \frac{5}{\lambda^2}$.
2. $E(X - 2Y) = -\frac{1}{\lambda}$, and $Var(X - 2Y) = \frac{5}{\lambda^2}$.
3. $E(XY) = \frac{1}{\lambda^2}$, and $Var(XY) == \frac{3}{\lambda^4}$.
4. $E(X^2) = \frac{2}{\lambda^2}$, and $Var(X^2) = \frac{20}{\lambda^4}$.
5. $E\left((X+Y)^2\right) = \frac{6}{\lambda^2}$, and $Var\left((X+Y)^2\right) = \frac{84}{\lambda^4}$.

6.2.9.

No: $E(X) = E(Y) = np = \frac{n}{2}$ and $E(XY) = \frac{n^2 - n}{4}$.

6.3.1.

Write $\widehat{X} = X - \mu_X$ and $\widehat{Y} = Y - \mu_Y$. Then evaluate $m_3(X + Y) = E\left(\left(\widehat{X} + \widehat{Y}\right)^3\right)$ using the independence of X and Y, from which the independence of $\widehat{X}$ and $\widehat{Y}$ follows.

6.3.3.

$\psi_Y(t) = \psi_X(at) e^{bt}$.

6.3.5.

$\psi_{X-\mu}(t) = E\left(e^{t(X-\mu)}\right)$, and so, by the result of Exercise 6.3.3, $\psi_{X-\mu}(t) = \psi_X(t)\,e^{-\mu t} = (pe^t + q)^n\,e^{-npt} = (pe^{qt} + qe^{-pt})^n$. Now, $Var(X)$ is the second moment of $X - \mu$. Use the expression above and Theorem 6.3.1 to compute $Var(X)$ as $\psi''_{X-\mu}(0)$.

6.3.9.

Use the appropriate definitions and the binomial theorem.

6.3.11.

Let X_1, X_2, X_3 denote the points showing on the three dice, respectively, and let $S = X_1 + X_2 + X_3$. Find $G_{X_i}(s)$ for any i, and then $G_S(s) = G_{X_i}^3(s)$. The coefficients of s^k here are the required probabilities, and so $p_3 = \frac{1}{216}$, $p_4 = \frac{3}{216} = \frac{1}{72}$, and $p_5 = \frac{6}{216} = \frac{1}{36}$.

6.4.1.

Write $Var(X + Y)$ as an expectation in terms of $X - \mu_X$ and $Y - \mu_Y$, and simplify.

6.4.3.

1. Write $Cov(U, V)$ in terms of $\widehat{X} = X - \mu_X$ and $\widehat{Y} = Y - \mu_Y$, and simplify.
2. Compute, for instance, $P(V = 0|U = 2)$ and $P(V = 0)$, and use Theorem 5.6.3.

6.4.5.

$Cov(X, Y) = \sum_{i=1}^m \sum_{j=1}^n p_{ij} x_i y_j - \sum_{i=1}^m p_i x_i \sum_{j=1}^n p_j y_j$.

6.4.7.

First show that $Cov(U, V) = acCov(X, Y)$, $\sigma_U = |a|\sigma_X$, and $\sigma_V = |c|\sigma_Y$.

6.4.9.

Use the result of Exercise 6.4.1 and Theorem 6.2.2.

6.5.1.

Use Definition 6.5.1, for discrete X and Y, and Theorem 6.1.3 with $g(Y) = E_Y(X)$.

6.5.3.

$E(X) = \frac{9}{4}$.

6.5.5.

$E_y(X) = \frac{1-y}{2}$ if $0 < y < 1$ and $E_y(X) = 0$ otherwise. $E_x(Y) = \frac{1-x}{2}$ if $0 < x < 1$ and $E_x(Y) = 0$ otherwise. $E(X) = E(Y) = \frac{1}{3}$.

6.5.7.

$E_z(X) = \begin{cases} \frac{z}{2} & \text{if } 0 < z < 2 \\ 0 & \text{otherwise} \end{cases}$ and $E_x(Z) = \begin{cases} x + \frac{1}{2} & \text{if } 0 < x < 1 \\ 0 & \text{otherwise.} \end{cases}$

6.5.9.

Use Definition 6.5.1 and Theorem 6.1.4.

6.5.11.

Use Theorem 6.5.1 and Definition 6.5.2.

6.5.13.

Show that for continuous (X, Y) with density $f(x, y)$, $E(Var_Y(X)) = \int_{-\infty}^{\infty} \int_{-\infty}^{\infty} [x - E_y(X)]^2$
$f(x, y) \, dx dy$ and $Var(X) = \int_{-\infty}^{\infty} \int_{-\infty}^{\infty} [x - E(X)]^2 f(x, y) \, dx dy$, and since $E_y(X) \neq E(X)$ in general, also $Var(X) \neq E(Var_Y(X))$ in general.

6.6.1.

1. Let $n = 2k + 1$ for $k = 1, 2, \ldots$. Then the median is $m = x_{k+1}$.
2. Let $n = 2k$ for $k = 1, 2, \ldots$. Then any number m such that $x_k < m < x_{k+1}$ is a median.

6.6.3.

The converse of Theorem 6.6.1 says: For m a median of a random variable X, $P(X < m) = \frac{1}{2}$ and $P(X > m) = \frac{1}{2}$ imply $P(X = m) = 0$. Show that this statement is true.

6.6.5.

Write $E(|X - c|)$ as two integrals without absolute values, and differentiate with respect to c using the fundamental theorem of calculus. Thus show that $E(|X - c|)$ has a critical point where $2F(c) - 1 = 0$ or where $F(c) = \frac{1}{2}$, that is, if c is a median m. Since we assumed that f is continuous and $f(x) > 0$, m is unique. Use the second derivative test to show that $E(|X - c|)$ has a minimum at $c = m$.

6.6.7.

For general X the 50th percentile is defined as the number $x_{.5} = \min\{x : F(x) \geq .5\}$. Show that this $x_{.5}$ satisfies the two conditions in the definition of the median.

6.6.9.

The quantile function is $F^{-1}(p) = 2\sqrt{p} - 1$ for $p \in (0, 1)$.

6.6.11.

Invert $p = x/4$ and $p = x/4 + 1/2$ separately. The graph of $F^{-1}(p)$ is the reflection of the graph in Fig. 5.8 across the $y = x$ line.

7.1.1.

1. $P(X(1) > 1) \approx 0.264$.
2. $P(X(2) > 2) \approx 0.323$.
3. $P(X_1(1) > 1$ and $X_2(1) > 1) \approx 0.0698$.
4. $P(X_1(1) = 2$ and $X_2(1) = 2 | X(2) = 4) \approx 0.375$.

7.1.3.

1. $P(X(1) > 2) \approx 0.323$.
2. $P(X(2) > 4) \approx 0.371$.
3. $P(X(1) > 4) \approx 0.053$.
4. $P(T_1 > 1) \approx 0.135$.
5. $P(T_2 > \frac{1}{2}) \approx 0.135$.

7.1.5.
$P(\text{even}) - P(\text{odd}) = e^{-2\lambda t}$. Also, $P(\text{even}) + P(\text{odd}) = 1$.

7.1.7.
Consider the instants $s - \Delta s < s < t <. t + \Delta t \le s' - \Delta s' < s' < t' < t' + \Delta t'$, and let T_1 and T_2 denote two distinct interarrival times. Compute $f_{T_1, T_2}(t - s, t' - s')$ as a limit, and in the last step, use part 2 of Theorem 7.1.7. If $t = s'$, the proof would be similar.

7.2.1.
Using the table, we obtain:

1. $P(Z < 2) \approx 0.9772$
2. $P(Z > 2) \approx 0.0228$
3. $P(Z = 2) = 0$
4. $P(Z < -2) \approx 0.0228$
5. $P(-2 < Z < 2) \approx 0.9544$
6. $P(|Z| > 2) \approx 0.0456$
7. $P(-2 < Z < 1) \approx 0.8185$
8. $z \approx 1.6448$
9. $z \approx 1.6448$
10. $z \approx 1.2815$

7.2.3.

1. Differentiate $\varphi(z) = \frac{1}{\sqrt{2\pi}} e^{-z^2/2}$ twice, and show that $\varphi''(z)$ changes sign at $z^2 = 1$, that is, at $z = \pm 1$.
2. Differentiate $f(x) = \frac{1}{\sqrt{2\pi}\sigma} e^{-(x-\mu)^2/2\sigma^2}$ twice, and show that $f''(x)$ changes sign at $\left(\frac{x-\mu}{\sigma}\right)^2 = 1$, that is, at $\frac{x-\mu}{\sigma} = \pm 1$.

7.2.7.
Compare $ce^{-(x+2)^2/24}$ with the general normal p.d.f. $\frac{1}{\sqrt{2\pi}\sigma} e^{-(x-\mu)^2/2\sigma^2}$.

7.2.9.
$P(|X_1 - X_2| > .5) \approx 0.27$.

7.2.11.
Solve $z = \Phi^{-1}(1-p)$ for p, to get the area of the tail to the right of z under the standard normal curve. Switch to the area of the corresponding left tail and solve the resulting equation for z.

7.3.1.
With the binomial: $P(X = 3) \approx 0.238$.
With the normal approximation: $P(2.5 < X \le 3.5) \approx 0.2312$.

7.3.3.
A single random number X is a uniform random variable with $\mu = \frac{1}{2}$ and $\sigma^2 = \frac{1}{12}$. By Corollary 7.3.2, the average $\overline{X}$ of $n = 100$ i.i.d. copies of X is approximately normal with $\mu_{\overline{X}} = \frac{1}{2}$ and $\sigma_{\overline{X}} = \sqrt{\frac{1}{1200}} \approx 0.029$. Thus,
$$P(0.49 < \overline{X} < 0.51) = P\left(\frac{0.49-0.5}{0.029} < \frac{\overline{X}-0.5}{0.029} < \frac{0.51-0.5}{0.029}\right)$$
$$\approx P(-0.345 < Z < 0.345) = 2\Phi(0.345) - 1 \approx 0.27.$$

7.3.5.

1. $n \ge 144$.
2. $n \ge 390$.

7.4.1.
$P(r$ successes before s failures$) = \sum_{k=r}^{r+s-1} \binom{k-1}{r-1} p^r q^{k-r}$.

7.4.5.
Use $P(X_r = k, X_{r+s} = l) = P(X_r = k)P(X_s = l - k)$.

7.4.7.
Differentiate the gamma density from Definition 7.4.2.

7.4.9.

1. Modify the proof in Example 7.4.2.
2. Use part 1 and $Var(T) = E(T^2) - [E(T)]^2$
3. Use the definition of $\psi(t)$ and again modify the proof in Example 7.4.2.

7.4.11.
Use mathematical induction on k.

7.4.13.
Compute $F_U(u)$ first and then differentiate. U turns out to be exponential with parameter $\lambda = \frac{1}{2\sigma^2}$. In particular, the χ_2^2 distribution is the same as the exponential with parameter $\frac{1}{2}$.

7.4.15.
Use Theorem 5.5.8

7.4.19.

Use Theorem 5.6.2 and Equation 5.142, but instead of evaluating the integral in the denominator, determine the appropriate coefficient for the numerator by noting that it being a power of p times a power of $1 - p$, the posterior density $f_{P|X}$ must be beta. Thus, $f_{P|X}$ is beta with parameters $k + r$ and $n - k + s$, and $c = \frac{1}{B(k+r,n-k+s)}$.

7.5.1.

Clearly, Y_1 and Y_2, as linear combinations of normals, are normal. To show that they are standard normal, compute their expectations and variances.

7.5.3.

Equate the coefficients of like powers in the exponents in Equation 7.148 and in the present problem.

7.5.5.

Use the result of Exercise 6.4.8 and Theorems 6.4.2, 7.5.1, and 7.5.4.

7.5.7.

First show that X_2 under the condition $X_1 = 80$ is normal with $\mu \approx 77$ and $\sigma \approx 8.57$. Hence $x_{.90} \approx 88$.

7.5.9.

If (X_1, X_2) is bivariate normal as given by Definition 7.5.1, then aX_1+bX_2 is a linear combination of the independent normals Z_1 and Z_2, plus a constant, and so Theorems 7.2.4 and 7.2.6 show that it is normal.

To prove the converse, assume that all linear combinations of X_1 and X_2 are normal, and choose two linear combinations, $T_1 = a_1X_1 + b_1X_2$ and $T_2 = a_2X_1 + b_2X_2$, such that $Cov\,(T_1, T_2) = 0$. Such a choice is always possible, since if $Cov\,(X_1, X_2) = 0$, then $T_1 = X_1$ and $T_2 = X_2$ will do, and otherwise the rotation from Exercise 7.5.5 achieves it. Next, proceed as in the proof of Theorem 7.5.1.

7.5.11.

$\mu_{U_1} = 0$, $\mu_{U_2} = 5$, $\sigma_{1,2} = 4.8$, $\sigma_{U_1}^2 = 44.2$, $\sigma_{U_2}^2 = 5.8$, $\sigma_{U_1,U_2} = -7$, and $\rho_{U_1,U_2} \approx -0.437$.

8.1.5.

a) Differentiate $\ln\,(L\,(\lambda)) = n \ln \lambda + (\lambda - 1)\sum \ln x_i$.
b) Express λ as a function of $E(X)$ and replace $E(X)$ by $\bar{x}_n$.

8.1.7.

Use Theorem 5.5.8 to find $f_Y(y)$ and the latter to compute $E\,(\Theta)$.

8.1.9.

$\hat{p} \approx 0.022$ and the required approximate confidence intervals are $(53.4\%, 60.6\%)$, $(52.7\%, 61.3\%)$, and $(51.3\%, 62.7\%)$.

8.2.1.
The P-value is about 0.294. This probability is high enough for us to accept the null hypothesis, that is, that the low average of this class is due to chance; these students may well come from a population with mean grade 66.

8.2.3.
Using a large-sample paired Z-test for the mean increase $\mu = \mu_2 - \mu_1$ of the weights, we find the approximate P-value to be 0.0002. Thus, we reject the null hypothesis: the diet is very likely to be effective; however, the improvement is slight, and the decision might hinge on other factors, like the price and availability of the new diet.

8.2.5.
For $H_0, \mu = 28$, the P-value is $\Phi(-11.4) \approx 2 \cdot 10^{-30}$, and for $H_0, \mu = 24$, it is $\Phi(-4.7) \approx 1.3 \cdot 10^{-6}$.

8.2.7.
$c = 0.2 \cdot \Phi^{-1}(0.975) \approx 0.2 \cdot 1.96 = 0.392$.

8.3.1.
b) If $\mu = 6.5$, then the drug has really reduced the duration of the cold from 7 to 6.5 days, and the test will correctly show with probability 0.841 that the drug works.

8.3.3.
The rejection region is $(-\infty, 26.5]$. The power function is given by $\pi(\mu) = P(\overline{X} \in C|\mu) = P(\overline{X} \le 26.5|\mu) \approx \Phi\left(\frac{26.5-\mu}{24}\right)$.

8.3.5.
Let X denote the number of nondefective chips. The rejection region is the set of integers $C = \{0, 1, 2, \ldots, 10\}$. The operating characteristic function is $1 - \pi(p) = P(X \in \overline{C}|p) = \binom{12}{0}p^{12}(1-p)^0 + \binom{12}{1}p^{11}(1-p)^1$.

8.4.1.
$862.26 < \mu < 1035.74$ and $41.8 \le \sigma < 201.4$.

8.4.3.
The P-value is $P(T \ge 0.995) \approx 0.2$, and so we accept the null hypothesis, the truth of the store's claim.

8.4.5.
Use the limit formula $\lim_{k \to \infty} \left(1 + \frac{x}{k}\right)^k = e^x$.

8.5.5.
Divide the interval into four equal parts (in order to have the expected numbers be at least five) and compute $\hat{\chi}^2$.

8.5.7.
Extend the given table to include the marginal frequencies. Hence, the expected frequencies under the assumption of independence can be obtained

by multiplying each row frequency with each column frequency and dividing by 88. Compute $\hat{\chi}^2$ and the number of degrees of freedom and use a χ^2 table.

8.6.1.
The P-value is about 0.0026.

8.6.3.
By the definition of $F_{m,n}$ and the independence of the chi-square variables involved,

$$E\left(F_{m,n}\right) = E\left(\frac{n\chi_m^2}{m\chi_n^2}\right) = \frac{n}{m}E\left(\chi_m^2\right)E\left(\frac{1}{\chi_n^2}\right).$$

Use the definition of the Γ function in the evaluation of $E\left(\frac{1}{\chi_n^2}\right)$.

8.7.1.
Use Equation 8.153.

8.7.5.
Use Equation 8.155.

8.8.1.

$$Cov\left(\widehat{B}, \widehat{C}\right) = \frac{\overline{x}\sigma^2}{n\widehat{\sigma}_1^2}.$$

8.8.3.
$SD\left(\widehat{B}\right) = 0.0028, SD\left(\widehat{C}\right) = 0.036.$

8.8.5.
$x_0 = 101.3, \widehat{\sigma} \approx 35.6.$

References

The following four references give more detailed introductions to probability at more or less the same or somewhat higher level than this book:

Gharahmani, Saeed: *Fundamentals of Probability,* 2nd ed. Prentice-Hall, 2000.

Larsen, Richard J. and Morris L. Marx: *An Introduction to Probability and its Applications,* Prentice-Hall, 1985.

Pitman, Jim: *Probability,* Springer, 1993.

Ross, Sheldon: *A First Course in Probability,* 3rd ed. Macmillan 1988.

The next reference is a great classic that gives a more advanced, deep treatment of discrete probability:

Feller, William: *An Introduction to Probability Theory and its Applications,* · *Vol. 1,* 3rd ed. Wiley, 1971.

The next reference gives an excellent elementary, nonmathematical introduction to the concepts of statistics:

Freedman, David, Robert Pisani, and Roger Purves: *Statistics,* 3rd ed. Norton, 1998.

The remaining references provide more detail on mathematical statistics and include good discussions of probability as well:

DeGroot, Morris H. and Mark J. Schervish: *Probability and Statistics,* 3rd ed. Addison-Wesley, 2002.

Larsen, Richard J. and Morris L. Marx: *An Introduction to Mathematical Statistics and its Applications,* Prentice-Hall, 1981.

Lindgren, Bernard W.: *Statistical Theory,* 3rd ed. Macmillan 1976.

© Springer International Publishing Switzerland 2016   379
G. Schay, *Introduction to Probability with Statistical Applications,*
DOI 10.1007/978-3-319-30620-9

Index

© Springer International Publishing Switzerland 2016
G. Schay, *Introduction to Probability with Statistical Applications*,
DOI 10.1007/978-3-319-30620-9

Printed in the United States
By Bookmasters